The theory of thermodynamics

Waldram, J. R.
The theory of thermodynamics / J.R. Waldram. -- Cambridge : Cambridge University Press, 1985.
xv, 336 p. : ill.
Bibliography: p. 316-318.
Includes index.
ISBN 0-521-24575-3 (cased)

1. Thermodynamics. I. Title

15 APR 88 12503853 CHCAxc GB85-16176

The theory of thermodynamics

J. R. WALDRAM

Lecturer in Physics, Cavendish Laboratory, Fellow of Pembroke College, Cambridge

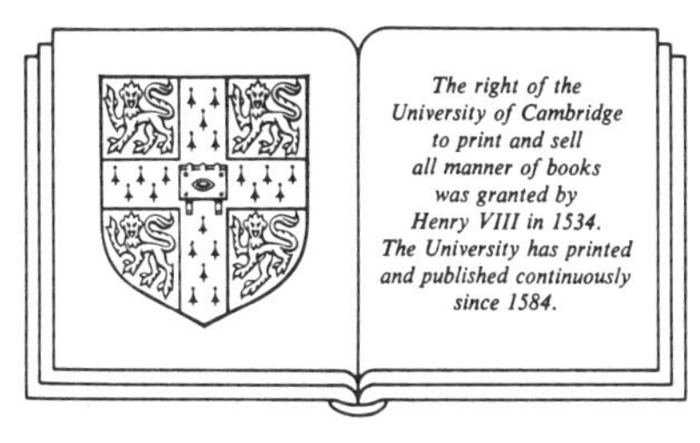

CAMBRIDGE UNIVERSITY PRESS

Cambridge

New York New Rochelle Melbourne Sydney

Published by the Press Syndicate of the University of Cambridge
The Pitt Building, Trumpington Street, Cambridge CB2 1RP
32 East 57th Street, New York, NY 10022, USA
10 Stamford Road, Oakleigh, Melbourne 3166, Australia

First published 1985
Reprinted 1987

Printed in Great Britain at the University Press, Cambridge

Library of Congress catalogue card number: 85-9544

British Library cataloguing in publication data

Waldram, J.R.
Theory of thermodynamics

1. Thermodynamics
I. Title
536′.71 QC311

ISBN 0 521 24575 3 hard covers
ISBN 0 521 28796 0 paperback

MP

For Harris Thorning and
John Ashmead

Contents

(Sections marked with an asterisk may be omitted at first reading.)

Preface

This is a textbook on the theory of heat for university students, primarily for students of physics and chemistry. The plan of the book is explained at the end of Chapter 1: this preface is more particularly concerned with the aims which the author hopes to achieve. In assessing existing texts and in organising the material I have had three points in mind. First, the *logic* of the theory of heat is notoriously obscure to students. In dealing with this branch of physics, they often feel themselves out of touch with the underlying principles, and, when faced with a calculation, are unsure of what approach to adopt. My first aim, then, has been to present the principles of the subject as coherently and concretely as possible. Second, it seems surprising that there is no single text on heat which a first-year student can buy and use in the knowledge that it will carry him through to his final year and beyond. This book is designed as an exposition of the theory of heat whose early parts may be understood by an able school leaver, but which is nevertheless a solid final-year text. Third, I am irritated, in texts at all levels, by much obscurity and exaggerated awe concerning the statistical principles upon which the theory of heat is based. I have therefore made an effort to present the starting assumptions of the subject clearly and in quite simple terms from the outset. This is not to deny that, as in other areas of physics, the fundamental notions turn out when rigorously examined to be of considerable difficulty and subtlety. Some of the difficulties and subtleties *are* explored in this book, but not until late in the treatment.

With these aims in view, it was important to work for a clear and compact presentation with relatively little illustrative material, so that my reader should quickly come to see what depends upon what. The chapters are laid out to display the logical structure of the subject which I hope that he or she will have accepted by the end of the course. The fundamental principles come first, and deductions follow. Moreover, the book is cross referenced using a system of marginal notes which makes it easy to locate particular topics, to trace arguments backwards and to anticipate later developments.

In using such a book, two points should be noted. Firstly, it is a book about *theory* – the sort of theory that experimenters should know and enjoy.

This means that it must be supplemented by illustrative material, which is provided in the text itself by a range of problems at the ends of the chapters: the student who tackles these problems energetically will not be short of illustrations of the ideas in action, but, unless he is quite unusually bright, he will need help and guidance in answering some of them. Secondly, although the book starts easily, in overall scope it is a final-year text. To make it more useful for elementary or specialised courses, I have arranged that certain subsets of ideas may be pursued in isolation. These subsets are identified under boldface headings in the index. In particular, there are two self-contained elementary courses, one on *classical thermodynamics*, which makes no use of the microscopic pictures of temperature and entropy, and one on *temperature, Boltzmann factors and elementary kinetic theory*, which does not involve the concept of entropy. There is also an abbreviated course on *chemical thermodynamics*. From Chapter 12 onwards the writing is aimed at students near the ends of their undergraduate careers, and a basic background in classical and quantum mechanics is assumed. In these later stages, and especially in the final chapter, I have not hesitated to expose the reader to advanced ideas, not normally regarded as undergraduate material, but without going into detail. At all stages of the book, sections marked with an asterisk may with advantage be omitted on first reading.

Argument concerning the true justification of statistical thermodynamics never ceases, and I shall not, I am sure, escape criticism from the experts for basing this book on Fermi's master equation. I have chosen to do so not because I believe the master equation to be the ideal logically fundamental starting point – as Chapter 19 shows, it is certainly not that – but because, from experience of teaching the subject, I find it the *only* starting point which is understandable at the beginning of the course, is not shrouded in mystery and does no violence to the essential physics. Moreover, it exhibits at the outset the intimate connection between kinetics and thermodynamics, giving the student confidence that he understands the nature of the argument, and it has for him a practical and familiar air, being directly related to the sorts of kinetic calculation which he is likely to have met in scattering or reaction theory. In the same spirit I have introduced statistical temperature (following recent practice in other books) by considering the physical properties of large heat baths rather than the mathematical properties of large ensembles, and defined statistical entropy from the outset as a measure of the logarithm of the number of quantum states over which the probability is spread, primarily because I know that students find definitions which are linked to an explicit physical picture the easiest to grasp.

It is a pleasure to thank Rex Godby, David Newson, Nick Safford, Charles Taylor and Chris Williams of Pembroke College, who took time from exam revision or research to read and comment on the various drafts, and my colleagues in the Cavendish Laboratory, Brian Pippard and Gil Lonzarich, who have reviewed the complete typescript, and, since they both know more about this subject than I do, have been well able to keep me on the rails. As is usual, they are not to be counted responsible for the outcome; and yet Professor Pippard must bear some sort of indirect responsibility, for he has been expounding physics to me for half my life, to my great pleasure, and I hope that he will continue in this task. Susan Cole and Jan Thulborn prepared the typescript, while at the University Press Simon Mitton, Sheila Shepherd and Jack Bowles guided the planning and organised the text for the printers.

A number of the problems at the ends of chapters are modified versions of Cambridge Tripos Examination questions or exercises which appeared originally in *Cavendish Problems in Classical Physics*, edited by A. B. Pippard and published by Cambridge University Press. I am grateful to the University, the University Press and Professor Pippard for permission to use this material, and to the American Physical Society, Dr B. J. Alder and Dr T. E. Wainwright for permission to reproduce Fig. 14.6. Where other figures and tables have been based on published work the sources are acknowledged in the text.

The happy tradition of dedicating even technical books is worth preserving. This one is respectfully offered to the two teachers who, at different stages, first made its subject intriguing to me. I hope that, in its turn, the book will prove useful to a new generation of students.

John Waldram,
Cavendish Laboratory,
March 1984

1

Introduction: thermodynamic systems seen from outside

1.1 Background ideas

This book is about thermodynamics, the study of heat. We shall start very near the beginning. I imagine that you, the reader, have some general knowledge of the subject. We all have an intuitive knowledge of hotness and coldness. Beyond that, I suppose that you are familiar with the idea of measuring hotness, or *temperature*, with a simple thermometer, some device whose properties are seen to change when it is placed in contact with a hotter or colder body. Given such a device – a mercury-in-glass thermometer, perhaps, in which the length of the mercury column increases with temperature – you know that we can set up an *empirical temperature scale* by assigning temperatures to two fixed points (as in the Celsius scale, where 0 °C is the temperature of melting ice and 100 °C is the temperature of boiling water), and then dividing the scale in some arbitrary way, by making temperature rise proportional to the increase in length of the mercury column, for instance.

Celsius temperature scale

You may have heard, too, of the *ideal gas scale of temperature*, defined by making the temperature proportional to the pressure exerted by a sample of gas of low density held in a container at fixed volume, and defining the triple-point temperature of water to be 273.16 K. (Ideal gas temperatures are given the symbol θ, and measured in units of *degrees kelvin*, or K. The value for the temperature of the triple-point of water is chosen so that this unit is equal in magnitude to the unit of the Celsius scale.) You may know that the temperature 0 K is known as *absolute zero*, and that this is thought to be the coldest possible temperature, but that there is no limit to how hot a body may be.

Ideal gas temperature scale

I assume also that you have met the idea that heat is a form of energy. We may, for instance, heat up a liquid by dissipating electrical energy in a heating coil. The heat required to do this may be measured electrically (in joules). The heat required to raise a body through unit temperature is called its *heat capacity* C and has units of J K^{-1}. Heat may also be needed to make a body *change phase* without changing the temperature – to melt ice, for instance. The heat required is known in this case as *latent heat* L.

Heat capacity

Latent heat

1.2 The approach to thermodynamics through statistical mechanics

In the eighteenth century some scientists thought that the phenomena of heat, like the phenomena of electricity, involved a new physical mechanism. Joseph Black, for instance, suggested that heat was a fluid known as 'caloric' which penetrated ordinary matter. At that time it seemed that the proper approach to the subject was to proceed empirically, to codify the experimental results into laws and to make deductions from those laws. We now know that this early approach was wrong in so far as it was looking for a new physical model. We have discovered, as Newton and Boyle had earlier suspected, that 'hot' bodies are simply bodies which contain a large amount of randomised energy, and that the atoms of which the bodies consist are perfectly ordinary atoms obeying the well-known laws of mechanics (or, rather, quantum mechanics). The laws of thermodynamics, therefore, though they first appeared as experimental laws, *ought to be deducible from the laws of mechanics.* Many textbooks still approach classical thermodynamics as an experimental, empirical subject, and the reasons are not wholly historical. The empirical approach to classical thermodynamics is logically very satisfying and, because it establishes rigorous links between thermodynamic facts without explaining the facts themselves in detail, it helps to emphasise the generality of the results in the student's mind. But, in my experience, it also has bad effects. Students, though they see the subject as beautiful, may also find it mystifying. It feels strange to be handling concepts such as temperature and entropy, which are clearly important, without having any idea of what they mean at the atomic level. Moreover, those students who do go on to study the mechanistic interpretation – 'statistical thermodynamics' – usually find it hard to bridge the gap between the two approaches. In this book, therefore, the approach is mechanistic – we *start* with the mechanical picture and derive the laws of thermodynamics from the mechanics. This means that the early part of the presentation may seem a long way from the everyday world of experiment. But I do not think that the approach is harder to understand than the traditional one, and I believe that in the long run it pays handsome dividends. In this introductory chapter, before plunging into the details of the mechanistic picture, we make some general comments on the nature of thermodynamic systems, how they behave as seen from outside, and what the tasks of the mechanistic theory will be.

What, then, is a thermodynamic system, and in what way does thermodynamics differ from mechanics (or quantum mechanics)? The answer is simple. There is no *fundamental* difference between a mechanical system and a thermodynamic one, for the name just expresses a difference of attitude. By a thermodynamic system we mean a system in which *there are so many relevant degrees of freedom that we cannot possibly keep track of all of them.* We think of the solar system of the Sun and nine planets as a mechanical system because, using Newton's laws of gravitation and mechanics we can in practice, knowing the starting positions and velocities of all the bodies in the system, make good

Thermodynamic systems

predictions of the future motions of all the bodies. We think, on the other hand, of a box of helium gas containing 10^{23} atoms as a thermodynamic system because, even if we had values for the starting positions and velocities for every atom, or the equivalent wave function for the gas, we could never in practice predict their future behaviours, partly because the calculations would be prohibitively long (even with the largest conceivable computer) and partly because we could never hope to have sufficiently accurate values of the starting conditions. The state of the helium gas is just as predictable *in principle* as the state of the solar system – we know the forces and the laws of motion. But in practice the prediction cannot be done because there are too many variables.

It follows that any theory of the gas has to be statistical in nature. We cannot speak of where a particular atom is, but we may speak of the *probability* of finding it in a given position and moving with a particular velocity. We shall examine the meaning of probability in thermodynamic systems more closely in the next chapter. Here we just note that our theory of the gas must proceed by the following route. Because we do not start with exact knowledge of the state of the gas, we have to assume, at some suitable starting point in time, a plausible or reasonable description of the system in terms of probabilities – known as the *initial probabilities*, the probabilities already present when we first meet the system. Then, from our knowledge of mechanics or quantum mechanics, which shows how the system would develop from each of its possible starting states, *we can find how the theoretical probabilities change with time*, and hence make predictions (which can themselves only be in terms of probabilities) of the properties of the system at future times. For instance, if the gas is known originally to have an uneven density, we represent this by a corresponding uneven probability distribution for the positions of the molecules, and, knowing how the molecules move, we can compute the probability distribution for their positions at later times. Knowing the theoretical distributions at later times, we can calculate how observable quantities, such as the pressure at a point in the gas, would be expected to behave on average. Because such predictions are essentially statistical, we can never make an *exact* prediction of the pressure. This is, however, a less serious limitation than might be expected. It turns out, as we shall see, that for many quantities the predicted statistical distributions are so sharply peaked that the statistical prediction is, in effect, an exact prediction for most practical purposes.

§2.1

The statistical method in outline

1.3 The thermodynamic state

What do we mean when we say that some system is in a definite *thermodynamic state*? First of all, to specify the state at some particular instant we must fix the *constraints*. A constraint means in thermodynamics some parameter of the system which can be held fixed or varied at will by the observer. The volume of a gas and the electric charge on a capacitor are examples of constraints. We usually treat a constraint as having a definite and

Constraints

exact value which can be fixed from outside the system (though this is an idealisation).

For a given set of constraints, the system still has access to a very large number of *microscopic states* – for a gas held at fixed volume, for instance, the molecules could be started at time zero in many different positions and with many different velocities. As time proceeds and collisions occur in the gas the microscopic state is continually changing: we say that it is subject to *thermal fluctuations*.

Microstates

Fluctuations

For a given set of constraints there are also many properties of the system which can be measured – the *thermodynamic observables*. Some observables may be thought of as responses to particular constraints. The pressure, for instance, is the system's response to the volume constraint, and the voltage between the capacitor plates is the system's response to the charge on them. Other measurable quantities such as the refractive index or the X-ray diffraction pattern are less obviously related to particular constraints. These are all observables.

Observables

Since the measured value of an observable must depend on the microscopic state of the system, the values of the observables will also be subject to thermal fluctuations. If sampled many times over a short period all observables have a *fluctuation distribution* (Fig. 1.1). The observable will have a mean value, but will sometimes lie above, and sometimes below, the mean. Where the observable involves the combined effect of a large number of particles (as with the pressure of a gas on a large area) its fluctuation

Fig. 1.1. The distribution of fluctuations in the pressure P of a gas on a very small area. The pressure is due to random bombardment by gas molecules and therefore wanders around the mean value. Such distributions are described by a *distribution function* $f(P)$, which is defined so that the probability of finding P in the narrow range between P and $P+\mathrm{d}P$ is equal to the corresponding area under the curve, $f(P)\,\mathrm{d}P$, which is shown shaded. The total area under the curve is unity. The mean pressure $\bar{P}$ is $\int Pf(P)\,\mathrm{d}P$.

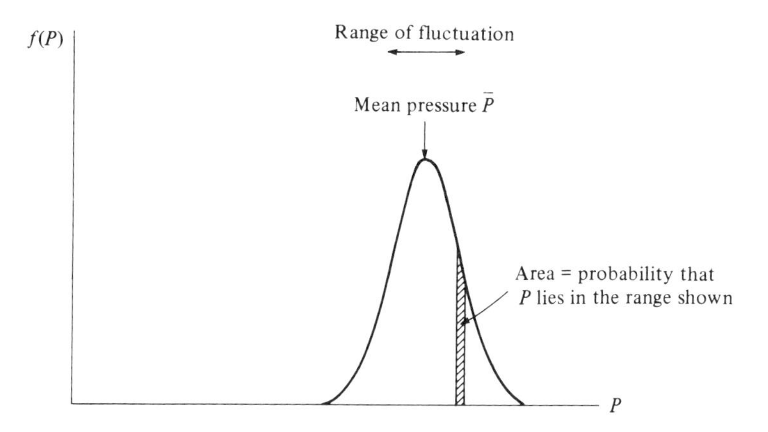

distribution will normally be narrow and the fluctuations fairly insignificant. If only a few particles are involved (as with the density measured over a small volume) the fluctuations may be relatively large. We shall regard the system as being in a definite thermodynamic state when its observables take definite (but not necessarily equilibrium) values. In statistical terms this means that *in a definite thermodynamic state we know the probabilities of the various micro-states sufficiently accurately to make good predictions of the values of the observables*, averaged over their fluctuations.

Thermal equilibrium

We are often interested in systems in *thermal equilibrium*. What does this mean? In essence, we say that a system has reached thermal equilibrium when its observables have ceased to change with time. For instance, a gas whose molecules are all at one end of the container is not in thermal equilibrium because the density at that end of the container – an observable – will be falling with time. But such a description of equilibrium has to be interpreted cautiously. Because of the fluctuations, we cannot define equilibrium by saying that the observables are strictly constant. Instead, we have to say that *the system is in thermal equilibrium when the statistical distributions of the observables, averaged over their fluctuations, are constant*. In a few cases the time to reach equilibrium may be so long that we risk misinterpreting the situation. A piece of glass, for instance, may appear to be in thermal equilibrium. Actually, it is not; the true equilibrium state is crystalline, but glass has no crystal structure. It takes many thousands of years for glass to crystallise, and in this case even the fluctuations in the arrangement of atoms are so slow that we do not notice them. Strictly speaking, to define the *full* thermodynamic equilibrium state for glass we ought to average over the fluctuations for millions of years, watching all the time to see when the distribution, averaged over the very slow fluctuations, has itself settled down to a statistical equilibrium state.

What fixes the nature of the equilibrium state?

Once the system has reached equilibrium, the observables all take definite mean values. It is important to ask what fixes these equilibrium values, apart from the constraints. (We know, for instance, that if we put a given quantity of gas into a given volume it does not always show the same pressure in equilibrium.) The answer given by experiment is a remarkable one. We shall see in the next section that the *internal energy*, the total energy inside the system, is a well-defined quantity. What experiment almost always shows is that *the nature of the equilibrium state is fixed entirely by the constraints and the value of the internal energy*. Although the system may have started from a great variety of different and macroscopically distinct non-equilibrium states, most of the details of these states have no effect on the nature of the equilibrium state. Evidently the randomising process which goes on as equilibrium is achieved is very effective, because all details of the starting state, except its energy, are lost. *All* the observables, such as the pressure, the temperature, the dielectric constant, the X-ray diffraction pattern, etc., are fixed once the constraints and the energy are fixed. In the equilibrium state the number of

§§2.5, 19.1

thermodynamic degrees of freedom is equal to the number of constraints plus one. If we fix the constraints and one other quantity, such as the temperature, this has the effect of fixing the energy and hence the values of all other observables. Thus the nature of the equilibrium state is fixed by a relatively small number of parameters.

1.4 Energy in thermodynamic systems

Modern physics has as its cornerstones the theories of quantum mechanics and relativity. The definition of energy and the fact that energy is conserved are built into the structures of these theories, and we shall not therefore spend long in discussing these two concepts here. The situation is straightforward. When forces act within a system of particles the net work done by these forces is equal to the increase in the *kinetic energy* ($\frac{1}{2}mv^2$ summed over all particles in the non-relativistic limit). It can also be shown that, for all the forces known to physics, the work done by the forces is equal to the decrease in the corresponding *field energies*. Field energy is a form of energy which we imagine to be associated with a particular configuration of fields (such as the electric and magnetic fields) and is usually thought of as being stored in space with a definite energy density. The mutual gravitational potential energy of two massive bodies, the mutual electrostatic energy of a pair of electric charges and the energy of a light wave are all examples of field energy. Clearly, then, the sum of the kinetic energies and the field energies – the total energy – is conserved. A thermodynamic system will have inside it a definite *internal energy*, E, which is the sum of whatever particle energies and field energies are regarded as being part of the system. (We note in passing that a thermodynamic system may consist entirely of field energy: the infra-red radiation in an evacuated oven is a thermodynamic system.)

Internal energy

Scientists before the mid-nineteenth century would have been very surprised at this conclusion. They were familiar with *dissipative forces*, forces such as friction or viscous drag, which seemed to destroy or dissipate energy. If we stir a viscous liquid with a rotating paddle we certainly do *external work*, w, on it. But we also return to our starting point, so the work done cannot be stored as potential energy, apparently. Has it been lost? In a famous series of experiments Joule and others showed that in such cases *we do not in fact return to our starting point because the thermodynamic state is altered* – typically, some part of the system is hotter because all dissipative processes generate heat. Joule was able to show that a definite amount of dissipative work was always associated with a given change of state, showing that the energy was not lost, but could be accounted for in the thermodynamic change. We now know what happens. *In a dissipative process the work done is converted into the energy of disorganised small-scale motions.* The energy is not lost. It is there on an atomic scale, though it cannot be recovered in any direct way.

Dissipative work

Energy may also enter a thermodynamic system in two other ways. First, we find that when two systems are placed in *thermal contact*, energy frequently flows spontaneously from one system to the other, without action by the observer. This can occur by various mechanisms, such as the transfer of vibrational energy between one solid and another whose surfaces are in contact, or the exchange of electromagnetic radiation. Such a spontaneous movement of energy is called a *heat flow*, and heat entering one body from another is written q. (The word 'heat' is simply a name for internal energy when we are thinking of the energy as randomised and free to move from one part of a system to another or between systems. The use of this special term can be rather misleading. There is no special part of the internal energy which can be identified as the 'heat', and it must not be thought that energies entering as work and as heat are stored differently inside the system.) Steps may be taken to prevent heat flow, by putting a body into a vacuum flask, for instance. Such a body is said to be *thermally isolated*. We shall refer to changes which occur inside thermally isolated systems as *adiabatic processes*. (Note carefully that authors differ over their use of this term, some reserving it for *reversible* processes in thermally isolated systems. We shall retain the earlier, more general sense.)

Heat flow

Adiabatic processes

We note finally that, because the system usually exerts a force on any constraint, work may be done and energy may enter the system when a constraint is altered. For instance, a gas responds to the volume constraint by exerting a pressure P, and, if the volume changes, the work $\mathrm{d}w$ done on the gas by the walls of the container is equal to $-P\,\mathrm{d}V$. There is an important difference between this type of work and the dissipative work discussed above. Dissipative processes such as stirring a viscous liquid or driving an electric current through a resistor are *irreversible*: once the work is done there is no way of recovering it by reversing the process. But imagine that we take a system, either alone or in thermal contact with a larger system which acts as a heat reservoir, and make a small and very slow frictionless change of constraint, such as an increase of volume, which we subsequently reverse. Since there is no friction we do no net work in this process, for the pressure remains at all times equal to the equilibrium pressure. Thus at the end of the process the constraints and the energy of the system (or system plus reservoir) return to their original values, and the state of the system is unchanged. The original slow and frictionless expansion is therefore *reversible*. We shall see later that if the gas is in contact with a heat reservoir during the expansion, heat will enter it from the reservoir, and leave it again if the gas is subsequently compressed. Evidently such heat flows which occur during a slow and frictionless change of constraint are also reversible.

Reversible work and heat flow in a slow, frictionless change of constraint

§ 1.3

§ 5.6

To summarise, then, energy may enter a thermodynamic system in three distinct ways:

(i) dissipative work, without change of constraint,
(ii) heat flow, and
(iii) reversible work, by slow, frictionless change of constraint.

Note that a *fast* movement of a constraint is not usually fully reversible, because the force involved may not be equal to the equilibrium force: in a sudden compression, for instance, the pressure may be greater than the equilibrium pressure. Such a process is best regarded as a combination of reversible and dissipative work. Evidently the change of internal energy of a system may always be written as

$$\mathrm{d}E=\mathrm{d}w+\mathrm{d}q, \tag{1.1}$$

First law of thermodynamics

where $\mathrm{d}w$ is the total work done on the system, either by reversible movement of the constraints or as dissipative work, and $\mathrm{d}q$ means the heat entering spontaneously from other systems. In thermodynamics, the principle of conservation of energy is known as the *first law*. Clearly, the first law can be expressed by Equation (1.1) combined with the *principle of conservation of heat*: heat leaving one body must reappear as heat gained by some other body or bodies.

Conservation of heat

1.5 Empirical temperature

Temperature is the thermodynamic parameter of which we have most intuitive knowledge: we are all aware of the sensations of hotness and coldness. But what does it mean to say that A is hotter than B? Certainly, temperature is connected with heat content and, generally speaking, adding heat to a system makes it hotter. But there are exceptions to this rule (adding latent heat to melt ice does not change the temperature), and the temperature change is not connected in any simple way with the amount of heat added. The real significance of temperature is that *it determines the direction of heat flow.* It is a remarkable fact that all equilibrium states of all systems can be placed in a sequence from 'cold' to 'hot' such that, when any two are placed in thermal contact (so that heat can pass between them) the heat *always* flows from the hotter body to the colder body, and *never* in the opposite direction. This is, of course, the basis of our physiological sense of hot and cold. A *hot* body is one from which, when we touch it, heat flows into us; a *cold* body is one which extracts heat from us. The temperature sensors in the skin respond to the change in heat density in the skin itself. We also make use of the same principle when we use a thermometer. When a thermometer is placed in contact with a warmer body, heat flows out of the body into the thermometer. In general, this makes the thermometer warmer and the body colder until they reach the same *temperature* – the same place in the sequence from cold to hot. We then (when we adopt an empirical temperature scale such as the mercury-in-glass or the perfect gas scale) use some visible parameter (such as the length of the mercury thread or the pressure of the constant volume gas thermometer) simply as a convenient label for the place on the sequence from cold to hot which the thermometer itself has reached when it is in the state concerned.

T fixes the direction of heat flow

Experience has shown physicists that the unwillingness of heat to flow from a colder to a hotter body applies not only to direct transfers of heat but also to the most elaborate indirect processes, and this conclusion is embodied in the *second law of thermodynamics*:

Second law of thermodynamics

> No process exists in which heat is transferred from a colder to a hotter body and the rest of the world is left unchanged at the end of the process.

The final condition is important. In a refrigerator, for instance, heat is made to pass from a colder to a hotter body at the expense of doing work, which also appears as heat in the hotter body.

1.6 The tasks of a mechanistic theory

Some of the tasks of a mechanistic theory of heat will by now be apparent. We shall need a formalism suitable for describing the statistical state of any thermodynamic system. The formalism will have to show how, starting from an arbitrary state, systems move towards thermal equilibrium, and how fast they do so. It will have to be able, for a given energy, to predict the mean values and distributions of all possible thermodynamic observables in the equilibrium state, and to show why this state is unique. It will have to provide some means for defining and calculating from first principles the temperature of an equilibrium state, and explain the extreme reluctance of heat to flow from colder to hotter bodies. We should also expect it to account for the magnitude and frequency spectrum of thermal fluctuations.

It may be useful to mention here that the theory divides itself into three parts (Fig. 1.2). The question of how the statistical state changes with time is the subject of *kinetics*. This is the most detailed branch of thermodynamics, and involves a knowledge not only of the possible states of the system but also the mechanism by which it moves from one state to another. It is a large subject, but we shall touch on its applications only lightly in this book, in Chapters 2, 5, 16, 17 and 18. The question of the nature of the equilibrium state and the distribution of fluctuations within it is the subject usually labelled *statistical thermodynamics*. It depends on kinetics (in the sense

Fig. 1.2. The logical structure of thermodynamics, and where to find its parts in this book.

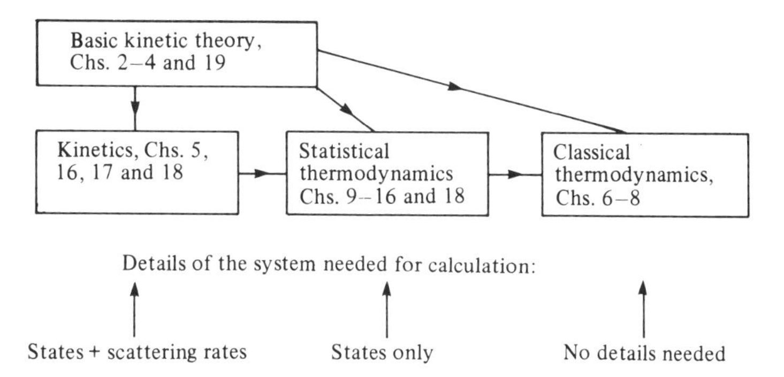

that only kinetics can show which states are in equilibrium) but it turns out not to involve the details of how the system gets from one state to another, and this makes it rather more general and less system-dependent than kinetics. Certain general results of statistical thermodynamics do not depend on the details of the system concerned at all. These general results constitute the subject known as *classical thermodynamics*, traditionally derived empirically, but in this book deduced from the concepts of statistical thermodynamics. Classical thermodynamic concepts and calculations appear in Chapters 6, 7 and 8 and are used also in later chapters. The statistical foundations and the fundamental concepts of temperature and entropy on which the whole subject depends are developed in Chapters 2, 3 and 4 and briefly reassessed in the final chapter.

2

The statistical foundations

In this chapter we shall explore the fundamental ideas concerning probability and the probabilities of jumps between quantum states upon which our main arguments will rest.

2.1 Probability

I shall assume that the idea of probability is already familiar. If you have not met it before it would be helpful to consult an elementary text on statistics before proceeding. Let us remind ourselves that the notion of probability is concerned with situations in which some experiment or *trial* can be repeated many times, and we are interested in how often some particular result or *outcome* occurs. The *probability* of outcome X means simply the fraction of trials in which X is expected to occur. For instance, we might set up an experiment in which we drop a dart from a fixed point, and record whether it falls within a circle marked on the floor beneath. If I say that the probability of falling within the circle is 0.12, I mean that I expect the dart to fall within the circle in 12% of the trials. At first sight this statement is rather vague, for if I make 100 trials, I do *not* mean that I expect the dart to fall in the circle *exactly* 12 times, but sometimes more and sometimes less. However, we may use the phenomenon known as *statistical convergence* to make the statement more precise. We usually find that if we make a *very large* number of trials the proportion of successes becomes more consistent: if, for instance, I make a *million* trials, and get a 12% success rate, I can be very confident that if I make a *further* million trials I shall again find a 12% success rate, or very close to it. This statistical convergence may be confirmed both experimentally and theoretically. Thus, more precisely, we may say that the *probability* of getting outcome X means *the fraction of trials in which X occurs when the number of trials is very large.* This fraction is normally consistent, and thus the probability is well defined.

Ensemble probability

In dealing with thermodynamics we frequently wish to discuss probabilities which are *changing with time*, and in that situation it is important to be clear about how the trials are imagined to be made. Suppose, for instance, that we have two identical cylinders, A and B. At time zero, A contains a single gas molecule, and B is empty. I then open a valve in a narrow tube connecting the cylinders. I might now assert that 'after one second the probability that the molecule will be found in B is 0.2'. What does a statement of this sort mean? Obviously, our repeated trials to find out whether the molecule is in B cannot be spread out in time after opening the valve on a single apparatus, for the probability is clearly changing with time. What we really mean is that if we set up a very large number of replicas of the apparatus, each with the molecule in A, and at time $t=0$ open each of the valves, then after one second we expect to find the molecule in B in 20% of the replicas. (Alternatively, we could say that, if we repeated the experiment a large number of times, then we expect to find the molecule in B one second after opening the valve in 20% of the repeats.) Such an imaginary set of replicas (or repeats) of the experiment was called by Gibbs an *ensemble*. In this book, when we speak of a 'probability' or 'mean value', it will normally be understood that the population over which the probability or mean value is to be calculated is a suitably chosen ensemble at a definite time after the start of the experiment. Using this understanding, we may speak of a 'probability' or 'mean value' which changes with time.

Of course, the ensemble does not exist. It is just a vivid way of saying what we mean when we use probabilities to describe the state of the system at a given moment. When we use a probability (or an ensemble) we are admitting that our detailed knowledge of the state of the system is incomplete. If we had known at $t=0$ *exactly* where the molecule was in A and how it was moving, we should have been able to calculate its exact trajectory, and would thus have *known* at any moment whether it was in A or in B: there would have been no need for probabilities. In practice we never have this sort of detailed knowledge of a thermodynamic system. The best we can do is to represent its state by a suitably chosen ensemble of systems (or the equivalent probabilities) which corresponds to what we know of the detailed microscopic states.

Let us see how this approach may be used to make statistical predictions. For instance, I might wish to *predict* the probability of finding the molecule in B after one second. To do this I must first determine or guess the *initial probabilities*, that is, the nature of the ensemble at $t=0$. For instance, if I know that at $t=0$ the molecule has speed c, but am otherwise ignorant of its microscopic state, it would be reasonable to assume that in the starting ensemble the molecule is uniformly distributed within the volume of A, and that the direction of its motion is also uniformly distributed. (These are sensible assumptions, because one can show that, when averaged over time, this is how the position and velocity would be distributed in a closed box; and we know that there is nothing special about the time at which we open the valve.) Then, for each state in the starting ensemble I could calculate the trajectory, and thus find in what fraction of cases the molecule would be in B after one second. In other words, I make an assumption about the initial probabilities, and then, using mechanical arguments, find how the ensemble,

Calculations are always based on some assumption about the initial ensemble

§ 19.2

and hence the probabilities, change with time. This will be our normal procedure in thermodynamic arguments. We shall invariably describe the thermodynamic state in terms of probabilities and, when we calculate the 'theoretical value' of a thermodynamic observable, we shall be calculating its mean value, averaged over whatever ensemble we have chosen.

2.2 Manipulations involving probability

When some object has a number of discrete possible states with label i, we define the probability p_i of finding it in state i by writing $p_i = N_i/N$, where N_i is the number of times the object is found in state i in a very large number of trials N. Since $\sum_i N_i = N$ we see that

$$\sum p_i = 1, \tag{2.1}$$

the sum of the probabilities for all possible states is unity.

The distribution function

Sometimes we wish to express the probability for some quantity which has a continuous range of possible values. In the dart experiment of §2.1, for instance, we might wish to express the probability that the dart falls at some particular distance r from the centre of the circle. In such a case, the chance that this distance takes some particular value r *exactly* is, of course, zero. What we really wish to express is the probability that the dart falls *in some narrow range*, between r and $r + \mathrm{d}r$ from the centre. The natural way to do this is to use $\mathrm{d}p/\mathrm{d}r$, the probability *density* per unit range of r, usually referred to as the *probability distribution function* for r, $f(r)$. This function is so defined that the probability $\mathrm{d}p$ that the dart will fall in the required narrow range is given by

$$\mathrm{d}p = f(r)\,\mathrm{d}r. \tag{2.2}$$

A typical distribution function is illustrated in Fig. 1.1. Once again, the sum of the probabilities for all possible ranges of r must be unity, so

$$\int \mathrm{d}p = \int_0^\infty f(r)\,\mathrm{d}r = 1, \tag{2.3}$$

the total area under the distribution curve must be unity.

We shall frequently be involved with *averages* or *mean values* of some quantity x taken over a large sample of size N. If the successive values found in the large sample are $x_1, x_2, \ldots, x_N$, the mean value $\bar{x}$ is defined as

Expressions for mean values

$$\bar{x} = \sum_{n=1}^{N} x_n/N. \tag{2.4}$$

This definition may be rewritten in terms of probabilities as

$$\bar{x} = \sum_i N_i x_i/N = \sum x_i p_i, \tag{2.5}$$

where x_i is the value of x in state i. If x is a continuously variable quantity then p_i has to be replaced by $\mathrm{d}p = f(x)\,\mathrm{d}x$, and we have instead

$$\bar{x} = \int x f(x)\, dx. \tag{2.6}$$

The averages of functions of x may be found in the same way. The mean value of x^2, for instance, written $\overline{x^2}$, is defined as $\sum x_n^2/N$, and by similar arguments we find that $\overline{x^2}$ is equal to $\sum x_i^2 p_i$ for a discrete variable and $\int x^2 f(x)\, dx$ for a continuous one.

Finally we need to mention the concept of *statistical independence*, which arises when we make more than one test at the same time. For instance, in the dart experiment we might draw a straight line passing through the centre of the circle on the floor, and, as well as measuring the probability of falling inside the circle, we could also measure the probability that the dart falls to the left of the line. We might, for instance, find that for *all* trials the dart fell to the left in 50% of cases. We should then say that the *absolute probability* of falling to the left of the line is 0.5. But we might alternatively measure the probability of falling to the left, using as our population only those cases in which the dart fell inside the circle: this type of probability is a *conditional probability*. If the conditional probability of falling to the left is the same as the corresponding absolute probability, we say that the outcome 'falling to the left' is *statistically independent of* or *uncorrelated with* the outcome 'falling inside'. In other words, the chance of falling to the left of the line is not affected by whether the dart falls inside or outside the circle. In such a situation we may be interested in the *joint probability* that the dart simultaneously falls inside the circle *and* to the left of the line. If these two outcomes are uncorrelated we may say that the fraction falling inside the circle is the corresponding absolute probability p_I, and *of these* the fraction falling to the left will be the corresponding absolute probability p_L. Thus the joint probability p_IL of falling both inside and to the left is given by

Statistical independence

Joint probabilities

$$p_\mathrm{IL} = p_\mathrm{I} p_\mathrm{L}. \tag{2.7}$$

In other words, the joint probability is equal to the *product* of the separate absolute probabilities so long as the two outcomes concerned are uncorrelated.

We shall use the manipulations summarised in this section quite frequently.

2.3 Quantum states and their distribution in energy

The statistical treatment of thermodynamics may be developed using either classical or quantum mechanics. In this book we shall use a quantum treatment at most points, partly because the real world obeys quantum mechanics and partly because it is in some ways easier to do so. For systems in the classical limit it is sometimes useful to know the classical version of the arguments, and that will be discussed, briefly, in Chapter 13.

Systems described by quantum mechanics do not usually have arbitrary internal energy: when isolated from the rest of the world they exist only in certain well-defined *energy states*. It is, for instance, well known that a

Energy states

one-dimensional harmonic oscillator of frequency ν has quantum states of energy

$$\varepsilon_n = (n + \tfrac{1}{2})h\nu, \tag{2.8}$$

Zero-point energy

where h is Planck's constant and n is an integer which is positive or zero. The lowest-lying state, with energy $\frac{1}{2}h\nu$, is called the *ground state*. The energy of the ground state is known as *zero-point energy*. Energy may only be added to such an oscillator in fixed amounts or *quanta* of $h\nu$: the integer n is the number of quanta which have been added in raising the energy above the ground state energy. A hydrogen atom, on the other hand, has a more complex set of energy states described by the three orbital quantum numbers n, l and m_z (plus others which describe the spin states of the electron and the proton). The energies are approximately proportional to $-1/n^2$, and are therefore not equally spaced in this case. One also finds that some distinct states of the hydrogen atom have exactly the same energy: such states are said to be *degenerate*.

Approximate energy states in large systems

If we turn to a complete thermodynamic system, such as a container of liquid or a lump of copper, it remains true that the system has some definite, and very large, number of energy states. In principle we could determine the nature of these states and their energies by analysing the quantum mechanics of the complete thermodynamic system. In practice, however, such a programme is completely impossible. In a large system the *exact* energy states are always exceedingly complicated, and we could never hope to find them and do practical calculations on them. Instead, we always have to work with *approximate* energy states, usually the true states for some simplified system of forces. For instance, we often describe a gas by using states in which the individual molecules each have a definite momentum. These *would* have been the true energy states if there had been no collisions. The forces which act between the molecules modify the situation: we may think of them as a *perturbation*, an extra force ignored in the simplified treatment. The presence of the perturbation leads to scattering. Thus the states of definite momentum are no longer completely stable, and the perturbation causes occasional *quantum jumps* in which a pair of molecules collide and exchange momentum, so altering the overall state of the gas.

In general, then, we shall describe a thermodynamic system in terms of some set of *approximate energy states*, with label i. We shall suppose that the approximation used is a good one, so that when the system is in state i we may write the energy as E_i without appreciable error. We shall always suppose, however, that there is some residual perturbation which causes quantum jumps between the different states i. Note that the states here described are states of the system *as a whole*, and not the states of the individual particles inside the system. (To emphasise this point we write E_i with a capital letter.) It is important to notice that, for a large system, the nature of the states i will depend on the constraints, such as the volume or the magnetic field. In particular, if we change the constraints on the system we shall usually change the energies E_i of the various states i.

For a small system, such as an oscillator or a hydrogen atom, we may identify the various quantum states and their energies without too much difficulty. But to understand the next two chapters it is important to realise that for a *large* system, such as a liquid or a lump of copper (containing, say, 10^{23} atoms), the situation is quite different in two important respects. The first difference is that the states are *very* much closer together in energy. For the oscillator the spacing between the states is $h\nu$ – perhaps 10^{-20} J for a small oscillator. For a typical large system the mean spacing between states is of order $10^{-10^{23}}$ J. This spacing is so *extremely* small that we cannot possibly enumerate and work with the individual states. The best we can do is to work with the *density of states in energy*, $g(E)$, which is dn/dE, the number of states per unit energy range. (The phrase 'density of states' is usually taken to mean the density of states in energy unless some other variable is specified.) For a large system the density of states may safely be taken to be a smooth function of energy on the atomic scale of energy.

The density of states

The second significant difference is that *the density of states rises very rapidly with energy in a large system*. For a *single* oscillator, the density of states is independent of energy and is equal to $1/h\nu$. But consider, as a rather artificial illustration, a system consisting of 100 identical oscillators. If it contains no quanta, there is only one possible state – each oscillator must be in its ground state. For one quantum there are 100 possible states, corresponding to the 100 possible ways of distributing the quantum amongst the 100 oscillators. For two quanta there are 5050 states (4950 with the two quanta in different oscillators and 100 with both in the same oscillator). It will be clear from Table 2.1 how rapidly the density of states g (which is $W/h\nu$, where W is the number of states for a fixed number of quanta) rises with the addition of a small number of quanta. The general formula for W when we have Q quanta in N oscillators is

$$W = \frac{(Q+N-1)!}{Q!\,(N-1)!}. \tag{2.9}$$

Problem 4.3

Table 2.1. *The number of ways, W, of arranging increasing numbers of quanta in a system of 100 identical oscillators*

Number of quanta	Number of arrangements
Q	W
0	1
1	100
2	5050
3	171 700
4	4 421 275
5	$\sim 10^{8.0}$
10	$\sim 10^{13.6}$
20	$\sim 10^{22.4}$

Note that each time we add a quantum g is *multiplied* by a factor $(Q+N)/(Q+1)$. If N and Q are large, this factor only changes slowly with Q, which means that g is rising roughly *exponentially* with energy. This is typical of large systems. Indeed, as we shall see in § 3.3, any system in which g did *not* rise roughly exponentially would have a negligible heat capacity, and we should not regard it as large in a thermodynamic sense. So *the density of states rises roughly exponentially with energy in all thermodynamically large systems.* This very rapid, nearly exponential, rise in g with energy means that it is usually more convenient to work with $\ln g$ rather than g itself. We show in Fig. 2.1 how $\ln g$ rises with E for our system of 100 oscillators. This system is still quite small and its density of states is not quite uniform on the atomic scale; nevertheless, the behaviour of $\ln g$ with increasing E is typical of large systems: $\ln g$ rises with increasing E, but the slope $\mathrm{d}(\ln g)/\mathrm{d}E$, which starts large for small E, gradually decreases. Notice the enormous magnitudes of W. You may like to confirm, using Equation (2.9) and Stirling's approximation, that for a system of 10^{23} identical oscillators containing 10^{23} quanta, g is of order $10^{10^{23}}$. (Stirling's approximation states that $n! = \sqrt{(2\pi)}n^{n+\frac{1}{2}}\mathrm{e}^{-n}$ to order $1/n$. When n is large we may approximate this further and write $\ln n! \simeq n(\ln n - 1)$.)

Stirling's approximation for $n!$

2.4 Quantum jump rates and the master equation

In practice we never know the quantum state of a large system precisely. A more practical description of it is to quote the *ensemble*

Fig. 2.1. Variation of $\ln g$ with energy for a system of 100 identical oscillators of frequency ν. Although quite small, this system is large enough for us to treat the states as essentially continuous in energy, and to treat $W/h\nu$ as the density of states in energy, g. For convenience the quantum energy $h\nu$ has been used as the unit of energy. Notice the enormous magnitudes of W.

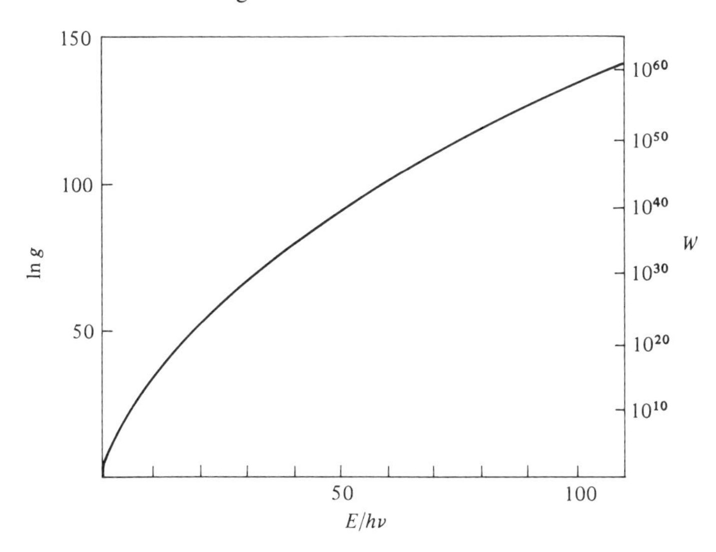

probabilities $\tilde{p}_i$ of finding the system in each of its various possible approximate energy states i. This description of the state of the system in terms of the $\tilde{p}_i$ values is the basic formalism of statistical thermodynamics, and when the $\tilde{p}_i$ values are known, we shall regard the system as being in a well-defined thermodynamic state. This situation is sometimes described by saying that *we define the thermodynamic macrostate by giving the probabilities of the various possible microstates*, or quantum states.

The microscopic probabilities $\tilde{p}_i$

§19.4

In general, the $\tilde{p}_i$ values will be changing with time – we use the superscript ~ as a reminder that the quantity so denoted may be altering, and does not necessarily correspond to the standard equilibrium condition of contact with a large heat reservoir at a definite temperature. The rate at which the $\tilde{p}_i$ values change with time is a matter of *quantum kinetics*, and to understand the kinetics it is helpful to start with the picture shown in Fig. 2.2 in mind. Here, A is some group of *very similar states*, spread over a small but finite range of energies δE, and B is another similar group. If the system is large we may assume that these groups of states are packed densely in energy, with densities of states g_A and g_B which we take to be uniform over the narrow energy range δE. Let the system be thermally isolated, and suppose that it is initially in some state i of group A. Since i is not an exact energy state we must expect the residual perturbation to cause quantum jumps out of i into the various states of group B.

The kinetic picture

There appears to be a difficulty about this picture. Since the system is isolated its energy will be conserved. Thus, although the states B are packed very densely, there is no guarantee that any one state in group B has *exactly* the energy of the initial state i, so at first sight it is not clear whether a quantum jump can occur at all. But to say this is to forget that the energies E_i are only approximate energies. In fact, using time-dependent perturbation theory we

The problem of energy uncertainty

Fig. 2.2. The accessibility convention. In reality, because of the uncertainty principle, quantum jumps from states in group A to states in group B are confined to a narrow range of energy of width $\Delta E \sim h/t$, which decreases with time. According to the accessibility convention we treat them as being spread evenly over group B. If $\tilde{p}_A$, g_A, g_B and the jump rates are independent of energy within the range δE this makes no difference to the rates of arrival in B and departure from A.

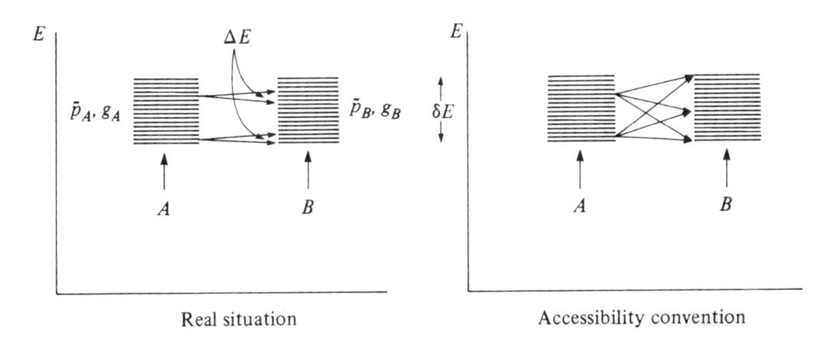

find that, if the system is initially in state i, the perturbation drives it into a *mixture* of states, in which we cannot say that the system lies in any particular state of group B, but there is some probability $\tilde{p}_j$ of finding it in several of them. This probability is concentrated in those states j of group B which are closest in energy to E_i, but is spread over a finite range of energies ΔE: thus, although the *exact* energy is conserved, a little blurring occurs in the *approximate* energies, and there is no difficulty in finding possible final states within group B. The value of ΔE depends on the circumstances. It is related to Heisenberg's energy uncertainty principle, and in a typical case would be less than 10^{-27} J: we shall simply assume that $\Delta E \ll \delta E$.

This slight blurring of the allowed final states makes it understandable that a transition from state i into one of the states of group B can occur, but also means that we do not know exactly *which* state of group B the system has entered. In fact the quantum calculation shows that the rate at which probability is transferred from state i into one or other of the states of group B, the *one-to-many jump rate* ρ_{BA}, is given by

Jump rates and the accessibility convention

§ 19.6

$$\rho_{BA} = 2\pi |H_{BA}|^2 g_B/\hbar. \tag{2.10}$$

This result is known as *Fermi's golden rule*. Here $\hbar$ is Planck's constant divided by 2π and H_{BA} is the *matrix element* for the transition: it is a complex number having dimensions of energy whose value depends on the details of the perturbation.

The difficulty of not knowing precisely the energy of the final state in group B may be circumvented by the following formal device. In practice we do not know in which state of group A the system lies at the outset: on the contrary, if δE is small the probability $\tilde{p}_i$ is likely to be spread *uniformly* over the states i of group A. When this is so the probability $\tilde{p}_j$ will be rising at the same rate for every state j of group B. It then makes no difference whether we treat the transitions out of any one state i as restricted to the real range ΔE or as spread evenly over all the states of group B: in either case the value of $\mathrm{d}\tilde{p}_j/\mathrm{d}t$ will be the same (Fig. 2.2). We therefore adopt the convenient fiction that the transitions from every state i of group A are spread uniformly over the states of group B which lie within the conventional *accessibility range* δE: this is the *accessibility convention*. It is, of course, not literally true, but it represents the rates of change of probability correctly so long as the probabilities are essentially uniform within the narrow range of energies δE. According to the accessibility convention we have a conventional *one-to-one jump rate* ν_{ji} from every state i in A into every state j in B given by

$$\nu_{ji} = \rho_{BA}/g_B \delta E = 2\pi |H_{BA}|^2/\hbar \delta E. \tag{2.11}$$

The principle of jump rate symmetry

At this point we meet a most significant fact. The quantum calculation shows that the matrix element H_{AB} for the *reverse* jumps from group B into group A is always equal to the complex conjugate H^*_{BA} of the forward matrix element. Thus $|H_{AB}|^2 = |H_{BA}|^2$, and we find that

$$\nu_{ji} = \nu_{ij}, \tag{2.12}$$

the forward and reverse one-to-one jump rates are equal. We shall refer to this important fact as the *principle of jump rate symmetry.*

We are now ready to write down a general expression for the rate at which any particular probability $\tilde{p}_i$ is changing under the influence of the perturbation. Consider jumps between the state i and *all* other states j of the system which lie within the conventional range of accessible energies δE (and not just those of group B). Using the accessibility convention we must write the rate at which $\tilde{p}_i$ is falling on account of jumps *out* of i into all these states j as $\sum_j \tilde{p}_i \nu_{ji}$. But we must also recognise that $\tilde{p}_i$ will be at the same time rising on account of jumps *into* i from all other states: the rate of rise for such processes will be equal to $\sum_j \tilde{p}_j \nu_{ij}$. If we combine these two rates we find using (2.12) that

The master equation

$$\mathrm{d}\tilde{p}_i/\mathrm{d}t = \sum_j \nu_{ij}(\tilde{p}_j - \tilde{p}_i). \tag{2.13}$$

This is Fermi's *master equation.* In using it we must remember that the sum on the right-hand side is taken over all states lying within the conventional accessibility range of energies δE, and that ν_{ij} is the conventional one-to-one jump rate, proportional to $1/\delta E$. Note that, as one would expect, a change in the purely conventional magnitude of δE leaves the value of $\mathrm{d}\tilde{p}_i/\mathrm{d}t$ unchanged, so long as δE is small.

The master equation is to be our chief statistical starting assumption. We had better pause for a moment therefore to comment on its range of applicability and validity. Note first that if a system is not thermally isolated, a jump may occur in which the system changes energy, while a compensating change occurs in the outside world. The above arguments, however, were limited to wholly internal quantum jumps between states having the same energy. *The master equation therefore applies only to thermally isolated systems.* It is not necessary, however, to assume that constraints such as the volume are fixed. A slow variation of the constraints will slowly change the state energies E_i, but it does not in itself cause jumps between the states, and those which occur are still governed by the master equation, so long as the system remains thermally isolated.

Comments on the master equation

Finally, note that although the content of the master equation seems simple and natural, it is by no means easy to derive it rigorously from first principles. Fermi's golden rule was originally developed to handle simple scattering processes, such as the scattering of an electron by an impurity atom in a metal. A feature of such processes is that they involve an element of *randomisation*: the electron may start in a state of definite momentum, but after the scattering the probability is spread over a wide range of momenta, so that we have, in some sense, lost precision in our knowledge of the state of the system. The master equation reflects this fact. It shows how the probabilities change as the randomisation occurs. And the master equation is certainly very successful. Kinetic calculations based upon it agree with experiment. In thermodynamics it shows correctly how systems approach equilibrium. And yet it is precisely this element of randomisation which is difficult to justify from first principles. The fact is that the element of randomisation present in the golden rule and the master equation is *not* present in the equations of quantum theory from which they were derived, for one can show that the final quantum state of a system, calculated according to Schrödinger's equation, actually contains exactly the same information about the system as the initial state. This distinction is related to the fact that the master equation, and the equations of thermodynamics which we shall derive from it, are not *time-reversible*: as we shall see, the equation predicts that systems approach equilibrium, and the quantity known as entropy always increases, in the forward time direction only. Schrödinger's equation, on the other hand, *is* time-reversible. We shall ask in detail how the element of randomisation gets into the master equation in the final chapter. Here we simply note two points. The first is a practical consideration. Not all perturbations have the randomising character of a scattering process. For instance, a weak and uniform electric field, though it certainly steadily *changes* the momentum of a free electron, does not *randomise* it in the way that a scattering centre does. The golden rule does not apply to such perturbations. In our general formulation we shall therefore have to assume that any such non-randomising forces have been already taken into account when the approximate energy states were calculated, so that the *residual* perturbation consists only of randomising fields for which the master equation is valid. Secondly, we observe that, since the randomising element is not present in Schrödinger's equation itself, it follows that we must have made some further, and crucial, assumption about the state of the world. What this crucial assumption is we shall see later. For the moment I invite you to accept the master equation as a simple and natural description of the quantum kinetics.

§19.2

2.5 The ergodic assumption

The actual values of the one-to-one jump rates ν_{ij}, though important for detailed kinetics, will not greatly concern us in this book: we can get most of the results which we need from the principle of jump rate symmetry alone. But we shall usually make one further general assumption about the jump rates. We shall normally assume that a closed system has access to *all* the microstates whose energies lie within the conventional accessibility range δE. In other words, we assume that non-zero jump rates exist which link every pair of states within the accessibility range of energy, either directly or indirectly. This is the *ergodic assumption*, the assumption that the system can go anywhere within the allowed energy band. We shall discuss the assumption in more detail later. Although it is usually effectively valid, there are special cases in which it fails.

§19.1

Two important consequences follow from the ergodic assumption. Imagine first that we have two identical, isolated systems, and that we start them in two different microstates i and j which have the same energy. We measure both systems periodically to find what state they have reached. According to the ergodic assumption if we wait for a long enough time we can be sure of finding the second system in state i eventually, and its statistical behaviour thereafter must be the same as that of the first system measured from time zero. Thus if we average over an extremely long time the statistical behaviours of the two systems must become identical. It follows that the equilibrium fluctuation distribution of any quantity, that is, its distribution averaged over a time long enough for the distribution itself to have become independent of time, must be the same for both systems. Thus we see that the ergodic assumption, if it is valid, provides a simple explanation of the important fact, noted earlier, that the nature of the equilibrium state of a system depends only on the constraints and the energy: the constraints and the energy fix the accessible microstates and, as we have just seen, it does not matter which microstate the system starts from.

The equilibrium distribution is independent of the initial microstate

§1.3

The second important consequence of the ergodic assumption is that for an ensemble of systems of well-defined energy in which thermal equilibrium is established, the distribution of any quantity averaged over the fluctuations in time of a single system is the same as its distribution averaged over the ensemble at a single time: *in equilibrium the fluctuation distribution and the ensemble distribution are identical.* To see this, consider the distribution for the quantity of interest averaged over all systems *and* all times. If we do the average over the systems first, we get the ensemble distribution. But since in equilibrium the ensemble distribution is time-independent, the subsequent average over time changes nothing: evidently we may identify our distribution as the ensemble distribution. On the other hand, if we first average over time for a single system, we get the fluctuation distribution. And, since the ergodic assumption shows, as we have just seen, that the equilibrium fluctuation distribution is the same for all systems, the subsequent ensemble average again changes nothing: evidently our distribution is *also* equal to the fluctuation distribution. We shall frequently use the fact that the two distributions are identical, remembering that it applies only in thermal equilibrium, and only when the ergodic assumption is valid.

The relation between fluctuation and ensemble distributions

2.6 The principle of equal equilibrium probability

In general, the state probabilities $\tilde{p}_i$ will be changing with time, but after a while they settle down to fixed equilibrium values. What is the nature of this equilibrium distribution in an isolated system? The principle of jump rate symmetry shows that if $\tilde{p}_i = \tilde{p}_j$ for any pair of states they will be in *dynamic equilibrium* – the mean rates for the forward and backward jumps will be equal. Extending this idea, we see that any distribution in which the $\tilde{p}$ values are the same for all mutually accessible states must be an equilibrium distribution – this is obvious from the master equation, for every term on the right is then zero. It is also true that any distribution in which the $\tilde{p}$ values are *not* the same for all mutually accessible states *cannot* be an equilibrium distribution. (To see this, note that, if the probabilities for a set of mutually accessible states are not all equal then the subset of states having the greatest probability, $\tilde{p}_{max}$, cannot be in equilibrium, for at least one of them must be linked with states of lower probability by allowed quantum jumps, and $\tilde{p}$ for this state must be falling.) It follows that

> When a thermally isolated system reaches equilibrium, the state probabilities $\tilde{p}_i$ for all the members of any set of mutually accessible states become equal.

We shall refer to this result as the *principle of equal equilibrium probability*. We emphasise again that it is a direct consequence of the principle of jump rate symmetry.

It is only the mutually accessible states whose probabilities become equal in equilibrium. If the probability is spread initially over a range of energies the $\tilde{p}$ values for states of different energy are *not* equalised. In practice, as we shall see, we do not usually know the energy of a large system to an accuracy better than, say, 10^{-10} J. It is, however, sometimes convenient to imagine that we know the energy much more accurately, to suppose, in fact, that the probability is finite within an accessibility band of width δE and zero outside. Then, using the accessibility convention, the principle of equal equilibrium probability and the ergodic hypothesis, we arrive at a very simple picture of the equilibrium state of an isolated system: *in equilibrium all the states within the accessibility band of width δE are equally probable, and all other states have zero probability*. An equilibrium distribution of this type, in an isolated system with a well-defined energy, was called by Gibbs a *microcanonical distribution*. This simple concept is often useful in discussions of the nature of equilibrium states.

§3.4

The microcanonical distribution

2.7* The principle of detailed balance

As we noted in the previous section, the principle of jump rate symmetry ensures that if $\tilde{p}_i$ is equal to $\tilde{p}_j$ the two states i and j will be in dynamic equilibrium – the average rate at which jumps occur from i to j will be equal to the average rate for the reverse jumps from j to i. Since, according to the principle of equal equilibrium probability, $\tilde{p}_i$ must be equal to $\tilde{p}_j$ in equilibrium, we deduce that

> When a thermally isolated system is in thermal equilibrium, the mean rate at which any quantum jump actually occurs is equal to the mean rate of the reverse jump.

This result is the *principle of detailed balance*. It is important to notice that if the quantum jump is from state j into state i, we mean by the 'reverse jump' a jump from state i into state j. There are situations where the phrase 'reverse jump' might be taken to mean something different. For instance, if a particle is scattered from a state of momentum $\mathbf{p}_1$ to a state of momentum $\mathbf{p}_2$, we might regard the reverse jump as being from the state of momentum $-\mathbf{p}_2$ into the state $-\mathbf{p}_1$ (i.e. the jump we should see if time ran backwards). We shall refer to such a jump as the 'time-reversed jump'. It turns out that the principle of quantum symmetry applies also to time-reversed jumps *provided the equations of motion of the system have internal time-reversal symmetry*, so we may add as a rider to the principle of detailed balance:

§ 17.7

> ... and if the equations of motion of the system have internal time-reversal symmetry is also equal to the mean rate for the time-reversed jump.

An important case for which the equations of the system do *not* have internal time-reversal symmetry arises when a charged system is subject to an external magnetic field.

3

Temperature

3.1 The meaning of temperature for large systems

We are now ready to take the important step of finding the statistical meaning of *temperature*, the quantity which fixes the direction of heat flow. We start with large systems.

Imagine two large systems A and B which are free to exchange heat, but sufficiently weakly interacting that their individual states are not modified by the contact. We may think of the combination of A with B as a *joint system*, and our first step must be to consider how the quantum states of this joint system may be enumerated. A convenient way of doing this is shown in Fig. 3.1. System A has its own states, with energies E_{Ai} which are marked by dashes along the x axis; for degenerate states we imagine a multiple dash. In the same way system B has states with energies E_{Bj}, marked on the y axis. The dots in the plane represent states of the joint system in which A is definitely in some state i, B is definitely in some state j, and the joint energy E_{ij} is equal to $E_{Ai}+E_{Bj}$. *The dots represent all the possible quantum states of the joint system.* Fig. 3.2 is an example of such a diagram for two large systems. Because the systems are large, the densities of states are extremely high, and it is not possible to show the individual dots. Instead, I have indicated on the axes the values of the densities of states g_A and g_B for the two systems, worked out for the case where system A consists of 5000 identical oscillators, and system B consists of 10 000 similar oscillators. The density of dots in the plane is equal to $g_A g_B$. (The illustration is chosen so that the systems are large enough to exhibit the characteristic behaviour of large systems, without making the numbers involved too difficult to write down, as they would have been for systems of everyday size containing, perhaps, 10^{23} oscillators.)

The joint states

§2.3

Now which of the dots in this diagram represent *accessible* states of the joint system? Let us assume that the joint system is thermally isolated and contains a definite energy. We shall use the *accessibility convention* and assume that the joint states within a narrow band of energy of width δE are mutually accessible. In the diagram this implies that dots lying between the sloping lines $E_A+E_B=E$ and $E_A+E_B=E+\delta E$ correspond to the accessible states, where E

§2.4

is the total energy in the system, which, for the sake of illustration, I have taken to be 15 000 quanta.

Suppose now that the system starts at some point such as a, with a definite energy division between system A and system B. In general, the joint system will not be in thermal equilibrium, and heat will flow between A and B, corresponding to a movement parallel to the sloping lines. *What determines the direction of movement?* The answer to this question depends on the values of the product $g_A g_B$ which are marked along the sloping lines. In the direction of state b the values of $g_A g_B$ increase *extraordinarily* rapidly (each of the marks between the lines corresponds to an increase in $g_A g_B$ by a factor of 10^{100}). In the direction of state b' the number of accessible states *decreases* equally rapidly. This effect is so strong that it seems certain to swamp all other effects, and make the probability of any significant movement towards b' totally negligible.

Heat flows so that $g_A g_B$ increases

We can explore this idea in more numerical detail for our model systems. If we assume that A and B are in relatively weak thermal contact, so that exchange of quanta within each system occurs much more frequently than exchange of quanta between the two systems, then one can show that in the neighbourhood of state a the probability that a quantum will jump in the 'right' direction (from A to B) is twice the probability of the reverse jump. (We omit the details of this calculation.) This makes it clear that *the flow of heat in the 'right' direction is a matter of statistics.* In a time so short that only one or two jumps have occurred the probability that heat will flow in the 'wrong' direction is quite high. However, if we wait long enough for 1500 jumps to have

Fig. 3.1. Joint states of a pair of systems A and B which are in thermal contact. Each dot represents a joint state. If the joint system is thermally isolated then, using the accessibility convention, only those dots which lie between the sloping lines represent accessible states.

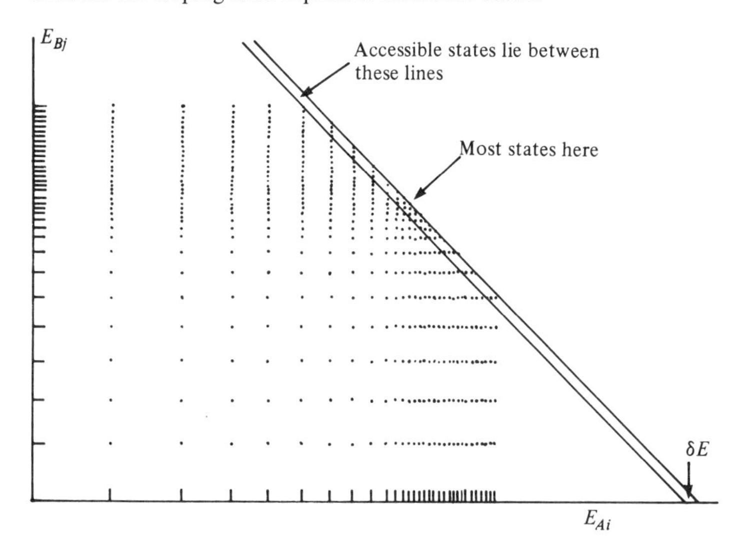

occurred, on average we expect 1000 to occur in the 'right' direction and 500 in the 'wrong' direction. We can also work out the standard deviations in these numbers, and we find (again omitting the details) that after 1500 jumps we expect E_A to take the value 9500 ± 36 quanta. The distribution of E_A after 1500 jumps is shown in Fig. 3.2 (curve *b*). Notice how narrow it is. There is not much uncertainty in the quantity of heat which we expect to flow from *A* into *B* in this time. There is, of course, a very small probability that after 1500 jumps the heat will have flowed the 'wrong' way, corresponding to the very small tail of curve *b* on the 'wrong' side of the starting state. For our model system this probability is about 10^{-38}. The probability that a *substantial* amount of heat would have flowed the 'wrong' way, say as far as state *b'*, is about 10^{-150}, a number so small that we can safely say that we should never see this event occur, even if we made the trial every microsecond during the whole history of the universe! Similar considerations apply to all pairs of large systems. *It is therefore safe to assume that when a substantial heat flow occurs between large*

Fig. 3.2. Heat flow between two systems. System *A* consists of 5000 oscillators and system *B* of 10 000 oscillators. The unit of energy is $h\nu$. The joint system starts in state *a*. After 1500 exchanges of quanta it reaches state *b*: the corresponding energy distribution is plotted normal to the sloping lines. After about 100 000 exchanges the system approaches the equilibrium distribution *c*, which is proportional to $g_A g_B$. The width of curve *c* represents the range of thermal fluctuations of E_A (or E_B) in equilibrium.

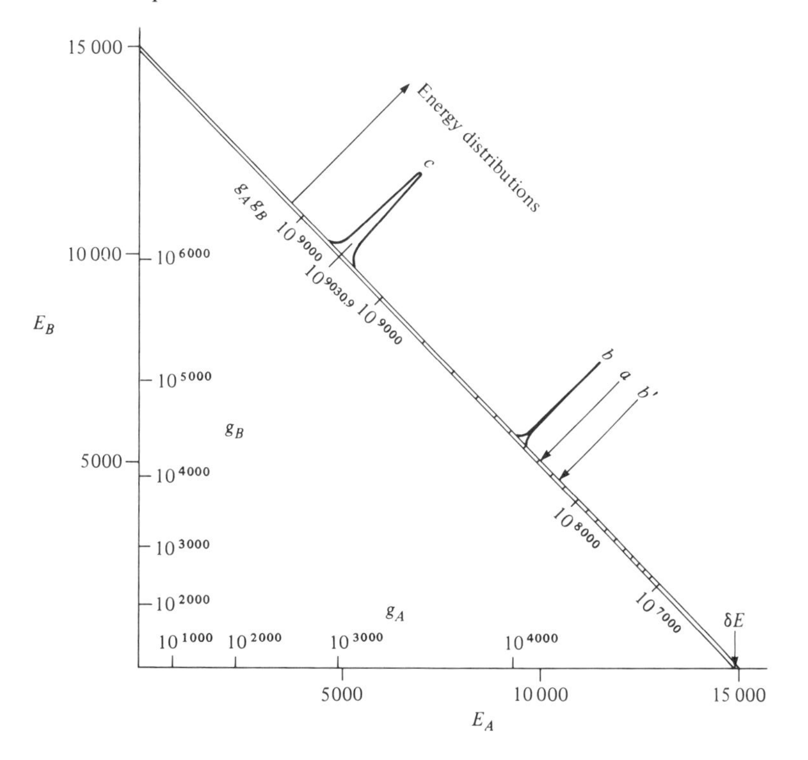

systems, in practice it always does so in the direction in which $g_A g_B$ increases.

We may use this conclusion to explore the statistical meaning of *temperature*. The heat always flows so that $\mathrm{d}(g_A g_B) > 0$, and thus

$$\mathrm{d}\ln(g_A g_B) = \mathrm{d}\ln g_A + \mathrm{d}\ln g_B \geqslant 0. \tag{3.1}$$

Since g_A and g_B are functions of E_A and E_B we may rewrite this condition as

$$(\partial \ln g_A/\partial E_A)\,\mathrm{d}E_A + (\partial \ln g_B/\partial E_B)\,\mathrm{d}E_B \geqslant 0$$

or

$$(\partial \ln g_A/\partial E_A - \partial \ln g_B/\partial E_B)\,\mathrm{d}E_A \geqslant 0, \tag{3.2}$$

since $\mathrm{d}E_B = \mathrm{d}E_A$ in any heat flow. (We have written $\partial \ln g/\partial E$ as a partial derivative to remind ourselves that the constraints are held constant, and no work is being done on the two systems.) This result shows that the sign of $\mathrm{d}E_A$ is connected with the sign of $(\partial \ln g_A/\partial E_A - \partial \ln g_B/\partial E_B)$. This is most significant: *it shows that $\partial \ln g/\partial E$ is a measure of the coldness of the system.* In words, what Equation (3.2) tells us is that if A is *colder* than B (the bracket is positive) heat cannot flow spontaneously from A into B ($\mathrm{d}E_A$ cannot be negative). In the same way, if B is colder than A heat cannot flow from B into A. The system with the larger value of $\partial \ln g/\partial E$ sucks the energy into itself. In all normal circumstances g increases with E, so the 'coldness' is a positive quantity. It is therefore natural to define the *statistical temperature* T as a positive measure of 'hotness' by writing

$\partial \ln g/\partial E$ is a measure of coldness

§ 3.5

Definition of T

▶ $$1/kT = \partial \ln g/\partial E, \tag{3.3}$$

where k is a constant. For the model system shown in Fig. 2.1 we see that the coldness $\partial \ln g/\partial E$ decreases as energy is added to the system, showing that the temperature increases. This is true of all large systems under normal circumstances. We shall see later that if we choose k to be *Boltzmann's constant*, equal to $1.38 \times 10^{-23}\ \mathrm{J\,K^{-1}}$, T will turn out to be identical with the *ideal gas scale temperature*. With this choice T has units of degrees kelvin, or K.

§ 2.3

§ 5.5

It may be objected that the above argument depended too much on the statistics of our particular model systems. Note, then, that for *any* pair of systems, if heat flows from A to B we have

$$\begin{aligned}\partial \ln(g_A g_B) &= (-\partial \ln g_A/\partial E_A + \partial \ln g_B/\partial E_B)\delta E_B \\ &= (-1/kT_A + 1/kT_B)\delta E_B,\end{aligned} \tag{3.4}$$

using the definition of T. Let us anticipate § 5.5 and assume that k is Boltzmann's constant and T the gas scale temperature. Suppose that T_A is 300 K, and T_B is 299.9 K – not a very large temperature difference. We can then easily find from (3.4) the following results:

δE_B	Factor by which $g_A g_B$ is multiplied
10^{-17} J	2.2
10^{-16} J	3×10^{3}
10^{-15} J	1.2×10^{35}
10^{-14} J	$\sim 10^{351}$

Substantial heat 'never' flows the wrong way

This makes it clear that the chance of seeing heat of order 10^{-17} J flow the 'wrong' way is quite high. But in a time long enough for a movement of 10^{-14} J to occur, the system has access to about 10^{702} times more states in the 'right' direction than it has in the 'wrong' direction. Once again, we may safely say that we would in practice 'never' see the system make the 1 in 10^{702} choice required for a heat flow of this size in the 'wrong' direction.

§ 1.5

The Clausius statement of the second law, which we quoted in the introduction, implies that heat *cannot* flow spontaneously from a colder to a hotter body. We have seen in this section that in one sense this statement is incorrect. The process is not mechanically impossible. It just happens that for large systems the *chance* of our ever seeing a measurable heat flow from colder to hotter, even when the temperature difference is small, is so extraordinarily minute that we may safely ignore it. *The second law is statistical in nature.*

3.2 Thermal equilibrium between two large systems

If we leave our two model systems in thermal contact heat will continue to flow so that $g_A g_B$ increases, until eventually $g_A g_B$ approaches its maximum value, at *c* in Fig. 3.2, where E_A is 5000 quanta. At this point $T_A = T_B$ and the two systems are in *thermal equilibrium.* Note that this occurs when the energy per oscillator is the same in both systems, as one would expect intuitively. The two model systems will be close to their final equilibrium state after about 100 000 quantum exchanges between them. After so many jumps the joint system has had a roughly equal chance of reaching *all* its accessible states, and, in accordance with the principle of equal equilibrium probability, *all* the dots between the sloping lines now have almost the same probability.

§ 2.6

At first sight one might jump to the conclusion that the probability distribution for E_A would have become very broad in this process. This is not so. Examination of the values of $g_A g_B$ shown in Fig. 3.2 shows that, although $g_A g_B$ is very large everywhere, it is nevertheless true that *the great majority of the accessible states lie very close to the maximum of* $g_A g_B$. In fact, since in equilibrium each dot between the sloping lines carries the same probability, the chance of finding the system in unit range of E_A must be just proportional to the density of dots in the plane, or, in other words, *the equilibrium distribution of* E_A *must be proportional to* $g_A g_B$. This distribution is shown in the diagram (curve *c*) and is narrow; in fact, $E_A = 5000 \pm 82$ quanta in equilibrium. Notice that the equilibrium distribution depends only on the densities of states g_A and g_B, and is quite independent of the scattering mechanism by which the equilibrium is achieved.

The equilibrium distribution of E_A is proportional to $g_A g_B$, and is narrow

The width of the equilibrium distribution represents the range of *equilibrium thermal fluctuations.* Within this range we expect to see the system moving sometimes in one direction and sometimes in the other, and we can no longer say that in practice it always moves so that $g_A g_B$ increases. It is therefore important to understand that, considered as a fraction of the total energy, the range of fluctuations in systems of everyday size is *much* smaller than its range

Heat flow in thermal fluctuations

for our relatively small model systems. We shall see how to calculate energy fluctuations later. In a pair of systems of everyday size the range of energy fluctuation will turn out to be of order 10^{-9} J, corresponding to a temperature fluctuation of 10^{-8} K. We are actually most likely to see heat flowing the 'wrong' way when close to equilibrium. Even in equilibrium, however, we see that only energy fluctuations smaller than about 10^{-8} J have any appreciable probability, and of course fluctuations as small as this would normally be quite unmeasurable. Thus even at equilibrium it remains true that the chance of ever seeing a *measurable* quantity of heat flow the 'wrong' way is completely negligible for large systems.

§3.4

3.3 Boltzmann's distribution and the meaning of temperature for small systems

We now ask an important question: *if a system is held at temperature T, what is the probability p_i of finding it in its* ith *quantum state, once equilibrium has been achieved*? The answer to this question is central to much of the later formal development of the subject. We shall use an argument which is equally valid for large and small systems (though the result is perhaps more often applied to small systems).

To hold the system of interest at temperature T we must place it in thermal contact with a much larger system, already at temperature T, which we shall refer to as the *reservoir*. We suppose that the *joint* system (the system of interest plus the reservoir) is thermally isolated. This is analogous to the joint system considered in §3.1, and may be treated in the same way (Fig. 3.3). The only difference is that the system of interest is not necessarily large, which we indicate by showing an appreciable spacing between its energy levels. But it remains true that the dots between the sloping lines represent the accessible states of the joint system, and that we know (by applying the principle of equal equilibrium probability to the joint system) that each of these accessible states carries the same probability once equilibrium has been established. In §3.1 we were interested in the probability of finding E_A in unit range of energy, which was proportional to $g_A g_B$. Here we are interested in the probability of finding the system of interest in *a particular one of its own states*, *i*. In the diagram the corresponding accessible joint states are represented by the N_i dots which lie in a vertical line above the energy E_i, and between the sloping lines. Since all the dots between the lines carry the same probability in equilibrium, it follows that p_i is proportional to the corresponding number of dots, N_i. But N_i is just equal to $g_R \delta E$, where g_R is the density of states for the *reservoir*. We thus reach the perhaps surprising conclusion that

p_i is proportional to the *reservoir*'s density of states

$$p_i \propto g_R, \tag{3.5}$$

where g_R is the density of states *not* for the system of interest but *for the reservoir with which it is in contact*.

Our problem, then, is to discover how g_R varies with energy. At this point we recall that the reservoir is supposed to be *much larger than the system of interest*. We mean by this that its heat capacity should be so large that its temperature does not change appreciably when it absorbs energies comparable with the energies in the system of interest. Now, we defined the temperature of a large body by the relation $1/kT = \partial(\ln g)/\partial E$. If the heat capacity is large, we mean by this that T does not change much with energy, or, in other words, that the slope $\partial(\ln g)/\partial E$ is nearly constant. (We showed in Fig. 2.1 how $\ln g$ varied with E for our relatively small system of 100 oscillators. The fall in slope with rising energy corresponds to the increase in temperature with energy. In a larger system the curvature of this plot would be very small because we would have to add a very large amount of heat to change the temperature by a given amount.) For the reservoir, therefore, we may write

g_R rises exponentially with energy

§2.3

$$\frac{\partial \ln g_R}{\partial E_R} = \frac{1}{kT} = \text{constant} \tag{3.6}$$

for energy changes on the scale of E_i. We deduce that the *density of states of the reservoir varies exponentially with energy in the energy range of interest*,

$$g_R \propto e^{E_R/kT}. \tag{3.7}$$

We may now combine this result with (3.5) and deduce that

$$p_i \propto e^{E_R/kT} \propto e^{-E_i/kT}, \tag{3.8}$$

since $E_R = E - E_i$ and E is constant. We find the constant of proportionality by

Fig. 3.3. The construction of Fig. 3.1 used to obtain Boltzmann's distribution.

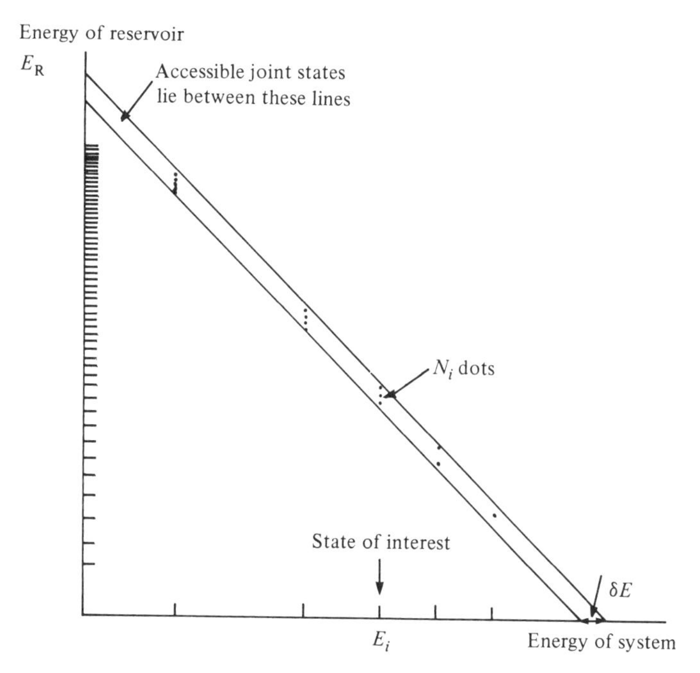

remembering that $\sum_i p_i = 1$, so that

The Boltzmann probability distribution

$$p_i = \frac{e^{-E_i/kT}}{Z}, \quad \text{where } Z = \sum_i e^{-E_i/kT}. \tag{3.9}$$

This very important result is *Boltzmann's distribution.* The normalisation constant Z is known as the *sum over states* or the *partition function.* The factor $\exp(-E_i/kT)$ is the *Boltzmann factor* for state i. The probability p_i is written without the superscript ~ to remind us that we are dealing with a system in equilibrium at temperature T. (Systems in equilibrium with a large reservoir at temperature T are said to be in their standard or *canonical* state.) We emphasise that the Boltzmann distribution is a consequence of the fact that the density of states of the *reservoir* varies exponentially with temperature, and that this in turn follows from our definition of T for large systems and the requirement that the reservoir should have a very large heat capacity.

The canonical state

The Boltzmann distribution is a natural extension of the principle of equal equilibrium probability. For a thermally isolated system in equilibrium all states having the same energy have the same probability, and this is also true for the Boltzmann distribution. A thermally isolated system, however, has no traffic between states of different energies, and without further information we can say nothing about how the probability varies with energy (which depends on the previous history of the system). By contrast, a system in the canonical state, in equilibrium with a reservoir, has access to all its possible states, and we know that the probability varies exponentially with the energy of the state, as described by the Boltzmann distribution. (This is not contrary to the principle of equal equilibrium probability because that principle applies to the *joint* states, and *not* to the states of the system of interest, which is not thermally isolated.) The variation of p_i with E_i for a typical small system at various temperatures is shown in Fig. 3.4. At $T=0$, the probability is concentrated in the ground state of the system. Note that as T becomes very large the probability becomes *equal* for all energies. (It is *not*, as students are apt to assume, higher for the higher energies.)

We can, in fact, use the Boltzmann distribution to *define* the temperatures of small systems of atomic size. There are difficulties in specifying the meaning of temperature for very small systems. We cannot write $1/kT = \partial(\ln g)/\partial E$, because the energy E for a small system is often comparable with the spacing between its energy levels, and we cannot regard $g(E)$ as a smooth function. Moreover, the energy fluctuations in small systems are relatively large, so that the direction of heat flow ceases to be well defined when a small system is involved. Small systems are, on the other hand, very frequently in the canonical state because they are almost always in thermal contact with a much larger reservoir. We shall take the view that the Boltzmann distribution *defines* the temperature of a small system which is in a canonical state, and in other situations regard the temperatures of small systems as undefined.

The meaning of T for small systems

3.4* The energy distribution for a large system at a given temperature

At this point in the argument the student is sometimes puzzled by the question of how the Boltzmann distribution can possibly apply to a *large* system held at temperature T. The Boltzmann distribution, after all, shows that the system is always more likely to be in its ground state than in any other, and that the probability falls off rapidly as the energy rises. But a large system such as a lump of copper is *never* found in or near its ground state at room temperature, but always very close to a particular energy which may be several joules above the ground state energy. How can this possibly be reconciled with the Boltzmann distribution, which shows that the probability for a state of this energy will be *smaller* than the ground state probability by a factor of about $\exp(10^{23})$?

The answer is that the *energy distribution must be carefully distinguished from the Boltzmann distribution.* The Boltzmann probability is a probability *per quantum state.* To get from it the probability *per unit range of energy* we must multiply by the density of states, so that the energy distribution

Fig. 3.4. The Boltzmann distribution applied to a small system. As T rises, Z rises, and the effect is that p_i falls at the lower energies and rises at the higher energies.

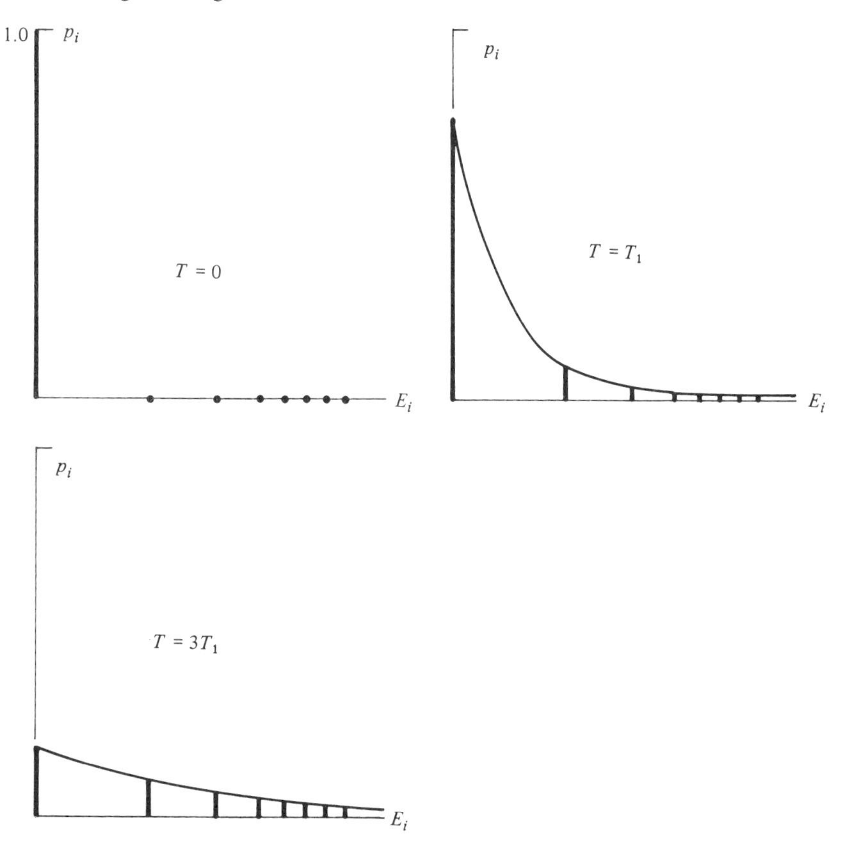

function $f(E)$ is given by the expression

$$f(E)\,\mathrm{d}E = p(E)g(E)\,\mathrm{d}E. \tag{3.10}$$

It is true that $p(E)$ falls rapidly as energy rises but, initially, $g(E)$ rises even more rapidly, so that the energy distribution has its peak far above the ground state energy, even though the individual states near the peak have extremely low probabilities.

The peak in the energy distribution occurs when $\partial(\ln f)/\partial E$ is zero. If the system is in contact with a reservoir at temperature T_0 we have

$$\ln f(E) = \frac{-E}{kT_0} + \ln g(E) + \text{const}, \tag{3.11}$$

thus

$$\frac{\partial \ln f}{\partial E} = \frac{-1}{kT_0} + \frac{1}{kT},$$

which is zero when $T = T_0$. So the most probable energy is the value for which $T = T_0$. To find how $f(E)$ varies near this maximum, we expand (3.11) in a Taylor series around the energy E_0 at which $T = T_0$. This gives on expansion to second order

$$\begin{aligned}\ln f(E) &= \frac{-\Delta E}{kT_0} + \left(\frac{\partial \ln g}{\partial E}\right)_{T=T_0} \Delta E + \frac{1}{2}\left(\frac{\partial^2 \ln g}{\partial E^2}\right)_{T=T_0} \Delta E^2 + \text{const} \\ &= \frac{-\Delta E}{kT_0} + \left(\frac{\Delta E}{kT}\right)_{T=T_0} - \frac{1}{2kT^2}\left(\frac{\partial T}{\partial E}\right)_{T=T_0} \Delta E^2 + \text{const},\end{aligned} \tag{3.12}$$

where $\Delta E = E - E_0$. Noting that the term in ΔE vanishes since $T = T_0$ and that $\partial E/\partial T = C$, the heat capacity, we find that

▶ $$f(E) \propto \mathrm{e}^{-\Delta E^2/2kT^2C}. \tag{3.13}$$

This is a *Gaussian distribution*, and on comparison with the standard form we see that the mean square energy fluctuation is given by

The magnitude of the energy fluctuation

▶ $$\overline{\Delta E^2} = kT^2C. \tag{3.14}$$

The Gaussian distribution

(The standard normalised form of the Gaussian distribution for a quantity x is $f(x) = (2\pi\sigma^2)^{-\frac{1}{2}} \exp(-\Delta x^2/2\sigma^2)$, where σ^2 is the mean square deviation, $\overline{\Delta x^2}$.) For a large system the fluctuations are small compared with the energy in the system and the approximation involved in terminating the expansion at the second order is very good.

How large are the energy fluctuations in practice? In many systems the internal energy is of order NkT, where N is the number of atoms in the system, and so C is of order Nk. Thus the root mean square (rms) fluctuation $(\overline{\Delta E^2})^{\frac{1}{2}}$ is of order $N^{\frac{1}{2}}kT$, while the *fractional* fluctuation $(\overline{\Delta E^2})^{\frac{1}{2}}/E$ is of order $N^{-\frac{1}{2}}$. This makes it clear why fluctuations in large systems are so inconspicuous. For a system of everyday size containing about 10^{22} atoms, the energy content might be 100 J at room temperature, but the energy fluctuation is only about 10^{-9} J.

For a very small system, on the other hand, in which the level spacing is comparable with kT, it is evident from the Boltzmann distribution that the energy fluctuation will be *comparable* with the energy in the system, both being of order kT.

3.5* Can heat capacity or temperature be negative?

Temperature fixes the direction of heat flow between bodies uniquely – heat can only flow from the hotter body to the colder body. But it does not follow that adding heat to a body necessarily makes it hotter, that the heat capacity has to be positive. Indeed, the addition of heat is *not* necessarily associated with a rise in temperature. For instance, adding latent heat to melt ice does not change the temperature at all, and if we add heat to a gas while simultaneously expanding it sufficiently rapidly, the gas will be cooling rather than warming. It is, however, true that *a canonical heat capacity cannot be negative.* A canonical heat capacity is one which is measured with the system in contact with a thermal reservoir whose temperature is slowly raised, and with the constraints held fixed so that no work is done on the system. In this arrangement the state energies E_i are fixed and their probabilities are given by the Boltzmann distribution. Raising T has the effect on the Boltzmann distribution shown in Fig. 3.4. The higher energies become *relatively* more probable as the temperature rises, and, since the total probability is conserved, it follows that the p_i values must rise at the higher energies and fall at the lower energies, as shown. This implies a positive heat capacity, because probability has been shifted from lower to higher energies, increasing the mean energy in the system. Clearly, *in the canonical state the heat capacity cannot be negative.*

Canonical heat capacity

§ 3.3

Negative temperature, however, can exist in certain rather artificial conditions. The crucial point to be understood about negative temperatures is that they describe systems which are not abnormally *cold*, but abnormally *hot*. This comes about in the following way. It is clear from our discussion of temperature for large bodies that it is $1/T$ (a measure of *coldness*) rather than T which is the significant quantity: Equation (3.2) shows that, whatever the sign of T, if $1/T_A > 1/T_B$, A is colder than B, and heat can only flow from B into A. With this in mind, consider a system in which $\ln g$ varies with energy as shown in Fig. 3.5. We see that as we add energy the *coldness* decreases smoothly from ∞ (at absolute zero) through 0 at the maximum of $\ln g$ to $-\infty$ at the highest possible energy. The states on the right-hand half of the diagram, where the temperature is negative, are *hotter* than those on the left. The temperature T, on the other hand, is discontinuous. It rises from zero to ∞ at the maximum, jumps to $-\infty$, and then rises to zero as we add energy. Note that the temperatures $+\infty$ and $-\infty$ are *identical*, but the temperature $+0$ (absolute zero) is quite different from the temperature -0. It is, perhaps, a pity that, for historical reasons, we use T as our measure rather than $1/T$. Note that, so long as it is understood that bodies at negative temperatures are *hotter* than all

Negative temperatures

§ 3.1

bodies at positive temperatures, the second law covers the case where one of the bodies is at a negative temperature.

The derivation of the Boltzmann distribution also still applies at negative temperatures, and the same continuity at $T=\pm\infty$ is apparent (Fig. 3.5). At positive temperatures $p(E)$ is a falling exponential; at $T=\pm\infty$ it becomes flat; and at negative temperatures it is a *rising* exponential. This comment helps to bring out the essential artificiality of the idea of negative temperature. *A system can only be in equilibrium at a negative temperature if it has an upper bound on its energy levels* – otherwise it would contain infinite energy.

§3.3

Do such systems exist? In fact, we know of no *complete* system which has an upper bound on its energy. There are, however, some situations in which *part* of the energy in a system may be regarded as residing in an independent subsystem. The most clear-cut example is the case of nuclear spins in a magnetic field, in copper, for example. The nuclear spin energy has a minimum (when all the nuclear magnetic moments are parallel to the field) and also a maximum (when all the moments lie anti-parallel to the field). At low temperatures it takes a very long time for the spins to come into thermal equilibrium with the solid lattice, so that we may regard the spin system as thermally isolated from the lattice. By a trick of spin resonance it is possible to

Fig. 3.5. Negative temperatures, which exist only in systems having an upper bound on their energies. The diagram shows $\ln g$ plotted against E for such a system. The 'hotness' increases steadily from left to right, but in the right-hand half of the diagram the temperature is negative, which means that the Boltzmann probability *rises* with increasing energy, as shown below.

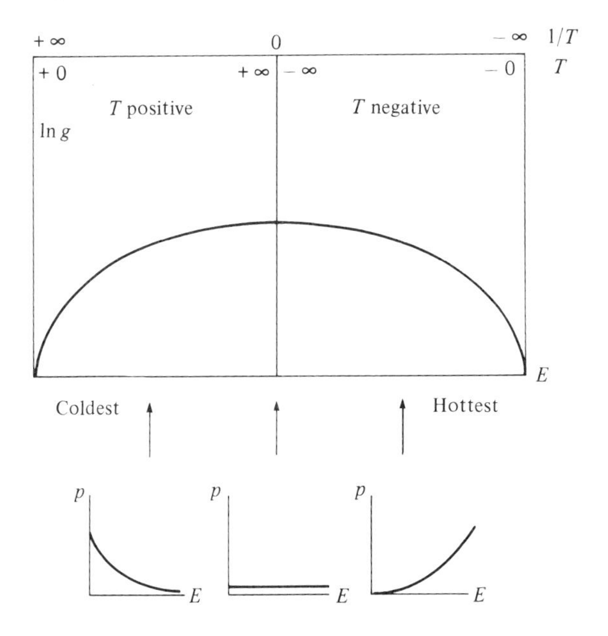

invert every spin with respect to the magnetic field. We may then use the curious, but quite realistic, description which says that the spins are at a very hot negative temperature, thermally isolated from the lattice and conduction electrons, which remain at a very cold positive temperature! The copper *as a whole* can never be at a negative temperature, however, because there is no upper bound on its energy.

Problems

3.1 According to (2.9), W for a system of N identical oscillators containing Q quanta is increased by a factor $(Q+N)/(Q+1)$ when a quantum of energy $h\nu$ is added. For a large system Q and N are both large, so this factor may be treated as approximately constant and equal to $1+N/Q$ for small changes of Q.

(i) Use the definition of T (3.3) to show that $kT = h\nu/\ln(1+N/Q)$ for this system.

(ii) If system A, consisting of 10^{23} oscillators for which $h\nu = 10^{-20}$ J, is in equilibrium with system B, consisting of 2×10^{23} similar oscillators, and the joint system contains 3×10^{3} J of energy over and above the ground state energy, find the number of quanta in A, the number of quanta in B and the temperature.

(iii) A single oscillator is in thermal equilibrium with system A. If p_n is the probability of finding n quanta in the single oscillator, explain carefully why $p_n \propto (1+N_A/Q_A)^{-n}$. Then compare this result with the Boltzmann distribution (3.9).

3.2 For a system of N free electrons it can be shown that, under certain conditions, the density of states for the *complete system* (not the individual electrons) is proportional to $\exp(NE/\varepsilon_0)^{\frac{1}{2}}$, where ε_0 is about 10^{-19} J and N is 10^{23}. Find T as a function of E, and hence find the heat capacity of the system at room temperature. Show that at room temperature the energy distribution is approximately Gaussian, and find the fractional energy fluctuation.

3.3 A small system has just two quantum states, of energy $E_1=0$ and $E_2=10^{-22}$ J. Find the probabilities p_1 and p_2 when the mean energy E in the system is $0.2E_2$ and $0.4E_2$. Assuming that these probabilities correspond to a Boltzmann distribution, deduce the corresponding temperatures.

3.4* Repeat problem 3.3 for the cases $E=0.5E_2$, $0.6E_2$, E_2, and comment on your answers.

4

Entropy

4.1 The Boltzmann–Gibbs statistical picture of entropy

We now introduce the important concept of *entropy*. Unlike temperature, entropy is a quantity of which we have no intuitive knowledge, so we begin by describing in general terms what it is, from a statistical point of view.

Entropy is connected with the fact, which we have already noted, that in most thermodynamic states the system is not in a definite quantum state, but is spread over a large number of states according to some probability distribution. The idea is to find some suitable definition of the *quantum multiplicity*, M, which is the effective number of quantum states over which the probability is spread, at any instant. We then define the entropy, $\tilde{S}$, as

$$\tilde{S}=k \ln M. \qquad (4.1)$$

If the probability is spread uniformly over the M states, this is equivalent to writing

$$\tilde{S}=-k \ln \tilde{p}, \qquad (4.2)$$

where $\tilde{p}=1/M$ is the probability that any particular state is occupied. (Boltzmann's constant appears in these definitions for reasons of convenience, as it did in the definition of temperature. Entropy has units of J K^{-1}.)

We note that, if the probability for system A is spread uniformly over M_A states and the probability for system B is spread uniformly over M_B states, then, taking the two systems together, the probability is spread uniformly over $M_A M_B$ joint states. Thus the joint entropy $\tilde{S}=k \ln (M_A M_B)=k \ln M_A+k \ln M_B=\tilde{S}_A+\tilde{S}_B$ – the entropy of the joint system is the *sum* of the entropies of the constituent systems. Entropy is evidently, in this case at least, an *additive* quantity. §4.3

It will be clear from the way in which it is defined that the entropy is a measure of *randomisation* or *disorder* or, more precisely, of our lack of detailed knowledge of the exact, microscopic state of the system. Consider again the system of 100 identical oscillators. If we happen to know that it contains no quanta, then there is only one possible quantum state – the ground state. In this case, $M=1$ and $\tilde{S}=0$. The system is *completely ordered*. If we happen to §2.3

know that it contains two quanta, and has reached statistical equilibrium, so that each state corresponding to two quanta has the same probability, then $M=5050$ and $\tilde{S}=k \ln 5050$. If we had not known the energy exactly, M and $\tilde{S}$ would have been even greater. In such situations, though the system may, at any instant, be in some particular state, *we do not know which one*. From our point of view, the system is now considerably *disordered*.

The Boltzmann–Gibbs definition of entropy

The meaning of $\tilde{S}$ should now be clear in the case where the probability is spread uniformly over the available states. But this is by no means the only possible situation. If the system has not yet reached equilibrium, or if the system is in thermal contact with a reservoir, $\tilde{p}_i$ will not be the same for all the states involved, and it is not clear how S should be defined. To cover such situations we shall adopt the definition

$$\blacktriangleright \quad \tilde{S} = -k \sum_i \tilde{p}_i \ln \tilde{p}_i. \tag{4.3}$$

This is a natural generalisation of (4.2). (It may be regarded as (4.2) averaged over the states i according to their probabilities.) This particular definition is chosen because it has clear, well-defined properties, as we shall see. It reduces to (4.2) or (4.1) in the case where $\tilde{p}$ is spread uniformly over M states. The quantity $H=\sum_i \tilde{p}_i \ln \tilde{p}_i$ was introduced into gas kinetics by Boltzmann and applied to general systems by Gibbs. Since $\tilde{S}=-kH$, $\tilde{S}$ is known as the *Boltzmann–Gibbs entropy*. In this book, however, we shall usually refer to it simply as the *entropy*. (Writers on this subject have adopted various different views on the best definition of entropy, but we shall stick to Equation (4.3) as our primary definition.)

4.2 The law of increase of entropy

It will be clear that the entropy in an isolated system will tend to increase as the probability spreads out over the accessible states and the system approaches equilibrium. For instance, if we place one oscillator, which we know to contain two quanta, in contact with 99 similar oscillators all in their ground states, then at the moment of contact we know the exact quantum state of the joint system, so that $\tilde{S}=0$. As time goes on, the quantum jumps will even out the probabilities among the 5050 accessible states until eventually $\tilde{S}$ reaches $k \ln 5050$.

The H-theorem

We can, in fact, prove quite generally that $\tilde{S}$ as defined by (4.3) can increase but never decrease in an isolated system, by making use of the master equation. By differentiating (4.3) we find that

$$\frac{d\tilde{S}}{dt} = -k \sum_i \ln \tilde{p}_i \frac{d\tilde{p}_i}{dt}, \tag{4.4}$$

where we have used the fact that $\sum d\tilde{p}_i/dt=0$, since $\sum \tilde{p}_i=1$. Now consider the effect of quantum jumps between two particular states, α and β. For an isolated system these jumps will make a contribution $\nu_{\alpha\beta}(\tilde{p}_\beta-\tilde{p}_\alpha)$ to $d\tilde{p}_\alpha/dt$ and a contribution $\nu_{\alpha\beta}(\tilde{p}_\alpha-\tilde{p}_\beta)$ to $d\tilde{p}_\beta/dt$, according to the master equation (2.13).

§2.4

They therefore make a contribution to $\mathrm{d}\tilde{S}/\mathrm{d}t$ equal to

$$k\nu_{\alpha\beta}(\ln\tilde{p}_\beta-\ln\tilde{p}_\alpha)(\tilde{p}_\beta-\tilde{p}_\alpha). \tag{4.5}$$

But the two brackets in this expression have the same sign, so *it cannot be negative*. Since the same applies to every such contribution to $\mathrm{d}\tilde{S}/\mathrm{d}t$, it follows that

▶ $$\Delta\tilde{S}\geqslant 0 \quad \text{in an isolated system.} \tag{4.6}$$

The statistical proof just given is known as the *H-theorem*. Like the master equation, this important result holds even if the constraints on the system are being altered. Note that in a *reversible* change we must have $\Delta\tilde{S}=0$ in an isolated system (for otherwise we should have $\Delta\tilde{S}<0$ in one direction of change or the other).

$\Delta\tilde{S}=0$ in a reversible change

The law of increase of entropy is one of the most significant results in physics. Amongst other things, it can be regarded as the point at which *the distinction between cause and effect* and the *arrow of time* enter physics. Although the above proof looks very simple it really involves some considerable subtleties, which we shall examine briefly in Chapter 19. For the time being we shall accept (4.6) at face value.

§§ 19.3, 19.6

4.3 The equilibrium entropy

The Boltzmann–Gibbs entropy $\tilde{S}$ is defined even when the system is not in equilibrium, and we use the superscript to remind us of that. Classical thermodynamics is concerned mainly with the equilibrium entropy. Let us assume for the moment that the system of interest is held at temperature T – we assume, in other words, that it is in an equilibrium canonical state in contact with a much larger reservoir. In this case we may write the equilibrium entropy as the *canonical entropy*

Canonical entropy

▶ $$S=-k\sum p_i\ln p_i, \tag{4.7}$$

where p_i is the equilibrium Boltzmann probability given by (3.9). It would be convenient if we could *always* identify the equilibrium entropy with the canonical entropy. Is this identification valid?

§ 3.3

For *small* systems, we are almost always concerned with an assembly of systems which are either in thermal contact with each other or some much larger system, or were recently in such thermal contact. In practice, then, small systems in equilibrium are always in a canonical state, and their entropy is indeed the canonical entropy just defined. *Large* systems are also frequently in a canonical state. Even when thermally isolated, a recent thermal contact will usually ensure that a large system has a Boltzmann distribution, and it can be shown that the slight blurring of energy which occurs when work is done on a large isolated system (due to fluctuations in pressure during a compression, for instance) tend to carry it towards a canonical distribution. But we cannot always make this assumption. For a thermally isolated system we can only assert that in equilibrium all states *of the same energy* have the same

§§2.6, 3.1

probability, and this leaves open a number of possibilities (Fig. 4.1). We may wish to consider equilibrium in a *microcanonical state*, in which the state probability is spread uniformly over a narrow band of energies of width δE, or we may wish to consider some more general distribution. These different distributions clearly have different equilibrium entropies $-k\sum \tilde{p}_i \ln \tilde{p}_i$. It can actually be shown that, for a given mean energy, the canonical entropy is the largest, which suggests that, in a sense, the canonical state represents a more complete equilibrium than the others. However, we can ignore such subtleties, for it turns out that *in large systems all such differences in equilibrium entropy are in any case negligibly small.* In any one of these distributions the typical value of $\tilde{p}_i$ will be of order $1/(g\Delta E)$, where ΔE is the width of the energy distribution. When we calculate the entropy as the mean value of $-k \ln \tilde{p}_i$, it becomes clear when we take the logarithm that a very rough estimate of $\tilde{p}_i$ is good enough. The reason is the enormous magnitude of g, which is typically of order $10^{10^{23}}\,\mathrm{J}^{-1}$ for a large system. This means that when we take the logarithm we could change the approximate value of $\tilde{p}_i$ by a large factor (say, 10^{100}) and still make no appreciable difference to $\ln \tilde{p}_i$. Indeed, g is so large that we can ignore $\ln \Delta E$ altogether! For this reason, we may safely write $S = -k\,\overline{\ln \tilde{p}_i} \simeq k \ln (g\Delta E) \simeq k \ln g$, or

The large system approximation

$$S = k \ln g \quad \text{for large systems.} \tag{4.8}$$

Fig. 4.1. Three different types of distribution for the state probability $\tilde{p}$ as a function of energy E for an isolated system in thermal equilibrium.

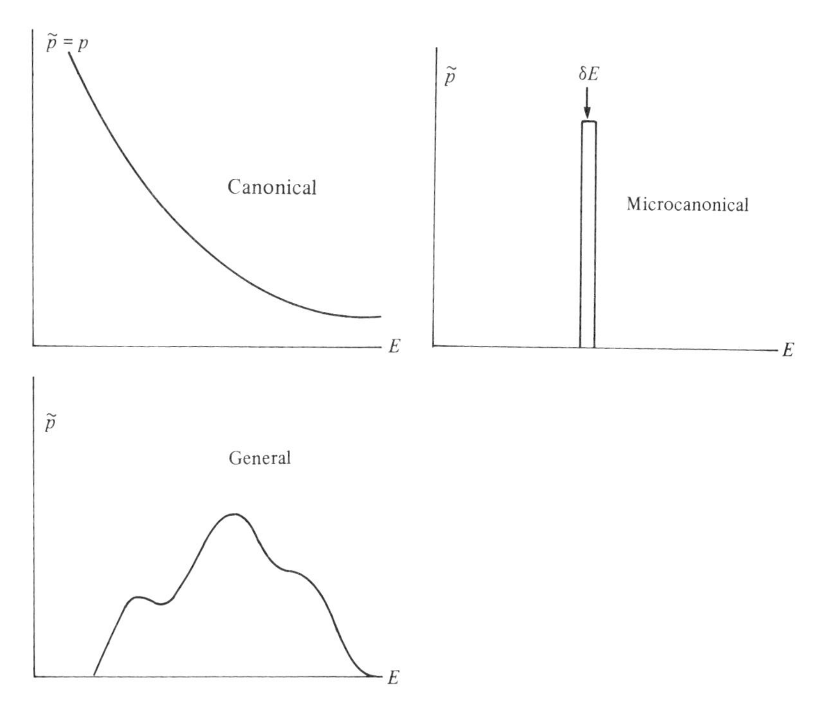

Normally such an expression would be unsatisfactory, since the value of S would depend on the units in which g was expressed. Here g is so large that *any* sensible units for g will give an accurate value for S. This simple formula for S in large systems is often very useful.

Since we may write the equilibrium entropy S as $k \ln g$ to an excellent approximation for *any* energy distribution in a large system, it is clear that *for both small and large systems we may safely treat the canonical entropy as the equilibrium entropy*, and we shall do so in our treatment of classical thermodynamics. For future reference, then, note first that the canonical entropy is *a well-defined function of state*, fixed when we fix the constraints and the temperature (or, equivalently, the mean energy in the system). Secondly it is *exactly additive*. If two systems are each in a canonical state they are statistically independent and the probability of the joint state ij is given by $p_{ij} = p_i p_j$. Consequently

§6.3

Properties of canonical entropy

§2.2

$$
\begin{aligned}
S &= -k \sum_{ij} p_{ij} \ln p_{ij} \\
&= -k \sum_{ij} p_i p_j (\ln p_i + \ln p_j) \\
&= -k \sum_i p_i \ln p_i - k \sum_j p_j \ln p_j \quad (\text{since } \textstyle\sum p_i = \sum p_j = 1) \\
&= S_A + S_B.
\end{aligned}
$$

Thirdly, of course, if we take a thermally isolated system in equilibrium, and upset the equilibrium by removing some internal constraint (by removing some internal thermal insulation or opening a valve, for instance), then, when the system has again reached thermal equilibrium, *the change in the equilibrium entropy ΔS which has occurred cannot be negative*. This is just a special case of the H-theorem (4.6).

Finally, we may conveniently mention here a special form of equilibrium entropy known as *configurational entropy*. Usually the states of a large system are spread out continuously in energy and we cannot identify any simple countable set of states as having the same energy. But situations do arise where the opposite is the case, where a large system has a definite, countable number W of different configurations all having exactly the same energy. Our system of 100 identical oscillators is a rather artificial example of this. Another example would be a crystal containing a definite number of vacancies, which could be arranged in a definite, finite number of configurations. In such a case, g must contain a factor W corresponding to the number of possible configurations and, since $S = k \ln g$, there must be a corresponding contribution to the entropy given by

Configurational entropy

▶ $$S_{\text{config}} = k \ln W. \tag{4.9}$$

This idea is often useful in detailed statistical calculations.

4.4 The connection between equilibrium entropy and reversible heat and work

We saw in §1.4 that a slow, frictionless change of constraint is a *reversible process*. One of the most important properties of the equilibrium entropy is that it is connected with the heat which enters the system in such a reversible process. We shall later give a macroscopic thermodynamic argument which leads to this result. Here, let us approach it by considering the meanings of heat and work in microscopic terms.

§6.9

For irreversible processes there is no clear microscopic distinction between heat and work, and, indeed, dissipative work has effects on the system which are indistinguishable from those of heat entering. For a reversible change, however, the situation is quite different. Let us first consider from the microscopic point of view a small reversible adiabatic change, in which no heat enters. In a slow movement of the constraints the individual quantum states slowly change, but remain identifiable. Now one can show using quantum mechanics that a sufficiently slow movement of the constraints does not, of itself, cause quantum jumps between one state and another. This is the *principle of adiabatic invariance*: when the constraints are slowly altered, the system stays initially in the same state. The energy of this state will, in general, change, and it is this change which represents the work done. The average reversible work done in such a small change of constraint is given by

§1.4

The principle of adiabatic invariance

$$\blacktriangleright \quad \mathrm{d}w_{\mathrm{rev}} = \sum \tilde{p}_i \, \mathrm{d}E_i. \tag{4.10}$$

We must not suppose, however, that a completed small change of this sort necessarily leaves the $\tilde{p}_i$ values unchanged at the end of the process. In general, such a change will alter the various state energies E_i by different amounts. This broadens the energy distribution somewhat, and normally leaves the system slightly out of thermal equilibrium. After the change has occurred, therefore, spontaneous quantum jumps inside the system will subsequently adjust the $\tilde{p}_i$ values until equilibrium is regained. These subsequent jumps conserve internal energy, however, and are not involved in the expression for $\mathrm{d}w_{\mathrm{rev}}$.

In a more general reversible change we may have heat entering the system as well. But this clearly does not affect the macroscopic expression for $\mathrm{d}w_{\mathrm{rev}}$ in terms of the response: that remains equal to $-P\,\mathrm{d}V$ for a reversible change of volume, for instance, whether heat enters or not. Thus, having identified $\mathrm{d}w_{\mathrm{rev}} = -P\,\mathrm{d}V$ as $\sum \tilde{p}_i \, \mathrm{d}E_i$ by considering a reversible adiabatic change, it is clear that the identification must hold also for *all* reversible changes. But the mean energy E is equal to $\sum \tilde{p}_i E_i$, and $\mathrm{d}E = \sum \tilde{p}_i \, \mathrm{d}E_i + \sum E_i \, \mathrm{d}\tilde{p}_i$. Since $\mathrm{d}E = \mathrm{d}w_{\mathrm{rev}} + \mathrm{d}q_{\mathrm{rev}}$, we see by subtraction that

$$\blacktriangleright \quad \mathrm{d}q_{\mathrm{rev}} = \sum E_i \, \mathrm{d}\tilde{p}_i. \tag{4.11}$$

It is evident that the clear macroscopic distinction between heat and work corresponds, in the case of reversible changes, to an equivalent microscopic distinction: *reversible work is the part of the energy change associated with changes of energy level at fixed occupation, while reversible heat flow is associated with changes in occupation at fixed energy level.*

Microscopic meanings of reversible work and heat

For a large system near to equilibrium, or small systems in contact with each other, we may always, without loss of precision, suppose that we are dealing with a canonical state, and so replace $\tilde{p}_i$ by p_i in the above relations. In this connection we observe that, in terms of canonical entropy, we may write

$$\begin{aligned} dS &= -k \sum dp_i \ln p_i \\ &= -k \sum dp_i(-E_i/kT - \ln Z) \\ &= \sum E_i \, dp_i/T, \end{aligned} \tag{4.12}$$

where we have used the Boltzmann distribution (3.9), and twice used the fact that $\sum dp_i = 0$. It follows that we can write, using (4.11),

Entropy and reversible heat flow

$$\blacktriangleright \quad dq_{\text{rev}} = T dS. \tag{4.13}$$

This unexpected relation between heat flow and entropy is extremely useful and important. Note that in general it applies only to *reversible* flows of heat.

4.5* The third law

The third law of thermodynamics is concerned with the behaviour of the entropy as the temperature tends to zero. It was discovered experimentally by Nernst that when a system is changed from one low temperature equilibrium state to another (when we change, for instance, liquid into solid helium by applying pressure, or convert separate samples of sodium and chlorine at, say, 4 K into sodium chloride at 4 K, by successively warming to a high temperature, allowing to react, and cooling the sodium chloride produced), then the net change of entropy is always very small. It seemed, therefore, that *the entropy of all states of all substances must be the same when sufficiently near absolute zero.* The theoretical basis for this is not hard to see. At room temperature the probability for a large system is spread over typically $10^{10^{23}}$ states, and the corresponding entropy is of order $10^{23}k$. It seems likely that near $T=0$ all systems are much more highly ordered, that the probability is spread over a very much smaller number of states, so that *the entropy is very small, or essentially zero.* It therefore seems reasonable to adopt, as an empirical law, the *Nernst theorem* or *third law of thermodynamics*:

> The equilibrium entropies of all systems and the entropy changes in all reversible isothermal processes tend toward zero as the temperature approaches zero.

It is, however, by no means obvious that there are no exceptions to this law. As a potential counter-example, consider the question of nuclear spin. In solid copper, for instance, the copper nuclei have a non-zero spin quantum number $I=\frac{3}{2}$. Each nucleus has, in zero magnetic field, a degeneracy of $2I+1$, corresponding to different orientations of the nuclear spin. The nuclear spins are essentially independent of everything else in the metal, and have a configurational entropy of $Nk \ln (2I+1)$, which is large. This entropy shows little sign of decreasing down to temperatures as low as 10^{-4} K.

This example illustrates an important point. If it were possible to treat systems at low temperatures as assemblies of small, identical parts (and this looks at first sight like a good model for the nuclear spins in copper), then the third law would hinge on the question of whether the ground states of the separate parts were degenerate: if the ground states are degenerate it seems that the law must fail. But to argue in this way is to ignore the fact that there are always some forces, however weak, which couple the separate parts together – if there were no such interactions the parts could not exchange heat. At the very lowest temperatures these couplings *always* become important, and this means that we must be concerned with the *collective motions* of the system as a whole. In a solid lattice, for instance, even at relatively high temperatures, it is not a good approximation to treat the different atomic vibrations as independent – we have to consider instead the vibrational modes of the solid as a whole. Even in the case of the nuclear spins the very weak coupling to the spins of the conduction electrons leads to *spin wave modes* which become important at temperatures below 10^{-5} K.

Importance of collective motions

§ 10.4

The question of the behaviour of the entropy at the lowest temperatures therefore has to be approached rather differently. The point is that the system will always have *some* very long wavelength modes, which have very closely spaced energy levels. Such modes will retain an appreciable entropy at the lowest temperatures, while other modes lose essentially all their entropy at higher temperatures. The theoretical discussion of the third law in this situation is a matter of finding how the *proportion* of modes which retain appreciable entropy falls as the temperature is lowered. For any given system it is usually easy to see that the proportion of excited modes does drop smoothly to zero as T is lowered. However, it has not so far proved possible to show that S tends to zero faster than any particular function of T for *all* systems, and in this sense the precise content of the third law remains a little obscure.

§ 6.9

The law is nevertheless useful, and is supported by experiment. In making use of it in measurements of entropy two points need to be watched. Firstly, as we have just seen, the law gives no hint of how close to $T=0$ we have to be before any particular form of entropy clearly approaches zero, and *the user must be alert to the possibility of unquenched entropies* at surprisingly low temperatures. Secondly, there is the distinct, but related, point that *the third law only applies to true equilibrium entropies.* At the lowest temperatures systems may easily be frozen into non-equilibrium states. For instance, a glass at low temperatures is not in true equilibrium, and the heat which has to be added, apparently reversibly, in raising it to the melting point cannot validly be used to measure the entropy at the melting point in the usual way.

Problems

4.1 Use the definition of canonical entropy (4.7) to find the values of S for the small system of problem 3.3.

4.2 Find the entropy at room temperature of the system of free electrons of problem 3.2, using the large system approximation (4.8). Confirm that the approximation is valid by showing that scaling g by a factor of 10^{100} has a negligible effect on the value of S.

4.3 Prove Equation (2.9). [One way of doing this is to show that W is equal to the number of distinct ways of arranging Q indistinguishable quanta and $N-1$ indistinguishable partitions in a row.]

In problem 3.1(ii) suppose that, before thermal contact is made, system A contains 2×10^3 J and system B contains 10^3 J of energy, in addition to the ground state energy. Use Stirling's approximation (§ 2.3) to find the total configurational entropy (4.9) (i) before contact, and (ii) after equilibrium is achieved.

4.4 In problem 4.3, 100 quanta flow from A into B immediately after contact is made. By what factor does $W=W_A W_B$ change? What is the corresponding change of entropy for each system? Check that $\mathrm{d}q=T\mathrm{d}S$ for each system.

4.5 A small system has two states only, of energy $E_1=0$ and $E_2=10^{-22}$ J, and is held at 5 K while the constraints on the system are changed slightly. This has the effect of increasing E_2 to 1.01×10^{-22} J, while E_1 is not affected. Find the work done on the system and the heat leaving it, using (4.10) and (4.13).

5

Elementary theory of the ideal monatomic gas

At this stage we need an illustration to show that the formal theory explored so far does produce intelligible and useful results when applied to a simple system. I have chosen as illustration the ideal gas because its properties are likely to be familiar, and because it will provide useful illustrations later in the book. Moreover, we introduced Boltzmann's constant k into our definition of temperature with the promise that this would make our statistical temperature T agree with the gas scale temperature θ, and we shall take the opportunity to establish that connection in this chapter. It must be said, however, that the full treatment of the gas problem involves some complications which we shall deal with later, in Chapter 12, but side step for the moment. For that reason I shall keep the discussion here fairly elementary. We shall limit the discussion to monatomic gases.

5.1 The model of the gas

A monatomic gas is a system of N free atoms of mass m moving around at random inside a container. Such a system will always expand to fill the container available. When the density is low enough, the gas is said to be *ideal*. From the theoretical point of view this means that collisions between the atoms are very rare, and for most of the time the atoms behave as though their neighbours were not present. The collisions serve to maintain thermal equilibrium by occasionally causing the atoms to jump from one state into another, but apart from this we can for most purposes ignore them. If we think of the atoms as classical particles of mass m we may imagine them moving in straight lines at some speed c which has velocity components u_x, u_y and u_z in a suitable co-ordinate system: c^2 will be equal to $u_x^2+u_y^2+u_z^2$. In thermal equilibrium at temperature T the velocity will be randomised and equally likely to lie in any direction. Thus the mean square values $\overline{u_x^2}$, $\overline{u_y^2}$ and $\overline{u_z^2}$ will be equal, and equal to $\frac{1}{3}\overline{c^2}$. We expect that the value of $\overline{u_x^2}$ will be related in some way to the temperature and, more specifically, that at temperature T there will be some definite probability distribution $f_u(u_x)$ for the velocity component u_x, which we should like to know. (The other components u_y and u_z will have the

The ideal gas

same distribution as u_x.) There will also be a definite probability distribution $f_c(c)$ for the speed c. Because the atoms are very far apart, we can assume that any potential energy due to interatomic forces is negligible. Thus the internal energy E of the gas is just the sum of the kinetic energies of the atoms: if there are N atoms we may write $E=\frac{1}{2}Nm\overline{c^2}=\frac{3}{2}Nm\overline{u_x^2}$. The atoms of the gas will collide periodically with the walls of the container. When they do so they deliver an impulse to the wall: it is the sum of these random impulses which constitute the pressure of the gas.

Internal energy in terms of mean square velocity

We have in earlier chapters written our statistical theory in quantum language, so we need to understand the possible energy states of the atoms in the gas. Consider an idealised rectangular container of sides L_x, L_y and L_z, with perfectly smooth elastic walls, containing just one atom. What states can this atom have? In *classical* mechanics the atom will have a trajectory of the type shown in Fig. 5.1(*a*). The atom is continuously bouncing off the walls, but we notice that, although u_x is periodically reversed in sign, its modulus $|u_x|$ is constant in this motion, and the same is true of $|u_y|$ and $|u_z|$. In *quantum* mechanics the atom is described by a wave function. When the wave function is confined inside the rectangular container, it can be shown that it takes the form of a three-dimensional standing wave

The quantum states of the atoms

$$\psi=\psi_0 \sin\left(\frac{2\pi x}{\lambda_x}\right)\sin\left(\frac{2\pi y}{\lambda_y}\right)\sin\left(\frac{2\pi z}{\lambda_z}\right)e^{-i\omega t}. \tag{5.1}$$

Since the standing wave has to fit inside the container (Fig. 5.1(*b*)), the length

Fig. 5.1. States of a single atom in a perfect rectangular enclosure. (*a*) shows a classical trajectory; note that $|u_x|$ and $|u_y|$ are conserved. (*b*) shows the equivalent wave function: here $|u_x|$ and $|u_y|$ are *quantised*.

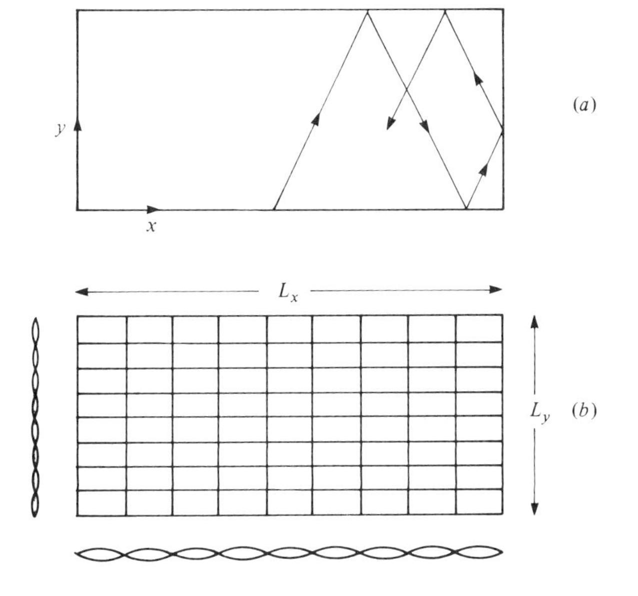

L_x of the container has to be an integral number of half-wavelengths, so that $\lambda_x = 2L_x/n_x$, where n_x is an integer. (Similar considerations apply to λ_y and λ_z.) Now in quantum mechanics we have *de Broglie's relation*, $p_x = h/\lambda_x$, which for a running wave function of the form

Quantisation of velocity and energy

$$\psi = \exp 2\pi i(x/\lambda_x + y/\lambda_y + z/\lambda_z) \exp(-i\omega t)$$

relates the momentum component p_x of the atom with the wavelength of the wave function through Planck's constant, h. If we analyse (5.1) into a mixture of such running waves we find that we must identify the x component of velocity u_x, with $\pm h/\lambda_x m$. Because only certain values of λ_x are possible, u_x is *quantised* – it is limited to the equally spaced values

$$u_x = \pm(h/2mL_x)n_x. \tag{5.2}$$

The spacing $h/2mL_x$ between the allowed values is, however, very small, of order $10^{-7}\ \mathrm{ms}^{-1}$ for a typical atom in a container of everyday size. By comparison, an atom at room temperature has a velocity of order $10^2\ \mathrm{ms}^{-1}$, so n_x is normally a very large number, and the quantisation is on so fine a scale that it is not very noticeable. Since the energy of the atom ε is equal to $\frac{1}{2}m(u_x^2 + u_y^2 + u_z^2)$, the energy is also quantised. The possible values are given by

$$\varepsilon = \frac{h^2}{8m}\left(\frac{n_x^2}{L_x^2} + \frac{n_y^2}{L_y^2} + \frac{n_z^2}{L_z^2}\right). \tag{5.3}$$

The quantum states of the atom, then, are labelled by the set of three integers (n_x, n_y, n_z). They have quantised velocities given by (5.2) and quantised energies given by (5.3).

5.2 The one-component velocity distribution and the internal energy

We can find the probability distribution for u_x very easily by applying the Boltzmann distribution to the single atom inside our rectangular enclosure. We first find the distribution of u_x for *particular* quantised values of u_y and u_z – that is, we hold n_y and n_z fixed, but allow n_x to vary. Let us find, subject to this limitation, the probability that the x component of velocity lies in the narrow range between u_x and $u_x + \mathrm{d}u_x$. The allowed u_x values are very closely and uniformly spaced. Thus, *the number of quantum states lying in the range of interest will be proportional to* $\mathrm{d}u_x$. Now, the Boltzmann distribution tells us that the probability of finding the atom in any single quantum state is proportional to the Boltzmann factor, $\exp(-\varepsilon/kT)$. Here ε is equal to $\frac{1}{2}m(u_x^2 + u_y^2 + u_z^2)$. Since we are holding u_y and u_z fixed, we may say that the *state probability for each of the states in the range is proportional to* $\exp(-\frac{1}{2}mu_x^2/kT)$. Thus the probability of finding u_x in the required range, which is the number of states in the range multiplied by the state probability, is given by

$$f_u(u_x)\,\mathrm{d}u_x = A\,\mathrm{e}^{-\frac{1}{2}mu_x^2/kT}\,\mathrm{d}u_x, \tag{5.4}$$

where A is a constant of proportionality. To find the constant we remember that the distribution must be *normalised*, that $\int_{-\infty}^{\infty} f_u(u_x)\,\mathrm{d}u_x = 1$. Since this distribution is Gaussian, A is most easily determined by comparison with the standard normalised form of the Gaussian, and we find that

§3.4

$$f_u(u_x)=\left(\frac{m}{2\pi kT}\right)^{\frac{1}{2}} e^{-mu_x^2/2kT}$$

and (5.5)

$$\overline{u_x^2}=kT/m.$$

This is *Maxwell's one-component velocity distribution.* So far, we have derived it for fixed values of u_y and u_z. But since u_y and u_z do not appear in the distribution it is clear that *the distribution of u_x is actually independent of u_y and u_z*. In fact, since it applies for every possible set of values for u_y and u_z it applies equally well when we include all possible states of the atom in the calculation. The distribution is plotted in Fig. 5.2(*a*). It is symmetrical, and $\overline{u_x}=0$, as one would expect. It is often useful in calculating the detailed kinetics of phenomena in gases.

The distribution for u_x is independent of u_y and u_z

We noted in §5.1 that the internal energy of the gas is given by the relation $E=\frac{3}{2}Nm\overline{u_x^2}$. Having found the value of $\overline{u_x^2}$, we may now write

$$E=\tfrac{3}{2}NkT \tag{5.6}$$

for the internal energy of a monatomic gas at temperature T. Some of the implications of this important result are set out in §5.5.

5.3 The speed distribution

To help us in finding the distribution of the speed, c, we shall introduce the idea of *velocity space*, a concept which will be developed further

Velocity space

Fig. 5.2. (*a*) shows the one-component velocity distribution $f_u(u_x)$ and the speed distribution $f_c(c)$ for atoms of mass m at temperature T, plotted on the same scales. (*b*) shows velocity space. In finding the speed distribution we need to determine the probability that the *representative point P* lies within a spherical shell of radius c and thickness dc.

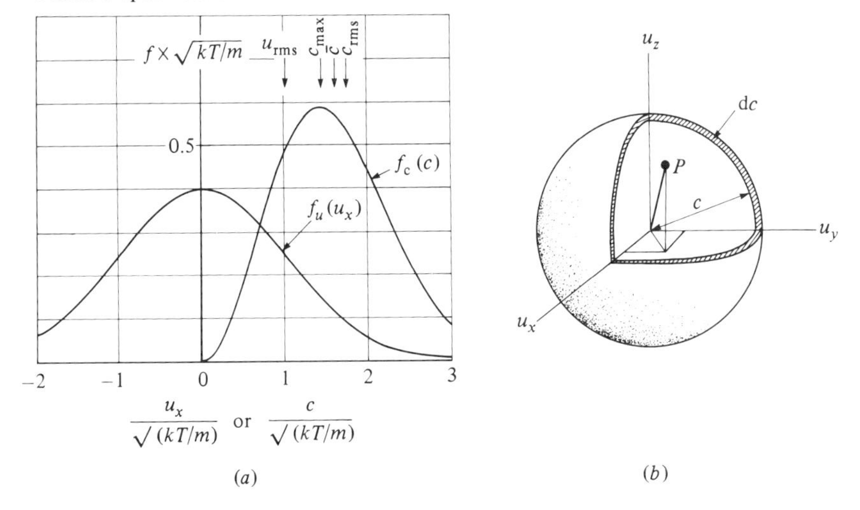

later in the book. For a single atom, velocity space is a three-dimensional space in which we plot u_x, u_y and u_z along the three axes (Fig. 5.2(*b*)). In velocity space we introduce the *representative point* whose co-ordinates correspond to the velocity vector of the atom, and whose distance from the origin represents the speed, c. We wish to find the probability that the speed lies in the range between c and $c+\mathrm{d}c$. This means that, in velocity space, we wish to find the probability that the representative point lies within a spherical shell of radius c and thickness $\mathrm{d}c$.

Because the probability distributions for u_x, u_y and u_z are independent, as we noted in § 5.2, we may write $\mathrm{d}p$, the probability that u_x lies in a narrow range of width $\mathrm{d}u_x$, while u_y and u_z lie *simultaneously* in narrow ranges of width $\mathrm{d}u_y$ and $\mathrm{d}u_z$, as a product,

$$\mathrm{d}p=(f_u(u_x)\,\mathrm{d}u_x)(f_u(u_y)\,\mathrm{d}u_y)(f_u(u_z)\,\mathrm{d}u_z)$$

or

$$\mathrm{d}p/\mathrm{d}V_v=\left(\frac{m}{2\pi kT}\right)^{\frac{3}{2}}\mathrm{e}^{-mc^2/2kT}, \tag{5.7}$$

where $\mathrm{d}V_v$ stands for $\mathrm{d}u_x\,\mathrm{d}u_y\,\mathrm{d}u_z$, an element of volume in velocity space. *This relation shows us the probability* $\mathrm{d}p/\mathrm{d}V_v$ *of finding the representative point per unit volume in velocity space.* Now the volume of the spherical shell which corresponds to speeds between c and $c+\mathrm{d}c$ is $4\pi c^2\,\mathrm{d}c$. The probability that c lies in this range is given by this volume multiplied by the probability per unit volume, so

$$\blacktriangleright \qquad f_c(c)\,\mathrm{d}c=\left(\frac{m}{2\pi kT}\right)^{\frac{3}{2}}\mathrm{e}^{-mc^2/2kT}4\pi c^2\,\mathrm{d}c. \tag{5.8}$$

This is the *Maxwell–Boltzmann speed distribution.* It is shown in Fig. 5.2(*a*). Note that it is not Gaussian, and not symmetrical. It has been confirmed by experiment to high accuracy and is very useful in discussing kinetic problems in gases.

5.4 The equation of state

The *equation of state* for a substance means the expression which gives the pressure as a function of volume and temperature. We shall now derive the equation of state for a monatomic gas from first principles.

The first step is to relate the pressure to the velocities of the atoms. We do this first by the *kinetic argument.* Think of the single atom alone in the container, as shown in Fig. 5.1. In the classical picture, $|u_x|$ is conserved, and the atom collides regularly with the right-hand wall at a frequency equal to $|u_x|/(2L_x)$. Each time it does so, it delivers an impulse to the wall equal to its change in momentum, which is $2m|u_x|$, normal to the wall. The mean force on the wall is the impulse delivered in unit time, which is mu_x^2/L_x, corresponding to a pressure of mu_x^2/V.

It is instructive to see that we can reach the same result using the quantum picture. If we slowly move the right-hand wall to the right the atom will not jump into a new state (with new values of n_x, n_y and n_z), but the energy of the original state will gradually change because L_x is increasing. In fact, §4.4

$$\begin{aligned} d\varepsilon &= d\left[\frac{h^2}{8m}\left(\frac{n_x^2}{L_x^2}+\frac{n_y^2}{L_y^2}+\frac{n_z^2}{L_z^2}\right)\right] \\ &= -\frac{h^2 n_x^2}{4mL_x^3}\, dL_x \\ &= -(mu_x^2/L_x)\, dL_x. \end{aligned}$$

The fall in $d\varepsilon$ represents work done on the moving wall, so the force on the wall must be mu_x^2/L_x. This is the same as we found by the kinetic argument, and again corresponds to a pressure mu_x^2/V.

When there are many atoms present we suppose that, when the density is low and the gas is ideal, each atom will exert its own pressure independently of the others, except that each atom will be colliding occasionally, and we must therefore replace u_x^2 by its average value, $\overline{u_x^2} = \frac{1}{3}\overline{c^2}$. For a gas containing N atoms, the total pressure will be given by $\frac{1}{3}Nm\overline{c^2}/V$, or §12.6

▶ $$P = \tfrac{1}{3}nm\overline{c^2}, \tag{5.9}$$

where n is the number density, N/V. This is the *kinetic theory expression for the pressure.* But we found in § 5.2 how the velocity is related to the temperature: we found that $\overline{u_x^2} = \frac{1}{3}\overline{c^2} = kT/m$. In terms of temperature, then, we have

▶ $$PV = NkT. \tag{5.10}$$

This is the *equation of state of the ideal monatomic gas.*

The equation of state shows how P depends on V, N and T, and each of these relations is of some interest. We see first that, at constant temperature, P is proportional to $1/V$, that is, P is proportional to the density (Fig. 5.3). This is *Boyle's law*, valid for all gases at low density, and discovered experimentally in about 1660. Rather more surprisingly, we see that *the formula for P depends only on the number of atoms present and not on any property of the atom itself* – it is quite independent of the mass or size or hardness of the atoms, for instance. Thus a given number of atoms of gas held at fixed pressure and temperature always occupies the same volume. Experiment shows that 1 mole of any gas (which contains N_A atoms, where N_A is *Avogadro's number*, 6.02×10^{23}) at standard atmospheric pressure, which is 1.013×10^5 N m^{-2}, and at the temperature of melting ice, always occupies 0.0224 m^3.

5.5 Gas scale temperature, the energy kT and Boltzmann's constant

Gas scale temperature θ is defined by making θ proportional to the pressure of a suitable gas held at constant volume and low density; the constant of proportionality is fixed by choosing the triple-point temperature of water to be 273.16 K. Statistical temperature, on the other hand, was defined by writing $1/kT = \partial \ln g/\partial E$ for large systems. Starting from this definition we have been able to deduce that $PV = NkT$ for an ideal gas. This §3.1

shows that P is proportional to the *statistical* temperature at fixed volume, so it is clear that T must be proportional to θ. The two can be made equal by choosing the scaling constant k appropriately. In this connection it is clear from the definition of statistical temperature that *kT is an energy characteristic of the temperature.* Indeed, it would have been possible and in some ways more natural to use kT as a 'temperature' having units of joules. Our derivation of the equation of state has shown us that kT is equal to PV/N for a gas. Using the experimental data given above, we find that PV/N has the value of 3.77×10^{-21} J at the temperature of melting ice. It follows that if we wish to make the gas scale and the statistical temperature scale agree, so that T is to be 273.15 K at this temperature, we must choose for k the value 1.38×10^{-23} J K^{-1}. This is *Boltzmann's constant*. It is perhaps best thought of as a scaling factor which links gas scale temperature, measured in the quite arbitrary units of degrees kelvin, with the more natural energy measure of temperature, kT. Because the degree kelvin is defined in an arbitrary way, this scaling factor has to be found experimentally and cannot be deduced from first principles.

From now on we shall treat gas scale and statistical temperatures as identical.

Fig. 5.3. Types of expansion for a monatomic gas. In *isothermal* expansion T is constant and PV is also constant (Boyle's law). The work done by the gas is the area under the curve, and is supplied by heat entering the gas from its surroundings. In *adiabatic* expansion no heat enters, S is constant, and PV^{γ} is also constant. The gas does work at the expense of its internal energy, and so is cooled. In *Joule* expansion the gas moves irreversibly from A to B, but no heat enters and no external work is done.

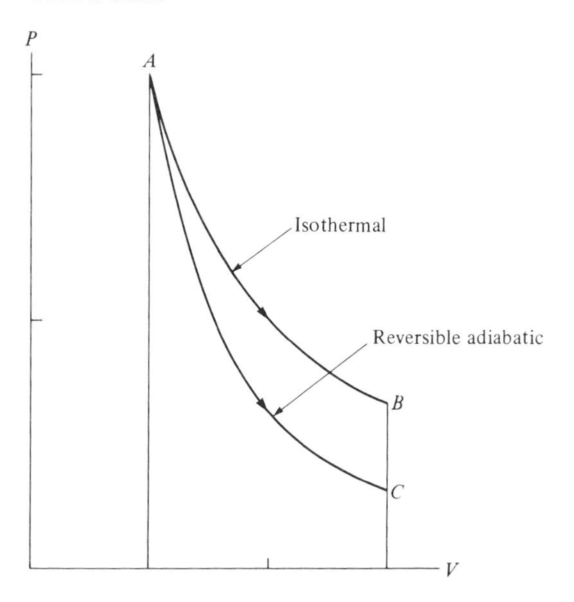

5.6 Properties related to heat entering the gas

Knowing that the internal energy of the ideal monatomic gas is given by $E=\frac{3}{2}NkT$, and that its equation of state is $PV=NkT$, we may deduce a number of useful properties, which we shall summarise here quite briefly.

If we heat the gas at constant volume we do no work on it, so any changes in E must be solely due to the heat entering the gas. It follows that the *heat capacity at constant volume* C_V must be given by

The heat capacities C_V and C_P

$$C_V=\tfrac{3}{2}Nk. \tag{5.11}$$

If, on the other hand, we heat the gas at constant pressure, it will expand and do work on the container equal to $P\,\mathrm{d}V$, which we can write as $Nk\,\mathrm{d}T$, using the equation of state. So we may write the heat entering the gas as $\mathrm{d}q=\mathrm{d}E+P\,\mathrm{d}V=\frac{3}{2}Nk\,\mathrm{d}T+Nk\,\mathrm{d}T$. Thus $(\mathrm{d}q/\mathrm{d}T)_P$, the *heat capacity at constant pressure* C_P, is $\frac{5}{2}Nk$, or

$$C_P=C_V+Nk. \tag{5.12}$$

If the gas expands, but is held at constant temperature (an *isothermal expansion*), it once again does work on its surroundings. But its internal energy does not change at constant T, so we are forced to the remarkable conclusion that *all* the work done by the gas in an isothermal expansion is supplied by *heat* which enters the gas from the surroundings. The gas simply acts as a channel through which the heat is converted directly into work. If, on the other hand, the gas expands reversibly while thermally insulated, the work which it does must be accounted for by a fall in internal energy, so $P\,\mathrm{d}V=-\frac{3}{2}Nk\,\mathrm{d}T$. Using the equation of state, the last term may be written as $-\frac{3}{2}(P\,\mathrm{d}V+V\mathrm{d}P)$. Thus $\frac{5}{2}\,\mathrm{d}V/V+\frac{3}{2}\,\mathrm{d}P/P=0$, and on integrating this equation we find that

Reversible isothermal expansion

Reversible adiabatic expansion

$$PV^{\gamma}=\text{constant for the reversible adiabatic,} \tag{5.13}$$

where $\gamma=\frac{5}{3}=C_P/C_V$. For an isothermal expansion, of course, $PV=$constant; the pressure falls *faster* in an adiabatic expansion because the gas is being cooled (Fig. 5.3). A thermally insulated gas may also be allowed to expand irreversibly through a valve into an insulated, evacuated container (a *Joule expansion*). In this process the gas only does work *on itself*. Its internal energy is therefore unchanged, and we deduce that the gas does *not* cool in a Joule expansion – its temperature is unchanged. (Its end point is actually the same as it would have been after an equivalent reversible isothermal expansion.)

Joule expansion

Finally, consider the entropy S of the gas. We may conveniently discuss changes in entropy by remembering that $\mathrm{d}S=\mathrm{d}q_{\mathrm{rev}}/T$. For changes at fixed temperature, as we have already noted, the heat entering is equal to the work done by the gas, so

§4.4

$$\mathrm{d}S=\frac{\mathrm{d}q_{\mathrm{rev}}}{T}=\frac{P\,\mathrm{d}V}{T}=\frac{Nk\,\mathrm{d}V}{V},$$

using the equation of state in the last step. At constant volume, on the other hand,

Entropy of a monatomic gas

$$dS = \frac{dq_{rev}}{T} = C_V \frac{dT}{T} = \frac{3}{2} Nk \frac{dT}{T}.$$

Integrating these equations for changes in both T and V we find that $S = Nk \ln (VT^{\frac{3}{2}}) + \text{const}$, which may be written as

▶ $$S = Nk \ln (\alpha V T^{\frac{3}{2}}), \quad (5.14)$$

where α is an unknown quantity independent of V and T. This expression for the entropy of the ideal monatomic gas is often useful. (We shall calculate the entropy again, using the statistical picture of entropy, in Chapter 12, and the value and significance of the quantity α will be discussed at that stage.)

§ 12.6

Problems

5.1 Starting from the distributions $f_u(u_x)$ (5.5) and $f_c(c)$ (5.8), show that $u_{rms} = (kT/m)^{\frac{1}{2}}$, $c_{max} = (2kT/m)^{\frac{1}{2}}$, $\bar{c} = (8kT/\pi m)^{\frac{1}{2}}$ and $c_{rms} = (3kT/m)^{\frac{1}{2}}$. [$\int_0^\infty s^n e^{-s} ds = n!$. Where n is non-integral and greater than -1, $n!$ is *defined* by this integral. When n is half-integral $n!$ is equal to $\sqrt{\pi}\, \frac{1}{2} \times \frac{3}{2} \times \frac{5}{2} \ldots n$.]

5.2 It may be shown that the number of atoms of a gas which strike area A of the walls of its container in unit time, and have speeds lying between c and $c + dc$, is equal to $\frac{1}{4} Ancf_c(c)\, dc$. A gas leaks from its container through a hole of area A into a vacuum. The hole is much smaller than the free path of the atoms between collisions, and this ensures that the distribution of atoms arriving at the hole is not substantially different from the equilibrium distribution. By integrating over all velocities obtain expressions for the total number of atoms and the total kinetic energy which leave the gas in unit time. Hence show that the mean kinetic energy of the atoms leaving the gas is $2kT$. Why, in general terms, is this greater than the mean kinetic energy of the atoms inside the gas?

5.3 An ideal gas of molecules has the same equation of state $PV = NkT$ as an ideal monatomic gas (why?), but because it has internal motions its heat capacity at constant volume C_V is greater than $\frac{3}{2}Nk$. If we continue to write C_P/C_V as γ, show that PV^γ, $TV^{\gamma-1}$ and $P^{1/\gamma - 1}T$ are all constant during an adiabatic expansion of *any* ideal gas.

5.4 The formula for the velocity of sound in a fluid is $(K/\rho)^{\frac{1}{2}}$, where K is the appropriate *bulk modulus*, $-V dP/dV$, and ρ is the density. In a gas a sound wave always passes so quickly that the compressions and rarefactions are effectively adiabatic. Use the results of problem 5.3 to show that the velocity of sound in a gas cannot exceed $(\sqrt{5/3})c_{rms}$, and comment on this result. Find the velocities of sound in argon (40 amu) and helium (4 amu) at room temperature. Why does the velocity of sound in the atmosphere fall with height?

5.5 An ideal monatomic gas at 300 K contained in a cylinder of negligible heat capacity expands slowly through a small valve into an identical evacuated cylinder. The two cylinders are thermally isolated from each other. Explain why the gas which remains in the first cylinder when the pressures are equalised must have expanded adiabatically, and show that the energy must then be divided equally between the cylinders. Hence find the final temperature in each cylinder.

6

The basic principles of classical thermodynamics

In this and the following two chapters we turn to *classical thermodynamics*, the part of the theory of heat in which we treat the thermodynamic system as a 'black box' and ignore the details of its internal structure. Classical thermodynamics is based on three laws. They are often introduced empirically, as experimental laws. In this book I prefer to treat them as consequences of the laws of quantum mechanics, obtainable from the theory of statistical thermodynamics which we examined in Chapters 2, 3 and 4. Nevertheless, in order to exhibit classical thermodynamics as a separate discipline, unconcerned with the microscopic details of systems, I shall not in this chapter make any use of the *microscopic* interpretations of temperature and entropy developed earlier. Instead I shall take as our starting point certain *macroscopic* statements about thermodynamic systems. If you have read only Chapter 1, you will be able to follow this chapter if you are prepared to accept these statements as plausible. If you have followed Chapters 2, 3 and 4 you will have some idea of how these statements may be established from first principles, and you will recognise that we are at various points repeating in thermodynamic language arguments which we set out earlier in microscopic terms.

6.1 General assumptions

We shall make certain general assumptions about thermodynamic systems which we have already discussed in Chapter 1, and set out again briefly here. We shall assume that the system is subject to *constraints*, under the observer's control. The most obvious type of constraint is the volume V of a fluid, which we shall use as a prototype, but there are many others, such as the electric and magnetic fields, the length and twist of a wire, the area of a soap film, etc. In thermal equilibrium the system generates an *equilibrium response* to each constraint, which determines the *reversible work* $\mathrm{d}w_{\mathrm{rev}}$ done when the constraint is changed. For the volume the response is the equilibrium pressure P, and $\mathrm{d}w_{\mathrm{rev}} = -P\,\mathrm{d}V$. For an external magnetic field of strength $\mathscr{B}_{\mathrm{E}}$ the response is the magnetic moment M and $\mathrm{d}w_{\mathrm{rev}} = -M\,\mathrm{d}\mathscr{B}_{\mathrm{E}}$. For the length L of a wire, the response is the tension F, and $\mathrm{d}w_{\mathrm{rev}} = F\,\mathrm{d}L$. We shall assume that the

§ 1.3

§ 1.4

§ 15.2

§ 1.3

system has a definite *internal energy* E, and that the nature of the equilibrium state is fixed once E and the constraints are fixed. Thus all quantities which take definite values in a given equilibrium state, the quantities known as the *functions of state*, will be functions of E and the constraints. The pressure, for instance, depends on E, V and any other constraints there may be, such as the electric and magnetic fields. If, however, a particular constraint, such as a magnetic field, is never varied it ceases to have any formal significance and we need never mention that P does in fact depend on such quantities, writing it simply as $P(E, V)$. The functions of state will show *thermal fluctuations*, but we shall usually assume that the system is large enough to make them negligibly small. Functions of state will normally be represented by a capital letter.

§ 1.3

In a homogeneous system, such as a quantity of uniform gas or solid, it is sometimes useful to distinguish between functions of state which are *extensive* variables, quantities such as the internal energy and the volume which are proportional to the quantity of the substance present, and those which are *intensive* variables, quantities such as the pressure, magnetic field or temperature which measure the strength of some parameter which is the same at all points. One must remember, however, that thermodynamic arguments may also be applied to inhomogeneous systems where the distinction between extensive and intensive variables is less clear.

Extensive and intensive variables

6.2 The first law

We shall take over from quantum mechanics the idea that energy is conserved, which we discussed in detail in Chapter 1. This means that we may always write

§ 1.4

▶ $$\mathrm{d}E = \mathrm{d}w + \mathrm{d}q, \tag{6.1}$$

which expresses the fact that the work dw done *on* the system (whether by reversible movement of the constraints or as dissipative work), and the heat dq *entering* the system from other bodies, both contribute to a well-defined *internal energy* E. In such processes we understand that some *other* system is doing the work or losing heat equal to the heat entering the system of interest, so that, taking all bodies together, energy is conserved.

6.3 The second law: entropy

The second law may be stated in several different ways which appear to be different in content but can be shown to be logically equivalent. We shall state it first as follows:

> There exists an additive function of state known as the *equilibrium entropy* S, which can never decrease in a thermally isolated system.

We shall refer to this as the *entropy statement of the second law*. If you have followed Chapter 4 you will understand that the entropy is a measure of the disorder in the system; the second law expresses the fact that a thermally isolated system cannot spontaneously regain order which has been lost. If you have not followed Chapter 4, you are invited to accept the second law as a plausible idea which may in fact be demonstrated using the principles of statistical thermodynamics. In classical thermodynamics we do not discuss the *nature* of entropy. We shall see in due course, however, that it can be *measured*, and has units of $\mathrm{J\,K^{-1}}$. When we say that the entropy is 'additive' we mean that the joint entropy of two systems is the sum of their separate entropies.

As it stands, the second law applies only to thermally isolated systems. In general, the system of interest will not be thermally isolated – it may be able to exchange heat with its 'surroundings'. These 'surroundings' may almost always, without affecting the essential physics, be idealised as large bodies disconnected from the outside world, so that we may treat the *joint* system, of system plus 'surroundings', as thermally isolated. We may then apply the second law to the *joint* system, and we observe that

▶ $$\Delta S_{\mathrm{tot}} \geqslant 0, \tag{6.2}$$

where S_{tot} is the total entropy of the system plus its surroundings. This principle is known as *Clausius' inequality*. It applies to all changes of state for which the surroundings may be suitably idealised.

Clausius' inequality

We saw in § 1.4 that certain thermodynamic changes may be regarded as *reversible*. In a reversible change we cannot have $\Delta S_{\mathrm{tot}} > 0$, for by reversing the change we should have $\Delta S_{\mathrm{tot}} < 0$, in contradiction of Clausius' inequality. This leaves only one possibility: we must have

Conservation of entropy in reversible changes

▶ $$\Delta S_{\mathrm{tot}} = 0 \quad \text{for all reversible changes.} \tag{6.3}$$

For the special case of an adiabatic change, in which no heat enters the system from the surroundings, (6.2) and (6.3) apply to the entropy of the system alone. Thus, in particular,

▶ $$\Delta S = 0 \quad \text{for all reversible adiabatic changes.} \tag{6.4}$$

Evidently all reversible adiabatic changes are also *isentropic changes*, changes in which the internal entropy is conserved.

6.4 The second law: temperature

We shall make no assumptions about temperature, but shall derive its properties from the second law. Suppose that we have some process which transfers heat between two bodies A and B, leaving the surroundings and the constraints on the bodies unchanged. Since entropy is a function of state it depends on the energy of the body concerned and its constraints. As the constraints on both bodies are held fixed we may write, after a small heat transfer,

$$\begin{aligned} \mathrm{d}S_{\mathrm{tot}} &= \mathrm{d}S_A + \mathrm{d}S_B \\ &= (\partial S_A/\partial E_A)_V\,\mathrm{d}E_A + (\partial S_B/\partial E_B)_V\,\mathrm{d}E_B \\ &= [(\partial S_A/\partial E_A)_V - (\partial S_B/\partial E_B)_V]\,\mathrm{d}q_A \geqslant 0, \end{aligned} \tag{6.5}$$

For partial differentiation see § 8.1

using the second law. The suffix V is here used as a shorthand to show that *all*

Definition of thermodynamic temperature

the constraints are to be held constant in the differentiation, and we have written $dq_A = dE_A = -dE_B$. We now *define* a quantity T, to be known as the *thermodynamic temperature*, by writing

$$1/T = (\partial S/\partial E)_V. \tag{6.6}$$

In terms of T we may rewrite our result as

$$(1/T_A - 1/T_B)\, dq_A \geqslant 0. \tag{6.7}$$

This equation shows that dq_A cannot be positive if $1/T_B > 1/T_A$. If we refer to the quantity $1/T$ as the *coldness*, we may restate this result in words as follows:

> No process exists in which heat is transferred from a colder body to a less cold body while the constraints on the bodies and the state of the rest of the world are left unchanged.

This is the *temperature statement* or *Clausius statement* of the second law. It shows that there is a quantity, the thermodynamic temperature, which alone determines the direction of heat transfer between bodies and that the temperature sequence from cold to hot is unique.

(Readers who have followed Chapters 3 and 4 will realise that, since S can be written as $k \ln g$ for large systems, the above definition of thermodynamic temperature is equivalent to the definition of *statistical temperature* adopted in Chapter 3. It follows from Chapter 5 that T is also identical with the *ideal gas scale temperature*. I shall give an alternative demonstration of this identity later in this chapter.)

§4.3 §3.1 §5.5 §6.6

6.5 The second law: heat engines

Pure heat flow

In the above discussion it was understood that the bodies A and B were gaining or losing energy by *pure heat flow*. In other words, it was understood that their constraints were held fixed and that no dissipative work was done on them. A body which provides heat in this way is known as a *heat source* and one to which heat is so rejected is known as a *heat sink*. Pure heat flow may be irreversible. For pure heat flow we have $dS = (\partial S/\partial E)_V\, dE$ or

$$dq = T\,dS \quad \text{for pure heat flow,} \tag{6.8}$$

where dq is the heat entering the body, equal to dE. This important relation between heat flow and entropy will be generalised later.

§6.9

We may use this result to discuss the problem which first stimulated the development of thermodynamics, that of the *efficiency of heat engines*. A heat engine is a device which turns heat abstracted from a heat source into work. We at once notice a difficulty. According to the result just derived the removal of heat from a heat source *lowers* the entropy of the source. If this heat is converted into useful work, the work must go somewhere; but since it may be stored by doing *reversible* work (by compressing a gas, for instance) the process of storage need involve no further change of entropy. Thus if the *whole* of the heat is converted into work, the total entropy can be made to fall. This

§6.3

would be contrary to the entropy statement of the second law, so we deduce that

> No process exists in which heat is extracted from a source at a single temperature and converted entirely into useful work, leaving the rest of the world unchanged.

This is the *heat engine statement* or *Kelvin statement* of the second law. It tells us that a heat engine cannot have an efficiency of 100%, and that part of the heat absorbed must be rejected to a heat sink.

The simplest possible heat engine is a device which works in a cycle, and in one cycle takes heat q_1 from a source at a high temperature T_1, converts part of the heat into useful work w, and rejects waste heat q_2 to a heat sink at a lower temperature T_2. Such a device is known as a *two-port engine*. At the end of one cycle the energy and entropy of the device itself are unchanged, so by energy conservation we know that $w = q_1 - q_2$. We also know that

Two-port engine

$$\Delta S_{tot} = \Delta S_1 + \Delta S_2 = -q_1/T_1 + q_2/T_2 \geqslant 0,$$

using the second law. We may rewrite this result as a condition on the waste heat q_2,

$$q_2 \geqslant (T_2/T_1)q_1. \tag{6.9}$$

The *efficiency* of the engine η is naturally defined as w/q_1, the ratio of the work got out to the heat put in at the higher temperature, which we may write as $1 - q_2/q_1$. Thus the engine will be at its most efficient when q_2 is as small as possible, and (6.9) shows that this occurs wherever the cycle is reversible, so that $q_2 = (T_2/T_1)q_1$ and $\Delta S_{tot} = 0$. Thus

Efficiency

§6.3

> All reversible two-port engines working between the same pair of temperatures have the same efficiency and no irreversible engine working between the same pair of temperatures can have a greater efficiency.

Carnot's principle

This is *Carnot's principle*, first enunciated in 1824. A reversible two-port engine is known as a *Carnot engine*. For a Carnot engine we have

▶ $$\frac{q_1}{q_2} = \frac{T_1}{T_2} \tag{6.10}$$

and the *ideal thermodynamic efficiency* η_I is therefore given by the relation

Ideal thermodynamic efficiency

▶ $$\eta_I = 1 - \frac{T_2}{T_1}. \tag{6.11}$$

The fact that the ideal efficiency of a heat engine depends in this simple way on the source and sink temperatures represents a new and remarkable property of the thermodynamic temperature. Kelvin, in fact, suggested that the thermodynamic temperature should be *defined* in terms of the efficiency of Carnot engines. For this reason the thermodynamic temperature is sometimes known as *work-scale temperature*.

Our technological society relies heavily on practical heat engines, and many types have been built. Steam and gas turbines, internal combustion engines and solar cells, for instance, are all heat engines. We shall give an example of a simple Carnot heat engine using an ideal gas in the next section, but we shall not have space in this book to discuss the design of practical heat engines. We should, however, note one important fact. Practical heat engines use many different sources of heat, such as burning fuel, nuclear reactions or solar radiation, while the heat sink is usually the atmosphere or, in the case of a power station, a river or the sea. Not much can be done in practice to change the heat sink temperature T_2, but it may well be possible to change T_1, the temperature at which the heat is delivered to the heat engine. Our expression for η_1 shows that it is best if the fuel is burnt and the heat delivered at the *highest possible temperature*. This is the reason for using superheated steam in power stations, high compression ratios in petrol and diesel engines and the high efficiency of the gas turbine. Typical best efficiencies are as follows (in 1983):

Practical heat engines

	Temperature at which heat is delivered	Ideal thermo-dynamic efficiency	Practical efficiency
Power station steam turbine	810 K	60%	47%
Petrol engine, 10:1 compression	650 K → 2800 K	60%	27%

If a two-port heat engine is run backwards, we have a device in which we put work *w in*, heat q_2 is *removed* from the cold body and heat q_1 is *delivered* to the hot body. This is a *refrigerator* or *heat pump*. Heat pumps are sometimes used to heat houses in winter by pumping energy from the surroundings into the house. For a heat pump it is natural to define the figure of merit, or *ideal heating efficiency*, as q_1/w (the ratio of the heat put into the house to the work done), which is $T_1/(T_1 - T_2)$, the *inverse* of the ideal heat engine efficiency. *For a heat pump the ideal heating efficiency is greater than unity* and we want $T_1 - T_2$ to be as *small* as possible. Heat pumps have the potential to save much energy – it is obviously much better to use electrical energy to pump heat from the surroundings into the house than simply to dissipate it in electric radiators. It should also be better, in theory, to burn fuel in power stations and use the power produced to run heat pumps in houses than to burn the fuel directly in the houses. In practice this is not yet true. It is difficult to make a heat pump work close to its ideal efficiency, because that requires very efficient heat exchangers with large surface areas, which are expensive, and exceptionally efficient mechanical compressors.

Heat pumps

6.6 Thermodynamic and gas scale temperatures

In this section we shall show that thermodynamic temperature is the same as gas scale temperature. We shall also take the opportunity to examine an example of a *Carnot cycle.*

See also §5.5

We imagine an ideal gas contained in a cylinder. The gas starts in state A (Fig. 6.1), where it is placed in contact with a heat source at thermodynamic temperature T_1. We then allow it to expand *reversibly and isothermally* to B. In this process the gas does work on the piston and heat q_1 flows into the gas from the heat source. At B the source is removed, and the expansion continues to C as a *reversible adiabatic* expansion. The gas continues to do work on the piston, but no heat enters, so the gas cools, reaching at C the temperature of the heat sink T_2. It is then placed in contact with the heat sink and compressed reversibly and isothermally to D. In this process the piston does work on the gas and waste heat q_2 is transferred to the sink. Finally, the sink is removed and the cycle is completed by a reversible adiabatic compression back to state A. In the course of the cycle the gas does net work w equal to the area of the closed loop on the diagram. Since energy is conserved, w must be equal to $q_1 - q_2$.

§§5.6, 6.3

What do we mean when we say that the working substance is an 'ideal gas'? In classical thermodynamics we are not concerned with internal structure and must define the ideal gas by its experimental properties. It is found that all gases become 'ideal' at sufficiently low densities. This means, firstly, that *all ideal gases have the same equation of state*

Experimental properties of the ideal gas: see also §§5.4, 5.6

$$PV = R\theta, \tag{6.12}$$

which relates the pressure P to the volume V, the *gas scale temperature* θ and the constant R known as the *gas constant*. This equation summarises a number of ideas. It expresses *Boyle's law*, the fact that PV is constant at fixed temperature for an ideal gas. It also involves the meaning of the gas scale temperature, which may be defined to be proportional to the pressure of a

Fig. 6.1. Carnot cycle using an ideal monatomic gas. The work done by the gas in one cycle is the area of the loop, and is equal to $q_1 - q_2$.

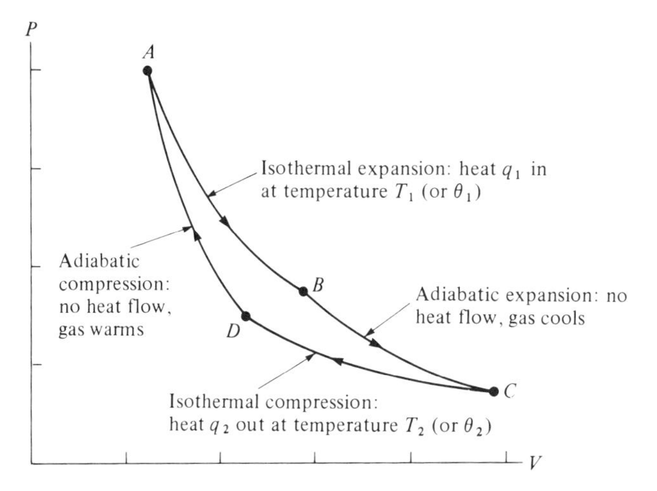

sample of helium held at a fixed low density, and set equal to 273.15 K at the temperature of melting ice. For helium, then, the fact that $P \propto \theta$ at fixed V is a matter of definition. The fact that $P \propto \theta$ at fixed V for *other* ideal gases and the fact that the gas constant R is the same for 1 mole of all gases are, however, matters of experiment. A second important property is that the *internal energy of an ideal gas is a function of temperature only, independent of the volume.* This may be demonstrated experimentally by showing that when an ideal gas is allowed to expand into an evacuated container through a thermally insulated valve its temperature does not change. Since no energy is entering the gas from outside as heat or work (for the constraints are not moving), it is clear that E must be the same for different volumes at the same temperature.

There are several ways of using these properties to show that the thermodynamic and gas scale temperatures agree. Perhaps the quickest is to observe that if we take the gas around the reversible cycle $ABCDA$ shown in Fig. 6.1 we have an example of a Carnot engine. At every stage of the cycle we may write

$$\frac{\mathrm{d}q}{\theta} = \frac{\mathrm{d}E - \mathrm{d}w}{\theta} = \frac{\mathrm{d}E}{\theta} + \frac{P\,\mathrm{d}V}{\theta}$$

or

$$\frac{\mathrm{d}q}{\theta} = \frac{\mathrm{d}E}{\theta} + \frac{R\,\mathrm{d}V}{V}, \tag{6.13}$$

using the equation of state. If we integrate this equation around the cycle we note that the first term on the right is a function of temperature only, while the second is a function of volume only. Since both θ and V return to their starting values after one cycle, *the loop integral of the right-hand side of the equation must be zero.* It follows that

$$\oint \frac{\mathrm{d}q}{\theta} = \frac{q_1}{\theta_1} - \frac{q_2}{\theta_2} = 0. \tag{6.14}$$

In § 6.5 we showed that $q_1/q_2 = T_1/T_2$ for a Carnot engine. Combining this with (6.14) we find that

▶ $$T_1/T_2 = \theta_1/\theta_2. \tag{6.15}$$

Units of S and T.

It follows that the thermodynamic temperature T is *proportional* to the gas scale temperature θ. In this chapter we have not said in what units S and hence T (which has units of energy/entropy) are to be written. Clearly, by choosing suitable units for S, *we may make T equal to θ.* From now on we shall suppose that this has been done, so that T is measured in units of K, and S in units of J K^{-1}.

6.7 The relation between pressure and entropy

In a simple situation where the only constraint to be varied is the volume, we think of S as a function of E and V. For general changes of both E and V we may therefore write $\mathrm{d}S$ in terms of its partial derivatives as

$$dS = (\partial S/\partial E)_V \, dE + (\partial S/\partial V)_E \, dV. \qquad (6.16)$$

(The essential facts concerning partial derivatives are set out in §8.1.) Now we have already seen that the temperature, which we can think of as the response of the system to its energy content, is related to $(\partial S/\partial E)_V$. It turns out that, in a similar way, the equilibrium pressure P, which is the response of the system to the volume constraint, is related to $(\partial S/\partial V)_E$. To see this, consider a small reversible adiabatic expansion. No heat enters the system, so $dE = dw = -P\,dV$; and, since the change is reversible, we have $dS = 0$. If we substitute these values in (6.16), and remember that $(\partial S/\partial E)_v = 1/T$ by definition, we find that

§6.3

▶ $$P = T(\partial S/\partial V)_E. \qquad (6.17)$$

If constraints other than the volume are varied, the corresponding responses may be found in an analogous way.

It is worth noting carefully that when we know the equilibrium entropy S of a system as a function of its internal energy E and constraints such as V, we may at once deduce the values of all the system responses such as T, P, etc. For this reason we shall refer to E, V, etc., as the *natural variables* in which S should be expressed. We could equally well have specified the state of a gas, for instance, by specifying other pairs of variables, such as T and V or E and P. But a little investigation shows that knowledge of S as a function of T and V, for instance, would give us less information: we might determine E from it by integration, but, unless we had some extra knowledge which would fix the constant of integration, we could not obtain E absolutely. We shall see later that other functions of state exist which have different sets of natural variables. It is often very helpful to remember which are the natural variables for any quantity of interest.

Natural variables for S

§7.1

6.8 The general expression for dE in terms of state functions

With the help of the argument just given we may now identify all the partial derivatives of S, and thus write dS in the general case as

$$dS = (\partial S/\partial E)_V \, dE + (\partial S/\partial V)_E \, dV + \text{other similar terms}$$
$$= dE/T + P\,dV/T \qquad + \text{other similar terms.}$$

This may be rearranged to read

▶ $$dE = T dS - P\,dV + \text{other similar terms.} \qquad (6.18)$$

This is the *general expression for* dE *in terms of functions of state*, and is the starting point of many calculations. Note that it depends only on our identification of the partial derivatives of S, and is therefore true for *any* small change of state whether it is achieved reversibly or irreversibly. If the change happens to be reversible, we can identify $-P\,dV$ as the reversible work done on the system in a small change of volume; and if other constraints are varied other similar terms which appear in dE can also be identified as reversible

See also §11.3

work terms. It follows from energy conservation that $T\mathrm{d}S$ must be identified as the heat which enters the system if the change is made reversibly, $\mathrm{d}q_{\mathrm{rev}}$.

6.9 Entropy, heat and the third law

See also § 4.4

It is important to understand that the argument just given showing that $\mathrm{d}q_{\mathrm{rev}} = T\mathrm{d}S$ is limited to *reversible* changes. For an irreversible change we can still write $\mathrm{d}E = T\mathrm{d}S - P\,\mathrm{d}V +$ similar terms and also $\mathrm{d}E = \mathrm{d}q + \mathrm{d}w$ – both of these relations are true for irreversible as well as reversible changes of state. But we have to remember that in the first relation P stands for the *equilibrium* pressure of the system. In an irreversible compression the pressure actually acting on the piston may be greater or less than P: for instance, in a *sudden* compression it is likely to be greater, while in compression by a *cold* piston it is likely to be less. Thus in an irreversible change of volume the work done may be greater or less than $-P\,\mathrm{d}V$. Similar considerations apply to the other work terms and consequently the heat $\mathrm{d}q$ entering the system may be less than or greater than $T\mathrm{d}S$. If $\mathrm{d}w$ is zero and the constraints are all held fixed, however, then of course $\mathrm{d}q$ must be equal to $T\mathrm{d}S$ even if the process is irreversible: this is the case of pure heat flow discussed earlier. So we find that

§ 6.5

▶ $$\mathrm{d}q = T\mathrm{d}S \quad \begin{cases} \text{(i) for } \textit{pure heat flow}, \text{ and} \\ \text{(ii) for } \textit{reversible changes}, \end{cases} \tag{6.19}$$

but there is no special relation between entropy change and heat flow for general irreversible changes.

Note finally that the result $\mathrm{d}q_{\mathrm{rev}} = T\mathrm{d}S$ is important because it allows us to *measure* changes in entropy by measuring the heat entering the system in a reversible process and the temperature at which it enters. Indeed, we can go further than this by making use of the *third law of thermodynamics*. This law states that, for all systems, S approaches zero as T approaches zero, showing that all systems in thermal equilibrium become perfectly ordered at absolute zero. (The theoretical basis for the law is discussed in § 4.5). Using the third law we may write the absolute entropy of any system as

Measurement of S and the third law

$$S = \int_{T=0} \mathrm{d}S = \int_{T=0} \mathrm{d}q_{\mathrm{rev}}/T. \tag{6.20}$$

§ 15.4

Although absolute zero is inaccessible, we may often get sufficiently close to it to compute the integral on the right-hand side from measurements of $\mathrm{d}q_{\mathrm{rev}}$ and T, made as the system is warmed from a very low temperature and brought to the state of interest. All heat entering the system reversibly must be taken into account, including latent heats as well as heat associated with changes of temperature and volume. The absolute entropies of many substances have been measured in this way, and may be found in tables of entropy.

Problems

6.1 An ideal gas, with $\gamma = 1.50$, undergoes a Carnot cycle, in which it is expanded at the source temperature of 240 °C from 10^{-3} m^3 at 10 atmospheres to 2.5×10^{-3} m^3 at 4 atmospheres. Between what limits of volume and pressure is it compressed at the sink temperature of 50 °C? Calculate the work done by the gas in the expansion at the source temperature, and the work done on the gas in the compression at the sink temperature. Hence find Q_1, Q_2 and η, and check that $Q_1/Q_2 = T_1/T_2$.

6.2 A building loses heat at the rate of 1 kW for each K by which the inside temperature exceeds the outside temperature. If 4 kW of electrical power is available to heat the building when the outside temperature is −5° C, find the maximum possible inside temperature if the power is (i) dissipated directly as heat, and (ii) used to drive a heat pump.

6.3 In an *idealised diesel engine cycle* (i) air at atmospheric pressure is compressed adiabatically from V_1 to V_2, (ii) the air is expanded to V_3 while a small amount of fuel is injected and burns at such a rate that P remains constant, (iii) the burnt gases expand adiabatically to V_1, and (iv) a valve opens, the pressure falls to atmospheric, and the burnt gases are swept out and replaced by fresh air. Sketch the P–V diagram for this cycle. The main effect of the fuel is to *heat* the air – its effect on the composition is small – so we may treat the burning as equivalent to supplying heat to the air from an outside source. On this assumption, if $1/r_c = V_1/V_2$ is the *compression ratio* and $1/r_e = V_1/V_3$ is the *expansion ratio*, find the work done on each of the strokes in terms of P_2V_2, r_c and r_e, and hence show that the ideal diesel efficiency η_D is given by

$$\eta_D = 1 - (r_e^\gamma - r_c^\gamma)/\gamma(r_e - r_c).$$

Taking $\gamma = 1.5$ for air and $r_c = \frac{1}{15}$, $r_e = \frac{1}{5}$ for a typical engine, calculate the ideal diesel efficiency. Show that, if very little fuel is burnt, η_D approaches the ideal thermodynamic efficiency. Why are large compression ratios preferred?

6.4 Calculate the maximum work obtainable from a Carnot engine which uses as source and sink (i) 10^6 kg of water at 100 °C and a lake at 10 °C, and (ii) 10^6 kg of water at 100 °C and 10^6 kg of water at 10 °C. [The specific heat capacity of water is 4.2×10^3 J kg^{-1} K^{-1}.]

6.5 Calculate the changes in total entropy of the system plus its surroundings produced by the following processes:

(a) A copper block of mass 0.4 kg and thermal capacity 150 J K^{-1} at 100 °C is placed in a lake at 10 °C.

(b) The same block at 10 °C is dropped from a height of 100 m into the lake.

(c) A 1 μF capacitor is connected by resistive leads to an ideal reversible cell of emf 100 V at 0 °C.

(d) The same capacitor is discharged through a resistor at 0 °C.

(e) One mole of an ideal monatomic gas at 0 °C is expanded reversibly and isothermally to twice its initial volume.

(f) The same expansion is carried out by opening a valve leading to an evacuated container of equal volume (a Joule expansion).

6.6 The entropies of hydrogen, oxygen and water, measured as described in § 6.9 at 25 °C and 1 atmosphere, are as follows. H_2 gas: 131 J K^{-1} mol^{-1}; O_2 gas: 205 J K^{-1} mol^{-1}; H_2O liquid: 70 J K^{-1} mol^{-1}. Find the heat given out if hydrogen and oxygen are made to combine *reversibly* to form 1 mole of water. If the reaction occurs irreversibly, will the heat given out be greater or less?

7

Energies in classical thermodynamics

A thermodynamic system has several different 'energies'. In this chapter I shall introduce the *Helmholtz free energy*, the *enthalpy*, and the *Gibbs function*, and try to give some idea of the role which these quantities play in the subject. The *availability*, a useful generalisation of the free energy, will appear in the last section. The more specialised problem of *field energies* in electric and magnetic systems will be deferred until we reach Chapter 15.

7.1 Thermodynamic potentials

§6.8

We saw in the last chapter that we may write $\mathrm{d}E$ as $T\mathrm{d}S - P\,\mathrm{d}V +$ other similar reversible work terms, and thus that $\mathrm{d}E$ is, in general, a sum of terms of the form $A\,\mathrm{d}B$, where A and B are functions of state. Using these terms as a guide, we may construct a set of quantities having dimensions of energy by subtracting from E one or more terms of the form AB. We have, for instance, the definitions

$$\begin{aligned} &\text{Helmholtz free energy} \quad && F \equiv E - TS, \\ &\text{Enthalpy} && H \equiv E + PV, \\ &\text{Gibbs function} && G \equiv E - TS + PV. \end{aligned} \tag{7.1}$$

§4.1

(The enthalpy H must not be confused with the Boltzmann H-function, which is related to the entropy.) Such quantities are known as *thermodynamic potentials*. The method of construction ensures that their differentials have simple forms. Using the expression for $\mathrm{d}E$ we find, for instance, that if the constraints other than the volume are held fixed

Differential forms and natural variables

$$\begin{aligned} \mathrm{d}F &= -S\,\mathrm{d}T - P\,\mathrm{d}V, \\ \mathrm{d}H &= T\mathrm{d}S + V\mathrm{d}P, \\ \mathrm{d}G &= -S\,\mathrm{d}T + V\mathrm{d}P. \end{aligned} \tag{7.2}$$

§6.7

These equations show that each of the thermodynamic potentials has its own set of *natural variables*, in the sense introduced earlier for the entropy. For the free energy, for instance, the natural variables are T and V. If we know F as a function of T and V we may at once obtain all other significant variables: S is

$-(\partial F/\partial T)_V$, P is $-(\partial F/\partial V)_T$, and knowing F, T, V, S and P we may write down E and the other thermodynamic potentials. The natural variables for each of the thermodynamic potentials are given in Table 7.1, which also lists other properties of the potentials which we shall meet later.

Other potentials

When we vary constraints other than the volume, other potentials become important. For instance, in dealing with the thermodynamics of soap films we must write $\mathrm{d}E$ as $T\mathrm{d}S+\Gamma\,\mathrm{d}\mathscr{A}$, where Γ is the surface tension, $\mathscr{A}$ is the area of the surface, and $\Gamma\,\mathrm{d}\mathscr{A}$ represents the reversible work done when we slowly stretch the film. In this situation we would still define F as $E-TS$, but we should then find that $\mathrm{d}F=-S\,\mathrm{d}T+\Gamma\,\mathrm{d}\mathscr{A}$. In place of the ordinary enthalpy and Gibbs function we should have to use the *surface enthalpy* H_s, defined as $E-\Gamma\mathscr{A}$ (so that $\mathrm{d}H_s=T\mathrm{d}S-\mathscr{A}\,\mathrm{d}\Gamma$), and the *surface Gibbs function* G_s defined as $E-TS-\Gamma\mathscr{A}$ (so that $\mathrm{d}G_s=-S\,\mathrm{d}T-\mathscr{A}\,\mathrm{d}\Gamma$). We shall meet several similar examples corresponding to other types of work term later in this book.

Potentials for inhomogeneous systems

We frequently wish to discuss the thermodynamic potentials for *inhomogeneous systems* in which the pressure or the temperature are different in different parts of the system, as in a liquid containing bubbles of vapour or a system of bodies in thermal isolation. In such a case we define the thermodynamic potential for the whole system as the sum of the thermodynamic potentials of its separate parts, defined according to (7.1) for each part separately.

7.2 Potential energy in thermodynamic systems: free energy

The idea of potential energy – that work done on a system may be thought of as stored inside it and recoverable – is often useful in mechanics. We are, for instance, used to the idea that when we bring two electric charges Z_1 and Z_2 from a great distance to positions a distance r apart we do work equal to $Z_1Z_2/(4\pi\varepsilon_0 r)$, and that this work is stored as electrostatic potential energy. We are also used to the idea that if we stretch a spring from its equilibrium length by an amount x, the tension in the spring will increase by αx, where α is the spring constant, and the stored energy will be $\frac{1}{2}\alpha x^2$. There is, however, an important distinction between these two cases. In the case of the charges, we do the work against one of the fundamental forces of physics, and the energy is definitely stored as field energy. Work done on the spring, however, is work done on a macroscopic thermodynamic system. Stretching a spring is somewhat like compressing a gas and it is not obvious where the energy of stretching has gone. Most of it will indeed have gone into the field energy of the interatomic forces acting within the spring, but some of it may have been converted into kinetic energy of the atoms, and some of it may have left the spring as heat flow. In what sense is it valid to speak of the 'potential energy of the spring', and how should we describe this energy in thermodynamic terms?

Obviously, we are hoping to find a 'potential energy' which is a well-defined function of the extension of the spring, and we at once meet a difficulty.

Unlike ε_0, the spring constant α is not a fundamental constant of physics and it varies with the temperature. Thus *there is no well-defined 'potential energy' for a thermodynamic system whose temperature is to be varied independently of the constraints.* Indeed, we could take the spring around a cycle in which we stretch it at a temperature where α is low, and allow it to relax at a different temperature where α is higher. In going around such a cycle we should continuously extract net work from the spring – we should have, in fact, a rather impractical heat engine, the work extracted coming from the heat sources used to change the temperature of the spring. This shows that the concept of 'potential energy' has to be treated with great caution for thermodynamic systems.

We may, however, speak of the 'potential energy' for a thermodynamic system which is *kept at fixed temperature*, and fortunately this often corresponds to cases of practical interest. The system may, in general, be a complicated one having a number of separate parts, and it may or may not be in mechanical equilibrium. But in every case we know that energy is conserved, and the work $\mathrm{d}w$ done on the system is equal to $\mathrm{d}E + \mathrm{d}q_\mathrm{R}$, where $\mathrm{d}q_\mathrm{R}$ is the heat entering the reservoir. We may write $\mathrm{d}q_\mathrm{R}$ as $T\mathrm{d}S_\mathrm{R}$, since the heat enters the reservoir as pure heat flow. We also know from the Clausius inequality that $\mathrm{d}S_\mathrm{R} \geqslant -\mathrm{d}S$. Thus

$$\mathrm{d}w = \mathrm{d}E + T\mathrm{d}S_\mathrm{R} \geqslant \mathrm{d}E - T\mathrm{d}S. \tag{7.3}$$

The free energy F is defined as $E - TS$. Since we are here considering a system, all parts of which are at the reservoir temperature both before and after any change, we may identify $\mathrm{d}E - T\mathrm{d}S$ as $\mathrm{d}F$. So we find that

▶ $$\mathrm{d}w \geqslant \mathrm{d}F \quad \text{for a system in contact with a heat reservoir,} \tag{7.4}$$

Free energy is potential energy at fixed T

where F is the Helmholtz free energy. This result shows that *it is the free energy which plays the role of 'potential energy' for systems held at constant temperature.* We interpret the equation as follows. In a *reversible* change, we have $\mathrm{d}w_\mathrm{rev} = \mathrm{d}F$ and we regard the reversible work as stored in the form of free energy. In irreversible processes the work put in may be greater than the increase in F, and the work recovered may be less than the fall in F. We interpret this as being due to the presence of *dissipative forces*, such as friction or viscous drag. Any work lost by dissipation reappears as heat transferred to the reservoir.

The reason why $\mathrm{d}w_\mathrm{rev}$ is related to F, equal to $E - TS$, rather than the internal energy E is that *there is almost always a heat flow associated with a reversible change of constraint at constant T*, and the term $-TS$ in F takes account of this: we have, by energy conservation, $\mathrm{d}w_\mathrm{rev} = \mathrm{d}E - \mathrm{d}q_\mathrm{rev} = \mathrm{d}E - T\mathrm{d}S = \mathrm{d}F$. In fact, if we remember that S is a measure of disorder we can often predict the direction of this heat flow. For instance, a liquid surface is more disordered than the bulk liquid, so stretching a soap film at constant temperature increases S, and heat must *enter* if the process is reversible. Thus the surface energy density is *greater* than the surface free energy density for a

Table 7.1. *Properties and significance of the thermodynamic potentials, with references to book sections. In the present chapter we consider only systems of fixed contents, so* $\mathrm{d}N=0$, *and we are concerned only with* E, H, F *and* G. *The symbol* $\mathrm{d}w_e$ *represents external work, that is, work other than that done by the pressure constraint. Each differential leads to corresponding Maxwell relations* (§§ *8.2, 11.3*). *New potentials corresponding to* H, G, Φ_H *and* Φ_G *must be introduced when constraints other than the volume are involved* (§ *7.1*)

Definition Differential Natural variables §§ **7.1**, **11.3**	Significance	Associated statistical computation	Associated inequality and fluctuation formula § **16.4**
Potentials significant for systems not in permanent contact with reservoirs			
Internal energy, E $\mathrm{d}E=T\mathrm{d}S-P\,\mathrm{d}V+\mu\,\mathrm{d}N$ S, V, N	Related to energy entering the system: $\mathrm{d}E=\mathrm{d}w+\mathrm{d}q$ § **1.4**	$S=k\ln g(E, V, N)$ gives $S(E, V, N)$ or $E(S, V, N)$ § **9.3**	$\mathrm{d}S\geqslant 0$ and $f\propto \mathrm{e}^{S/k}$ for all systems isolated from reservoirs § **6.3**
Enthalpy, $H=E+PV$ $\mathrm{d}H=T\mathrm{d}S+V\mathrm{d}P+\mu\,\mathrm{d}N$ S, P, N	Related to energy entering system at fixed P: $\mathrm{d}H=\mathrm{d}w_e+\mathrm{d}q$ $\Delta H=$process energy, latent heat, heat of reaction § **7.3**		As above
Potentials significant for systems in contact with a heat reservoir			
Free energy, $F=E-TS$ $\mathrm{d}F=-S\,\mathrm{d}T-P\,\mathrm{d}V+\mu\,\mathrm{d}N$ T, V, N	Effective potential energy for system held at fixed T § **7.2**	$Z=\mathrm{e}^{-F/kT}=\sum \mathrm{e}^{-E_i/kT}$ gives $F(T, V, N)$ § **9.2**	$\mathrm{d}w\geqslant \mathrm{d}F$, and $f\propto \mathrm{e}^{-F/kT}$ for fixed constraints, if all parts can exchange energy with heat reservoir at fixed T § **7.2**
Gibbs function, $G=F+PV$ $\mathrm{d}G=-S\,\mathrm{d}T+V\mathrm{d}P+\mu\,\mathrm{d}N$ T, P, N	Energy determining the direction of phase and chemical changes: $G=\sum \mu_i N_i$ §§ **7.4**, **11.3**		$\mathrm{d}w_e\geqslant \mathrm{d}G$ and $f\propto \mathrm{e}^{-G/kT}$ if all parts can exchange energy and volume with heat and pressure reservoir at fixed T and P § **7.4**

Potentials significant for systems in contact with a heat and particle reservoir

Grand potential, $\Phi_F = F - \mu N$ $d\Phi_F = -S\,dT - P\,dV - N\,d\mu$ T, V, μ	Effective potential energy for system held at fixed T and μ: $\Phi_F = -PV$ in homogeneous system §**11.3**	$\Xi = e^{-\Phi/kT} = \sum_G e^{-(E_i - \mu N_i)/kT}$ gives $\Phi(T, V, \mu)$ §**11.4**	$dw \geqslant d\Phi$, and $f \propto e^{-\Phi/kT}$ for fixed constraints, if all parts can exchange energy and particles with a reservoir at fixed T and μ
$\Phi_G = G - \mu N$ $d\Phi_G = -S\,dT + V\,dP - N\,d\mu$ T, P, μ	For Φ_G the three natural variables are all intensive. In a homogeneous system $\Phi_G = 0$. This corresponds to the fact that the three intensive variables are not independent: fixing any two fixes the third, but the quantity of material is then indeterminate. Thus we cannot usefully discuss fluctuations at constant T, μ and P. We observe, however, that $-S\,dT + V\,dP - N\,d\mu = 0$; the *Gibbs–Duhem relation* §**11.3**		

Other potentials

$\Phi_E = E - \mu N$ $d\Phi_E = T\,dS - P\,dV - N\,d\mu$ S, V, μ $\Phi_H = H - \mu N$ $d\Phi_H = T\,dS + V\,dP - N\,d\mu$ S, P, μ	These potentials would be important if the system could be connected to an ideal particle reservoir (§11.1) only. In real reservoirs particle exchange cannot be separated from ordinary heat exchange, however, so these potentials have no special significance §**11.3**

liquid. On the other hand, compressing a gas at constant temperature *decreases* the disorder, so heat must *leave* the gas, and the increase in internal energy is *less* than the increase in free energy. In fact, for an ideal gas, as we saw earlier, E does not change at all in an isothermal compression. In this case the *whole* of the free energy is stored in the form of heat transferred to the surroundings, corresponding to a change in the second term of F only. Note that this free energy stored as heat is recoverable: if we allow the gas to expand, the heat reenters the gas, and we get our work back.

§5.6

If a system having several parts all held at fixed temperature is mechanically isolated so that $\mathrm{d}w=0$, but is out of mechanical equilibrium internally, our result shows that $\mathrm{d}F \leqslant 0$. In other words, in the irreversible process of approaching the equilibrium state, F falls and at fixed temperature *the system reaches equilibrium when F is minimised*. This result agrees with the well-known principle of mechanics that equilibrium corresponds to the position of minimum potential energy. 'Minimising the free energy' in order to determine the equilibrium state of a complex system is a common procedure in theoretical physics.

F is minimised in equilibrium at fixed T

7.3 Energy in flow processes: enthalpy

In a *flow process* one or more substances flow steadily into some device where the process takes place, and one or more substances flow steadily out. Flow processes are common in technology. Compressors, turbines, internal combustion engines, gas liquefiers and many chemical plants involve them. In a flow process heat may be injected into or extracted from the substances which pass through and work may be done on them or by them. The net energy which enters the flow as heat or work is known as the *process energy*. How is the process energy related to the thermodynamic parameters of the substances which pass through?

Suppose that in a certain time a volume V_1 of input material of internal energy E_1 and at pressure P_1, enters the process box (Fig. 7.1). During this time the material immediately *behind* the volume V_1 in the flow is acting like a piston, pushing the volume V_1 into the process box, and it does work on the material entering the box equal to P_1V_1. Thus the total energy entering the box from the left is the original internal energy E_1 of the material entering plus the work P_1V_1 done by material behind it. We note that this is just the *enthalpy* H_1 of the material entering. If during the same time a volume V_2 of output material of internal energy E_2 emerges at pressure P_2, work P_2V_2 has to be done to push the material immediately *ahead* of the volume V_2 out of the way. Thus the total energy leaving the box during the same time is $E_2+P_2V_2$, equal to the enthalpy H_2 of the output material. Since energy is conserved, it is evident that *the process energy must be equal to the enthalpy change which occurs in the process, or*

Process energy equals enthalpy change

▶ $dH = dq_p + dw_p$ for flow processes, (7.5)

where dq_p is the *process heat* and dw_p is the *process work*. Contrast this result with the relation $dE = dq + dw$, which applies to a system of fixed contents. Equation (7.5) depends only on energy conservation and is equally valid for reversible and irreversible processes.

The fact that the process energy is equal to the enthalpy change makes enthalpy an important quantity for engineers. For instance, the electrical energy obtained from a well-insulated steam turbine is equal to the fall in enthalpy of the steam as it passes through and the power required to drive a well-insulated air compressor is equal to the rate of enthalpy increase in the air as it passes through. Internal combustion engines give out both heat and work, and the total process energy extracted is the enthalpy of the fuel plus the air taken in, less the enthalpy of the exhaust gases. Extensive tables of enthalpy as a function of pressure and temperature for many substances are available.

Enthalpy and C_P

It is worth noting that if a substance is heated as it passes along a tube at constant pressure, the heat added is equal to the enthalpy change, so that we can write for the heat capacity at constant pressure

▶ $C_P = (\partial H/\partial T)_P$, (7.6)

a result which we can also get directly from the expression for the differential $dH = T dS + V dP$.

Enthalpy is conserved in a Joule–Kelvin expansion

§ 14.4

Finally, if a substance is simply allowed to flow irreversibly through an insulated valve, so that no heat enters and no work is done, we see that the enthalpy change must be zero. Such a process is known as a *Joule–Kelvin expansion* and we shall meet it again in Chapter 14. For the moment we note that

▶ $\Delta H = 0$ in a Joule–Kelvin expansion. (7.7)

Fig. 7.1. Energy balance in a flow process. The process energy entering is equal to the change in enthalpy which occurs.

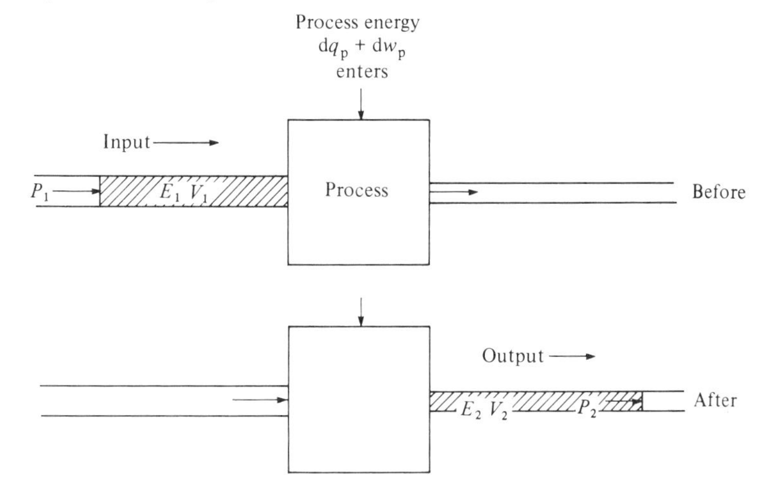

7.4 Energy in phase changes and chemical reactions: enthalpy and Gibbs function

In a phase change, such as the evaporation of a liquid, or a chemical reaction, one or more substances disappear and one or more new substances appear, with the absorption or emission of heat. If we imagine the phase change or chemical reaction occurring as a flow process at constant pressure in a chemical plant, it is clear from the argument of § 7.3 that the *latent heat* for the phase change must be equal to the *increase of enthalpy* which occurs, while the *heat of reaction*, the heat given out in the reaction, must be equal to the *enthalpy of the reagents* less the *enthalpy of the products.*

Enthalpy, latent heat and heat of reaction

When we consider *phase equilibrium* and *chemical equilibrium* we find that a different energy is important. We imagine a mixture of phases, or a mixture of reagents and products, held at fixed temperature and pressure. We saw in § 7.2 that for systems held at constant temperature we have $dw \geqslant dF$. If the system is also held at constant pressure P, but otherwise mechanically isolated, we may write dw as $-P\,dV$, the work done by the constraint, for both reversible and irreversible changes. If P is the same in all parts of the system, we may write $G = F + PV$ for the parts of the system added together, and it follows that

▶ $$dG \leqslant 0 \quad \text{for a system held throughout at fixed } T \text{ and } P \text{ and otherwise mechanically isolated.} \qquad (7.8)$$

Evidently G decreases as the system approaches equilibrium. If G is greater for the first phase than the second, the phase change can occur only in the *forward* direction; while if G for the reagents is greater than G for the products the chemical reaction can only proceed *forwards.* If the reverses are true the changes can only proceed in the *reverse* direction. If the phase change or the reaction involves no change in G, the system will be in phase or chemical *equilibrium.*

The Gibbs function fixes the direction of change

Thus it is ΔG which determines the *direction* of the change, and ΔH which determines the *heat evolved or absorbed* in the change, whether the change occurs reversibly or not. Note that there is no direct relation between these quantities, so a phase change or chemical reaction may either evolve or absorb heat, depending on whether ΔG and ΔH have the same or opposite signs.

Exothermic and endothermic reactions

We shall reexamine the question of the direction of chemical and phase changes when we discuss the quantity known as the *chemical potential* in Chapter 11.

§§ 11.2, 11.3, 12.7

7.5 Useful work and the general condition for equilibrium: availability

In § 6.5 we showed how a hot body may be used to drive a heat engine by pure heat flow, and so obtain useful work. In § 7.2 we showed how a body at fixed temperature can be regarded as having potential energy which may also be used to obtain useful work. Let us now generalise the discussion. If we have, for example, a boiler full of steam at 800 K, *what is the maximum useful work which can be extracted from it using any conceivable process*?

Note first that the answer will depend on the *surroundings*. If the surroundings are also at 800 K a Carnot engine using the steam as source and the surroundings as sink will have zero efficiency; and if the steam happens to be at atmospheric pressure, no *useful* work is obtained by letting it expand – all the work which it does is used up in pushing back the atmosphere. We shall therefore assume that the boiler has definite surroundings, which we may think of as a heat and pressure reservoir, having equilibrium temperature T_R and equilibrium pressure P_R.

We now use an extended version of the argument of §7.2. We are interested in a system which may have several parts at different temperatures which are not in equilibrium with the heat reservoir. We are also interested in the possibility of using heat engines which use different parts of the system and the reservoir as sources and sinks. But since we are supplying no external heat and the internal energy of any heat engine is unchanged at the end of its cycle, the principle of energy conservation still takes the simple form

$$\mathrm{d}w = \mathrm{d}E + T_R\,\mathrm{d}S_R \geqslant \mathrm{d}E - T_R\,\mathrm{d}S, \tag{7.9}$$

as in (7.3), except that we now have to distinguish the reservoir temperature from the various system temperatures. Here E and S are the total energy and entropy of the system, and $\mathrm{d}w$ represents the total work done, either directly on the system itself or on the heat engines. We write $\mathrm{d}w$ as $-P_R\,\mathrm{d}V + \mathrm{d}w_u$, where V is the total volume of the system, $-P_R\,\mathrm{d}V$ is the work done by the pressure reservoir, and $\mathrm{d}w_u$ represents all other forms of external work done on the system or the engines: we refer to $\mathrm{d}w_u$ as the *useful work input*. Thus

$$\mathrm{d}w_u \geqslant \mathrm{d}E - T_R\,\mathrm{d}S + P_R\,\mathrm{d}V, \quad \text{or}$$

▶ $$\mathrm{d}w_u \geqslant \mathrm{d}A, \tag{7.10}$$

where A is the *availability*, defined as $E - T_R S + P_R V$.

This result shows that the availability is a useful generalisation of the free energy. *Provided that the thermodynamic system can only interact with surroundings at a single temperature T_R and a single pressure P_R*, we have $\mathrm{d}w_u = \mathrm{d}A$ for a reversible change, so any reversible work input may properly be regarded as stored in the form of availability for *any* change of state. In any change of state the maximum useful work which can be extracted is the fall in the availability, and if the change is made reversibly this useful work *will* be extracted, whatever route is chosen. Note that for a simple system such as the boiler full of steam, which is homogeneous and has no internal constraints, $-\mathrm{d}A$ may be written as

$$-\mathrm{d}A = (T_R - T)\,\mathrm{d}S + (P - P_R)\,\mathrm{d}V. \tag{7.11}$$

The first term represents useful work obtainable from a Carnot engine which uses heat extracted reversibly from the system, while the second term represents useful work obtained directly when the system expands. This expression shows that once $T=T_R$ and $P=P_R$, the availability reaches an absolute minimum for the homogeneous system and no further useful work can be extracted.

Note also that if some more complex system is mechanically isolated, so that $dw_u=0$, then when some internal constraint is removed $dA \leqslant 0$, showing that the availability falls as the system moves irreversibly towards its new equilibrium state. *A is minimised in equilibrium, and this provides a useful general test for finding equilibrium states.* Note that this general condition for equilibrium contains two of our earlier results as special cases. For a system held at fixed volume, all parts of which are in thermal equilibrium with the surroundings, the condition $dA \leqslant 0$ implies that $dF \leqslant 0$ since $T_R = T$ and V is constant; while for a system all parts of which are in thermal *and* pressure equilibrium with the surroundings, the condition $dA \leqslant 0$ implies that $dG \leqslant 0$, since $T_R = T$ and $P_R = P$.

The general condition for equilibrium

§7.2

§7.4

We shall meet the availability again when we discuss the magnitudes of thermal fluctuations.

§16.4

Problems

7.1 With the help of (5.6) and (5.14), obtain an expression for F in terms of T and V for an ideal monatomic gas. Find from it the expression for P in terms of T and V.

Find also the corresponding expressions for S and E in terms of their natural variables, and confirm that they also lead to the correct expression for P.

7.2 Show that the extra free energy required to create area $\mathscr{A}$ of free surface in a liquid is equal to $\Gamma\mathscr{A}$, and that the free energy of a quantity of ideal gas kept at fixed temperature may be written as $-F_0 \ln(V/V_0)$, where F_0 and V_0 are constants.

A bubble of air of radius r exists in a liquid at temperature T_L and pressure P_L. Write down an expression for F for the bubble, and explain why $dF/dV = -P_L$ in equilibrium. By using this relation show that the pressure inside the bubble is equal to $P_L + 2\Gamma/r$. Check that this is consistent with a direct mechanical calculation of the excess pressure inside. Why is the bubble spherical?

7.3 A compressor (i) draws in air at 300 K and 1 atmosphere, (ii) compresses it reversibly and adiabatically to 2 atmospheres, and (iii) delivers the compressed air to a pipe. By considering the enthalpy change, show that the process work may be written as $\int V dP$ calculated for the adiabatic compression. Confirm this conclusion by considering the P–V diagram for the compressor. If the compressor consumes 200 W of useful power, what volume of air will it deliver per second? [Assume that $\gamma = 1.4$.]

7.4 When hydrogen and oxygen combine at atmospheric pressure and 25 °C to form 1 mole of liquid water, the heat given out is 2.86×10^5 J. Using the result of problem 6.6, determine the change in Gibbs function ΔG in the formation of water from its elements. Show that if ΔZ is the charge which circulates when 1 mole of water is electrolysed into gaseous hydrogen and oxygen, and ϕ is the voltage across the cell, then $\phi \Delta Z \geqslant -\Delta G$. Hence find the minimum value of ϕ.

7.5 Find the maximum useful work obtainable from 1 mole of an ideal monatomic gas at 100 °C and 10 atmospheres pressure, if it can interact only with the atmosphere at 20 °C, by first finding an expression for its availability, with the help of (5.14).

Choosing any convenient reversible process which brings the gas to atmospheric temperature and pressure (such as cooling at fixed V followed by isothermal expansion, or adiabatic expansion followed by warming at fixed P, etc.), show in detail how the maximum useful work could be obtained with the help of suitable Carnot engines.

8

Thermodynamic relations

A *thermodynamic relation* is an exact rule, such as the formula for the ideal efficiency of a Carnot engine, which may be obtained by general thermodynamic reasoning and which applies to a wide range of systems. The phrase is usually limited to rules which depend on the second or third laws. Thermodynamic relations frequently link quantities which are not, at first sight, closely related. In experiment they often make it possible to relate data obtained in quite different ways, or to replace a difficult measurement by an easier one. In theoretical work they can be used to deduce values of one quantity from calculations of another which may be more straightforward; and it is always wise to check the validity of a new theoretical model by making sure that it does not lead to nonsense when thermodynamic relations are applied to it.

§6.5

Thermodynamic relations may be found by a number of elementary methods, but general results are obtained most quickly and systematically by using the techniques of *partial differentiation*. The mathematical results which we shall use are set out in the first section below, but if the techniques are unfamiliar you will probably find it best to consult a suitable textbook before reading further.

There are very many thermodynamic relations, and even more possible routes for deriving them, and this makes the results and their derivations hard to remember. I would encourage you not only to read and understand this chapter but to practise deriving its results without referring to the text.

8.1 Partial derivatives

The basic theorem

The basic theorem of partial differentiation states that if $z(x, y)$ is a function of two variables x and y which may be varied independently, then when there are simultaneous arbitrary infinitesimal changes $\mathrm{d}x$ and $\mathrm{d}y$ in x and y the corresponding change in z may be written as

$$\mathrm{d}z = (\partial z/\partial x)_y\,\mathrm{d}x + (\partial z/\partial y)_x\,\mathrm{d}y, \tag{8.1}$$

where the symbol $(\partial z/\partial x)_y$ stands for the *partial derivative of z with respect to x at fixed y*, with a similar understanding for $(\partial z/\partial y)_x$. The partial derivative of z with respect to x at fixed y is obtained by differentiating z with respect to x in the usual way, treating y as a constant. The theorem may be extended to more variables. If z is a function of a *set* of variables x_i, then $\mathrm{d}z = \sum_i (\partial z/\partial x_i)\,\mathrm{d}x_i$, where it is understood that the derivative $\partial z/\partial x_i$ is taken with respect to x_i, holding *all* the other variables in the set constant.

§6.1

The basic theorem has obvious applications in thermodynamics because thermodynamic systems normally have two or more independent variables, corresponding to the internal energy and the various constraints. For a sample of gas, for instance, the nature of the equilibrium state depends on E and V and we think of the equilibrium pressure, temperature, entropy, etc., as functions of E and V, all having partial derivatives with respect to E and V. From the mathematical point of view, however, it is not *necessary* to use E and V as the independent variables, and it is often convenient to choose a different pair. For instance, we commonly think of the state as determined by fixing V and T. From that point of view we think of E, as well as P, S, etc., as functions of V and T, having partial derivatives with respect to V and T. Since there are a great many possible choices of independent variable *it is vital to be clear about which variables are being treated as independent, and what is held constant in any partial differentiation.*

Choice of independent variables in thermodynamics

In addition to the basic theorem, it is helpful to bear in mind three further mathematical rules of manipulation. First, note that the *chain rule for differentiation* holds in the usual way provided the *same* variables are held constant in each derivative:

The chain rule

$$\left(\frac{\partial z}{\partial x}\right)_y = \left(\frac{\partial z}{\partial u}\right)_y \left(\frac{\partial u}{\partial x}\right)_y. \tag{8.2}$$

Secondly, note that if, in the basic theorem (8.1), we set $\mathrm{d}z = 0$ and divide by $\mathrm{d}x$ we find that

The relation between the three derivatives in x, y, z

$$\left(\frac{\partial z}{\partial x}\right)_y = -\left(\frac{\partial z}{\partial y}\right)_x \left(\frac{\partial y}{\partial x}\right)_z. \tag{8.3}$$

This result is superficially similar in form to (8.2). But note that here the variables held constant are different in each term, and that we have, in fact, *a relation between the three possible partial derivatives which involve x, y and z.* (There are only three, because $(\partial x/\partial z)_y$ is just $[(\partial z/\partial x)_y]^{-1}$, etc.) Finally, it can be shown that, provided the derivatives in question are all continuous, *the partial differential operators commute*, so that

The commutation rule

$$\left(\frac{\partial}{\partial x}\right)_y \left(\frac{\partial z}{\partial y}\right)_x = \left(\frac{\partial}{\partial y}\right)_x \left(\frac{\partial z}{\partial x}\right)_y. \tag{8.4}$$

8.2 Maxwell relations

When the only constraint to be varied is the volume we write the basic energy relation as $\mathrm{d}E = T\,\mathrm{d}S - P\,\mathrm{d}V$. At the same time, if we choose to use S and

§6.8

V as our independent variables we have from (8.1) the relation $\mathrm{d}E = (\partial E/\partial S)_V\,\mathrm{d}S + (\partial E/\partial V)_S\,\mathrm{d}V$. On comparing these relations we find that we must identify $(\partial E/\partial S)_V$ as T, and $(\partial E/\partial V)_S$ as $-P$. Now according to (8.4) we know that $(\partial/\partial V)_S(\partial E/\partial S)_V = (\partial/\partial S)_V(\partial E/\partial V)_S$. Using the identifications just made, it follows that

$$\left(\frac{\partial T}{\partial V}\right)_S = -\left(\frac{\partial P}{\partial S}\right)_V. \tag{8.5}$$

A relation of this type is known as a *Maxwell relation*. Starting from the expressions for the differentials of the thermodynamic potentials, $\mathrm{d}F = -S\,\mathrm{d}T - P\,\mathrm{d}V$, $\mathrm{d}H = T\mathrm{d}S + V\mathrm{d}P$ and $\mathrm{d}G = -S\,\mathrm{d}T + V\mathrm{d}P$, and using the same method, we find the following three Maxwell relations in addition:

§7.1

$$\left(\frac{\partial S}{\partial V}\right)_T = \left(\frac{\partial P}{\partial T}\right)_V, \tag{8.6}$$

$$\left(\frac{\partial T}{\partial P}\right)_S = \left(\frac{\partial V}{\partial S}\right)_P, \tag{8.7}$$

$$\left(\frac{\partial S}{\partial P}\right)_T = -\left(\frac{\partial V}{\partial T}\right)_P. \tag{8.8}$$

To remember these results note that the pairs of variables, T, S and P, V always appear in diagonally related positions inside the brackets; that the suffixes and denominators change places on the two sides of the equation; and that wherever T and V appear in the same bracket the two sides of the equation have opposite signs: negaTiVe and posiTiVe.

How to remember Maxwell relations

It is important to understand, by tracing back the argument of Chapter 6, that the various identifications on which the Maxwell relations depend all stem from our original identification of $(\partial E/\partial S)_V$ as a temperature. We were only able to take this step by using the second law, and it follows that the Maxwell relations, unlike the relations for partial derivatives (8.1)–(8.4), are not purely mathematical in content: *when we use a Maxwell relation we are in effect using the second law.* Maxwell relations are prolific generators of thermodynamic relations, as we shall see.

§6.4

When constraints other than the volume are varied, new reversible work terms appear in the expression for $\mathrm{d}E$, and new thermodynamic potentials may have to be introduced. Wherever this happens, corresponding new Maxwell relations appear which are analogues of (8.5)–(8.8).

Problem 8.5

8.3 Examples of thermodynamic relations

8.3.1 *Vapour pressure and latent heat*

If an evacuated container is partly filled with a liquid, molecules of the liquid will escape from the free surface and form a *vapour* in the empty space above the liquid. This process will continue until the vapour density is high enough to make the rate at which molecules return to the liquid from the vapour equal to the rate at which molecules are evaporating. The pressure when this equilibrium density is reached is called the *equilibrium vapour pressure*, P_{vap}. It does not depend on the quantities of liquid and vapour present. So long as both *are* present, and the system has reached dynamic equilibrium, *the vapour pressure is a function of temperature only.*

Imagine then, a cylinder which contains a liquid with some of its vapour in equilibrium. Let us apply to this system the Maxwell relation (8.6), which requires that, for any thermodynamic system, $(\partial P/\partial T)_V = (\partial S/\partial V)_T$. If we hold our cylinder at constant volume, and make a small increase in the temperature, all that will happen is that enough liquid will evaporate to ensure that the pressure in the vapour reaches the new equilibrium vapour pressure. In this system, then, *we may identify* $(\partial P/\partial T)_V$ *as* $\mathrm{d}P_{vap}/\mathrm{d}T$. On the other hand, if we expand the cylinder at constant temperature the vapour pressure will stay constant, but liquid must evaporate to fill the extra space with vapour of the required density. If the mass of liquid which evaporates is $\mathrm{d}m$, then we know that $T\mathrm{d}S = \mathrm{d}q_{rev} = L\,\mathrm{d}m$, where L is the *specific latent heat of evaporation.* At the same time the volume change $\mathrm{d}V$ will be equal to $(V_v - V_l)\,\mathrm{d}m$ where V_v and V_l are the *specific volumes* (the volumes per unit mass, the inverses of the densities) for the vapour and liquid respectively. On combining these results we find that we may identify $(\partial S/\partial V)_T$ as $L/\{T(V_v - V_l)\}$. Using the Maxwell relation, then, we find that

The Clausius–Clapeyron relation

$$\frac{\mathrm{d}P_{vap}}{\mathrm{d}T} = \frac{L}{T(V_v - V_l)}. \tag{8.9}$$

This is the well-known *Clausius–Clapeyron equation* which relates the temperature derivative of the vapour pressure and the latent heat of evaporation. The argument may be extended to any phase change which involves a latent heat. An interesting case is that of melting ice. This process is unusual because water is *denser* than ice – the specific volume of the high-temperature phase is the smaller of the two. It follows from the Clausius–Clapeyron equation that $\mathrm{d}P/\mathrm{d}T$ taken along the melting curve is negative: in this case the melting temperature *falls* with pressure. This means that ice melts when enough pressure is applied, an unusual fact of some significance in the study of glaciers.

§ 18.1

8.3.2 *Relations involving C_P and C_V*

To a theoretician it is much easier to calculate the various contributions to the heat capacity of a solid at constant volume, when the lattice spacing is constant, than at constant pressure. But to the experimenter the heat capacity is much more easily *measured* at constant pressure because solids expand so powerfully on heating. Fortunately, when we have to compare theory with experiment, a thermodynamic relation comes to our rescue. Since $\mathrm{d}q_{rev} = T\mathrm{d}S$, any heat capacity may always be written in the form $T(\mathrm{d}S/\mathrm{d}T)$. Remembering that $\mathrm{d}S = (\partial S/\partial T)_V\,\mathrm{d}T + (\partial S/\partial V)_T\,\mathrm{d}V$ and multiplying both sides of this relation by $T/\mathrm{d}T$ while holding P constant, we find that

$$T(\partial S/\partial T)_P = T(\partial S/\partial T)_V + T(\partial S/\partial V)_T(\partial V/\partial T)_P,$$

or

$$C_P = C_V + T(\partial P/\partial T)_V(\partial V/\partial T)_P,$$

using (8.6). Using (8.3) also we deduce that

▶ $$C_P - C_V = -T(\partial P/\partial V)_T[(\partial V/\partial T)_P]^2. \quad (8.10)$$

The relation between the heat capacities C_P and C_V

The value of $C_P - C_V$ may easily be obtained from this formula by measuring the *compressibility*, which gives $(\partial P/\partial V)_T$, and the *volume expansion coefficient*, which gives $(\partial V/\partial T)_P$. For a solid $C_P - C_V$ turns out to be relatively small. The theoretical C_V may thus readily be compared with the experimental C_P, using experimental values of the compressibility and the expansion coefficient to calculate the small correction required.

When a substance is compressed reversibly and adiabatically the temperature rises, and consequently the pressure rises faster than it would have done at constant temperature. You may like to show that the ratio of the two rates is given by

▶ $$\frac{(\partial P/\partial V)_S}{(\partial P/\partial V)_T} = \frac{C_P}{C_V}. \quad (8.11)$$

The ratio of the adiabatic and isothermal bulk moduli

This result does not require a Maxwell relation; it may be derived using the mathematical relations (8.3) and (8.2) only, and depends on the first law alone.

See also problem 8.4

Our two results apply, of course, to any system. You may like to check that they agree with what we obtained earlier, using a theoretical model, for the ideal gas.

§5.6

8.3.3 *Expansion coefficients at low temperatures*

As an example of a thermodynamic relation which depends on both the second law and the third, consider the question of thermal expansion at low temperatures. The Maxwell relation (8.8) tells us that $(\partial V/\partial T)_P = -(\partial S/\partial P)_T$. But according to the third law the entropy change in all isothermal processes vanishes at low enough temperature, so the right-hand side of the Maxwell relation will be very small. Evidently then, the same must be true of the left-hand side: *all expansion coefficients must tend to zero at low temperatures*. This simple result seems much less trivial when we try to derive it directly from theoretical models of solids, for instance.

§6.9

8.3.4 *The energy density of equilibrium radiation*

When an empty container is held at a fixed temperature it becomes filled with isotropic electromagnetic radiation, characteristic of the temperature, which is known as *equilibrium radiation* or *black body radiation*, and has a definite energy per unit volume, E/V. We shall consider the detailed properties of equilibrium radiation later. Here we examine a simple argument which shows how its energy density varies with temperature.

§10.2

To carry through this argument we have to understand that radiation exerts a pressure on its container. One can find the magnitude of this pressure by calculating directly the electromagnetic forces acting on the walls when the radiation is reflected, but a quicker method is to treat the *photons* of which the radiation consists as particles whose mass is given by the relativistic formula $\varepsilon = mc^2$, where ε is the energy of the photon. We may apply to the photon gas the classical kinetic argument which we used for the ideal gas in § 5.4. As before, we find that $P = \frac{1}{3}n\overline{mc^2}$. For the photon gas we identify $\overline{mc^2}$ as $\bar{\varepsilon}$. So $P = \frac{1}{3}n\bar{\varepsilon}$, or

The photon gas

$$P = \tfrac{1}{3}E/V. \tag{8.12}$$

A quantum argument leading to the same result is given later.

§ 9.1.3

Now the energy density E/V may be written as $(\partial E/\partial V)_T$ for the container of radiation, since expanding the box at fixed temperature just adds more radiation of the same density. Dividing the general expression $dE = T dS - P dV$ by dV at fixed T shows that

$$E/V = (\partial E/\partial V)_T = T(\partial S/\partial V)_T - P$$
$$= T(\partial P/\partial T)_V - P,$$

using the Maxwell relation (8.6). Since $E/V = 3P$, we have

$$T(\partial P/\partial T)_V = 4P,$$

or

$$dP/P = 4\,dT/T$$

at fixed V. It follows that $P \propto T^4$, and, since $P \propto E/V$, we have

$$E/V \propto T^4. \tag{8.13}$$

The fact that the energy density is proportional to T^4 is closely related to *Stefan's law*, which states that the radiation emitted by a black body at temperature T is proportional to T^4.

§ 10.2

The ability of Maxwell relations to generate unsuspected theoretical relationships is worth noting. Few students have an intuition sufficiently developed to *expect* to find a connection between the pressure of radiation and the temperature dependence of its energy density.

8.3.5 *Heats of reaction from cell voltages*

Chemists frequently wish to know the *molar heat of a reaction*, the heat evolved when the equivalent weights of the reagents react chemically and irreversibly. It is, however, notoriously difficult to measure heats of reaction accurately. When the reaction of interest may also be made to take place reversibly in an electric cell there is a way round this difficulty. For instance, if zinc is placed in a solution of a copper salt, a reaction occurs in which zinc enters solution and solid copper appears. Normally this reaction occurs irreversibly at constant pressure and the heat of the reaction is equal to the fall in enthalpy $-\Delta H$ which occurs, as we noted in § 7.4. The *same* reaction can also be made to occur reversibly in a Daniell cell, which consists of zinc and copper rods immersed in acidified copper sulphate solution. Since electrical energy may be extracted from the cell the heat evolved will be different in this case, and we must write the fall in enthalpy which we wish to find as

$$-\Delta H = \text{process energy leaving per mole}$$
$$= -q_{rev} + \phi\Delta Z, \quad (8.14)$$

where ϕ is the emf of the cell, and ΔZ represents the charge which circulates when the equivalent weights react. At first sight we have gained nothing by considering the cell because we still have to find the heat, q_{rev}. However, since the reaction is reversible we may write q_{rev} as $T\Delta S$, and thus

$$-\Delta H = \left[-T\left(\frac{\partial S}{\partial Z}\right)_{T,P} + \phi\right]\Delta Z, \quad (8.15)$$

since the reaction takes place at constant T and P. In this situation we must allow for electrical work done on the cell and therefore write the general expression for dE as $T dS - P\, dV - \phi\, dZ$. We therefore have a Maxwell relation, the analogue of (8.6),

$$\left(\frac{\partial S}{\partial Z}\right)_{T,P} = \left(\frac{\partial \phi}{\partial T}\right)_{Z,P}, \quad (8.16)$$

which may be derived from the expression $dG = -S\, dT + V dP - \phi\, dZ$ using the relation $(\partial^2 G/\partial T\, \partial Z)_P = (\partial^2 G/\partial Z\, \partial T)_P$. Using this Maxwell relation we find that the heat of reaction in the *original* reaction, equal to $-\Delta H$, may be written as

The Gibbs–Helmholtz relation

▶ $$-\Delta H = [-T(\partial\phi/\partial T)_{Z,P} + \phi]\Delta Z. \quad (8.17)$$

This is the *Gibbs–Helmholtz* relation for the cell. Using it, chemists can determine the heat of reaction $-\Delta H$ from measurements of the emf of the cell and its temperature dependence *only* – which are very much easier than heat flow measurements.

Other examples of thermodynamic relations will appear later in the book.

Problems

8.1 For an ideal gas there is no temperature change on Joule expansion through a valve into an evacuated container. Explain why this remark is equivalent to the statement $(\partial T/\partial V)_E = 0$. Using (8.3), deduce that E is independent of V at fixed T. Using (8.4), deduce also that C_V, expressed as $(\partial E/\partial T)_V$, is independent of V at fixed T. §5.6

8.2 Knowing that $(\partial E/\partial V)_T = 0$ for an ideal gas, adapt the argument of §8.3.4 for the photon gas to show that at fixed volume $P \propto T$ for an ideal gas. [This shows that the thermodynamic and ideal gas temperature scales are the same, and is an alternative to the argument of §6.6.]

8.3 Use the expression for dH (7.2) with a Maxwell relation to show that $(\partial H/\partial P)_T = 0$ for an ideal gas obeying the equation of state $PV = RT$. Deduce that for an ideal gas (i) there is no temperature change on Joule–Kelvin expansion, and (ii) C_P is independent of P at fixed T. §7.3

8.4 Use a Maxwell relation and (8.2) to show that for any substance the rate of change of P with T in a reversible adiabatic compression is given by $(\partial T/\partial P)_S = (T/C_P)(\partial V/\partial T)_P$. Find an equivalent expression for the rate of change of V with T, and check that both results are valid for an ideal monatomic gas.

§ 5.6

8.5 The *surface entropy density* of a liquid surface is the extra entropy which enters the system when unit area of new surface is created. Show that there is a Maxwell relation of the form $(\partial S/\partial \mathscr{A})_T = -(\partial \Gamma/\partial T)_{\mathscr{A}}$ for a liquid surface. The surface tension of liquid argon is well fitted by the formula $\Gamma = \Gamma_0(1 - T/T_c)^{1.28}$ where $\Gamma_0 = 3.8 \times 10^{-2}$ N m^{-1} and T_c is the critical temperature, equal to 151 K. Find the surface entropy at the triple point where $T=$ 83 K, and deduce an approximate value for the surface entropy per surface atom in units of k. [Density of liquid argon at the triple point $= 1.4 \times 10^3$ kg m^{-3}; atomic mass = 40 amu.]

8.6 Use the Maxwell relation of the previous problem to find a formula for the liquid *surface energy density* $(\partial E/\partial \mathscr{A})_T$ in terms of Γ and $\mathrm{d}\Gamma/\mathrm{d}T$. The surface tension of water changes linearly with temperature from 0.075 N m^{-1} at 5 °C to 0.070 N m^{-1} at 35 °C. Find the surface energy density at 20 °C. Why is the surface energy density not equal to the work done in creating the surface?

8.7 Estimate roughly the reduction in melting point of ice due to the pressure exerted on it by a skater who is standing still. [Specific latent heat of melting of ice $= 3.3 \times 10^5$ J kg^{-1}.]

9

Statistical calculation of thermodynamic quantities

We turn in this chapter to statistical thermodynamics and in particular to the calculation from first principles of specific thermodynamic quantities in particular systems. We shall suppose that some quantum analysis has been done, and that we know a suitable set of approximate energy states for the system of interest. Even when this is so, it is not always easy to decide how to proceed in calculating the equilibrium functions of state. In this chapter we shall describe and illustrate three standard methods of calculation, and give some idea of when they are likely to be useful. In the following three chapters we shall extend these methods in various ways. In Chapters 13 and 14 we shall see how similar calculations may be done from the classical point of view.

It is perhaps worth emphasising that *the methods to be described can only be carried through straightforwardly when the approximate energy states involved are easily described and enumerated.* In real life we often have to deal instead with systems of interacting particles whose states are difficult to describe, or unknown. Recently the methods of *quantum field theory* have been extended by Nambu and others to include thermal averaging, and this powerful technique sometimes makes it possible to calculate thermal averages directly from the system Hamiltonian *without finding the quantum states.* These methods are beyond the scope of this book, but you will find references to them in the end papers.

Nambu formalism

9.1 First method: direct application of the Boltzmann distribution

Many quantities of interest have average values which may be written down at once as a function of temperature by averaging over states of known energy using the Boltzmann distribution. For instance, the mean value of the internal energy itself is just given by $E=\sum_i p_i E_i$ where p_i is the Boltzmann probability $\exp(-E_i/kT)/\sum_i \exp(-E_i/kT)$. In the same way the mean magnetic moment $\bar{m}$ for a paramagnetic ion in a field $\mathscr{B}$ at temperature T may be written as $\sum_i p_i m_i$, where m_i is the component of magnetic moment parallel to $\mathscr{B}$ when the ion is in state i. The entropy S may be written as $-k\sum_i p_i \ln p_i$, using the expression for the canonical entropy. The pressure P may be found

§15.2

§4.3

§4.4

using the relation $\mathrm{d}w_{\mathrm{rev}}=\sum p_i\,\mathrm{d}E_i$: since $\mathrm{d}w_{\mathrm{rev}}=-P\,\mathrm{d}V$ we see that $P=-\sum p_i(\mathrm{d}E_i/\mathrm{d}V)$. Other responses to a constraint can be found in the same way. Let us summarise these results:

$$\begin{aligned} &\text{Energy} && E=\sum p_iE_i, \\ &\text{Magnetic moment} && \bar{m}=\sum p_im_i, \\ &\text{Entropy} && S=-k\sum p_i\ln p_i, \\ &\text{Pressure} && P=-\sum p_i(\mathrm{d}E_i/\mathrm{d}V). \end{aligned} \tag{9.1}$$

This simple approach should always be considered first when only a single function of state is required for a small system at temperature T. We now consider some examples of its use in more detail.

9.1.1 *Energy and entropy of a harmonic oscillator*

§2.3

Quantum theory shows that a harmonic oscillator of frequency ν has a ladder of states with energies $\varepsilon_n=(n+\frac{1}{2})h\nu$ where n is a non-negative integer. Let us calculate the mean energy in the oscillator at temperature T as $\bar{\varepsilon}=\sum_n p_n\varepsilon_n$. We find that

$$\bar{\varepsilon}=\frac{\sum_n \mathrm{e}^{(n+\frac{1}{2})h\nu/kT}(n+\frac{1}{2})h\nu}{\sum_n \mathrm{e}^{-(n+\frac{1}{2})h\nu/kT}}. \tag{9.2}$$

This result may be reduced to a more convenient form. We note that, if we write β for the quantity $1/kT$, the denominator Z is equal to $\sum_n \exp[-(n+\frac{1}{2})h\nu\beta]$, while the numerator is equal to $-\partial Z/\partial\beta$, and thus $\bar{\varepsilon}=-(\partial Z/\partial\beta)/Z$. Now Z is the sum of a geometric series, and may be written as $\mathrm{e}^{-\frac{1}{2}h\nu\beta}(1-\mathrm{e}^{-h\nu\beta})^{-1}$. If we insert this value into the expression for $\bar{\varepsilon}$, we find after a little algebra that

Planck's formula for the thermal energy in an oscillator

$$\bar{\varepsilon}=\tfrac{1}{2}h\nu+\frac{h\nu}{\mathrm{e}^{h\nu/kT}-1}. \tag{9.3}$$

This is *Planck's formula* for the mean energy in an oscillator at temperature T.

The variation of $\bar{\varepsilon}$ with temperature is shown in Fig. 9.1 with the corresponding heat capacity, $C=\mathrm{d}\bar{\varepsilon}/\mathrm{d}T$. At high temperatures where $kT\gg h\nu$, $\bar{\varepsilon}$ approaches the value kT. This is the *classical limit*. We shall see in Chapter 13 that when kT is much greater than the spacing between the states in energy we may use *classical statistical mechanics* and that an energy of kT is what we should expect for an oscillator according to the *classical equipartition theorem*. At low temperatures, where $kT\ll h\nu$, we reach a situation where the system is almost certain to be in its ground state (with energy $\frac{1}{2}h\nu$) and the probability of being in even the first excited state is extremely small. In this limit, the energy is greater than the classical value but *the heat capacity is very small because the spacing between the levels makes the probability of any excitation very small*. We say that the classical heat capacity is *quenched* at low temperatures where $kT\ll h\nu$. As we shall see, many physical systems contain oscillators and Planck's formula is often useful in calculations of heat capacity.

The oscillator is in the classical limit if $kT\gg h\nu$: see §§13.1, 13.5

The oscillator's heat capacity is quenched if $kT\ll h\nu$

The entropy in the oscillator at temperature T may be found in the same way. We have

$$S = -k \sum p_n \ln p_n$$
$$= k \sum p_n(\varepsilon_n/kT + \ln Z)$$
$$= \bar{\varepsilon}/T + k \ln Z,$$

or

$$S = \frac{h\nu/T}{e^{h\nu/kT} - 1} - k \ln (1 - e^{-h\nu/kT}). \tag{9.4}$$

You may like to check that at low temperatures $S \to 0$, in accordance with the third law, while at high temperatures $S \to k \ln (e\, kT/h\nu)$. The high temperature limit may be interpreted as telling us that, at high temperatures, the effective number of states over which the probability is spread is $e\, kT/h\nu$. This is reasonable – the characteristic energy width of the Boltzmann distribution is

§4.1

Fig. 9.1. Thermal predictions for a harmonic oscillator of frequency ν, as a function of temperature. Upper curve: internal energy, $\bar{\varepsilon}$. Lower curve: heat capacity, C.

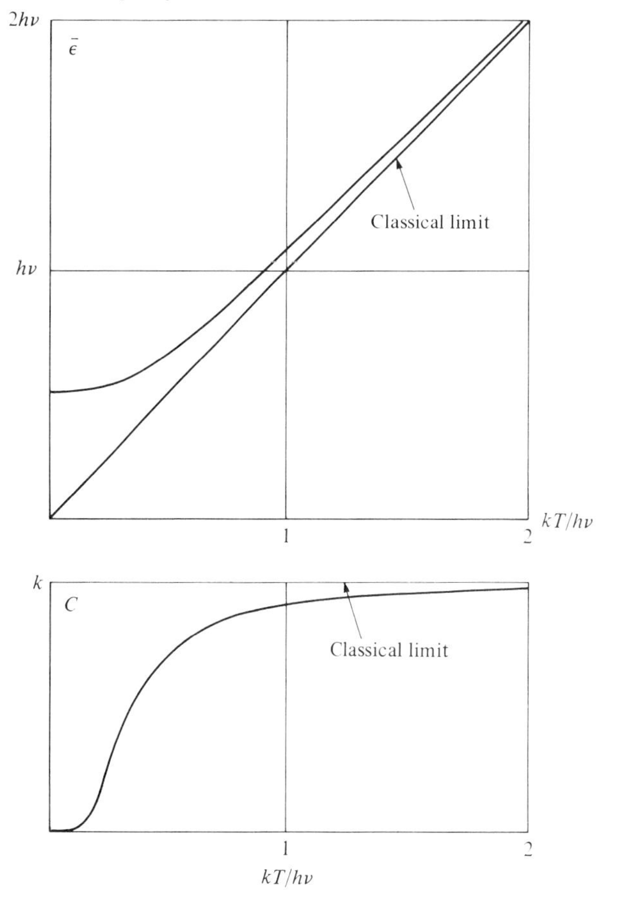

kT, so we expect the number of states having appreciable occupation at high temperature to be of order $kT/h\nu$.

9.1.2 *Energy of vibration and rotation in an asymmetrical diatomic gas*

In Chapter 5 we showed that a monatomic gas should have an atomic heat capacity of $\frac{3}{2}k$ at constant volume, due to the kinetic energy of translation of the atoms. This result agrees with experiment. For a simple diatomic gas such as carbon monoxide the molecular heat capacity at constant volume is normally about $\frac{5}{2}k$, but rises quite sharply around 1000 K and approaches $\frac{7}{2}k$ before it finally dissociates. This difference is due to the *energy of internal motions*, which we now discuss.

§5.2

We note first that even the monatomic gas may have internal motions. In a monatomic gas, however, internal motions are motions of the *electrons* – if the atom is 'rotating' we mean that the electrons are rotating around the nucleus, and if the atom is 'vibrating' we mean that its electrons are in excited states. It turns out that the *energy spacing between the lowest electron levels in an atom is almost always large compared with kT* – typically the spacing is about 10^{-19} J, while kT at room temperature is about 4×10^{-21} J. Thus the Boltzmann probability for any excited electron state is totally negligible at all ordinary temperatures, and we may ignore electronic excitation when we calculate the heat capacity. The electronic heat capacity is *quenched* at all ordinary temperatures.

The electronic excitations are always quenched

In a molecule the orbiting electrons move much faster than the nuclei. We may therefore safely ignore the motion of the nuclei in calculating the electronic state of the molecule: this is the *Born–Oppenheimer approximation*. We may also safely assume that the electrons remain in their ground state at all normal temperatures, as in the monatomic gas. If we apply the Born–Oppenheimer approximation to a diatomic molecule such as carbon monoxide and calculate the energy of the molecule as a function of the distance between the nuclei, we find a sharp minimum. Evidently the electrons tend to hold the nuclei a fixed distance apart and the molecule behaves like a fairly rigid dumb-bell. If we now consider the relatively slow motions of the two nuclei in this situation, we find that the molecule has *six independent modes of motion* (Fig. 9.2) corresponding to its six spatial degrees of freedom – it takes six numbers, in three dimensions, to specify where the two nuclei are. These six independent modes are: *three modes corresponding to translation of the centre of mass* in the x, y and z directions; *one mode of vibration parallel to the line of centres*; and *two modes of rotation* about the centre of mass. (Note that rotation *around* the line of centres is an electronic excitation which is quenched at ordinary temperatures.)

The Born–Oppenheimer approximation

The six nearly independent modes of motion

Each of these modes of motion has its own quantised energy, which may be separated from the others if the molecule is rigid enough; and this is usually a good approximation. The argument which we used in §5.2 to determine the quantisation of the *translational* energy of the monatomic gas

Translational energy

applies just as well to the centre of mass motion of the diatomic gas, and thus the translational contribution to the molecular heat capacity is $\frac{3}{2}k$ at all temperatures. The *vibration* is not a perfect harmonic oscillation, but is quite close to one, as may be seen by measuring the infra-red vibration spectrum of the molecule. For carbon monoxide, for instance, such measurements show that the quantum of vibration $h\nu$ is equal to 4.24×10^{-20} J. We may use Planck's formula to estimate the energy in the oscillator. At room temperature $kT/h\nu \sim 0.1$, and inspection of the heat capacity curve in Fig. 9.1 shows that the vibrational heat capacity will be almost completely quenched; we have to go to 1000 K before the vibrational heat capacity becomes important.

Vibrational energy

Fig. 9.2. Six modes of motion of the nuclei of a diatomic molecule. (Rotation about the line of centres is not included, because it does not involve movement of the nuclei, but only the electrons, and is quenched at ordinary temperatures.)

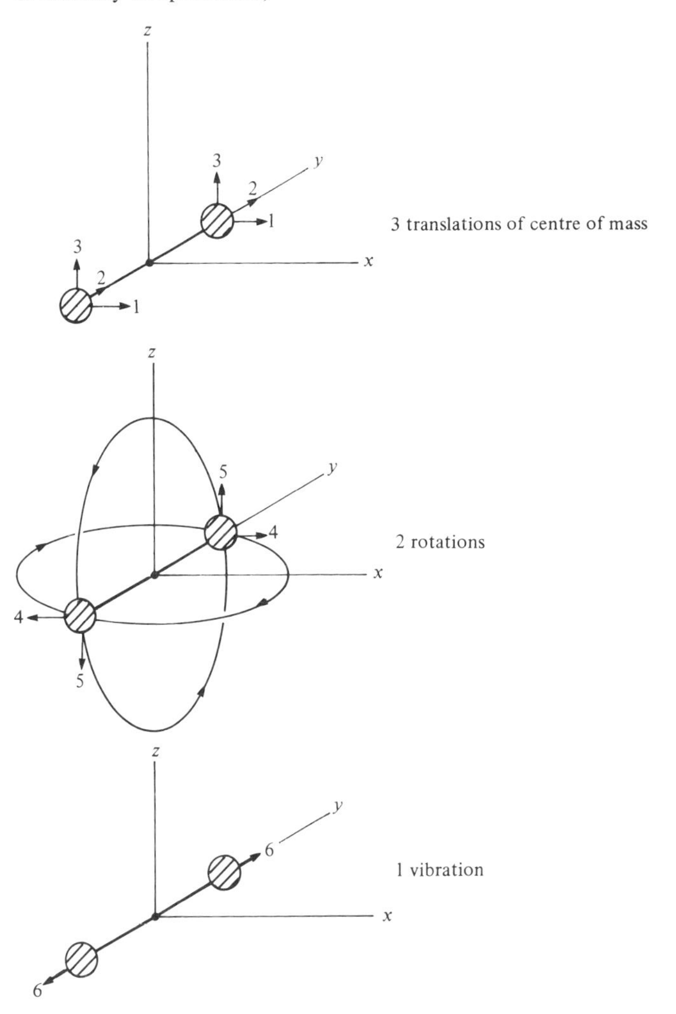

The *rotational* energy requires a new calculation. If we treat the molecule as a *rigid rotator*, we may write the rotational energy as $L^2/2I$, where L is the angular momentum and I is the moment of inertia. The quantum theory of the rigid diatomic molecule shows that L is quantised, and that L^2 takes the values $\hbar^2 K(K+1)$ where $\hbar = h/2\pi$ and K is a non-negative integer. Thus the possible rotational energies are

Rotational energy

$$\varepsilon_K = \hbar^2 K(K+1)/2I. \tag{9.5}$$

The states of fixed K are, however, *degenerate* – for an asymmetrical molecule such as CO there are $2K+1$ different rotation states for each K-value. (For a symmetrical molecule such as H_2 the degeneracy depends in a more complicated way on the nuclear spin state.) Thus writing the mean rotational energy $\bar{\varepsilon}_K$ as $\sum_K (2K+1) p_K E_K$ we find that

Problem 12.9

$$\bar{\varepsilon}_K = \frac{\sum_K (2K+1) \exp[-\hbar^2 K(K+1)/2IkT] \hbar^2 K(K+1)/2I}{\sum_K (2K+1) \exp[-\hbar^2 K(K+1)/2IkT]}. \tag{9.6}$$

This expression cannot be reduced to a simple formula, but its value is easily computed; $\bar{\varepsilon}_K$ and the corresponding heat capacity are shown in Fig. 9.3. As in the case of the harmonic oscillator, at high temperatures $\bar{\varepsilon}_K$ approaches kT, the value according to the classical equipartition theorem, while at low temperatures, where the excitation energy $\hbar^2/I$ for the first excited state is much greater than kT, the rotational heat capacity is quenched. For CO the characteristic energy $\hbar^2/2I$ may be measured using observations of the infra-red vibration–rotation spectrum: it is 3.8×10^{-23} J. This is very much *less* than kT at room temperature, so at ordinary temperatures the rotational heat capacity of CO is close to the classical value of k per molecule. In fact, we should have to go down to about 1 K before serious quenching of the rotational heat capacity occurred – and, of course, CO would be solid at such temperatures.

Problem 9.4

Putting these results together, we see that the usual molecular heat capacity at constant volume for CO of $\frac{5}{2}k$ is accounted for by contributions of $\frac{3}{2}k$ from the translational energy and k from unquenched rotational energy. Above 1000 K, the vibrational heat capacity ceases to be quenched, and the heat capacity approaches $\frac{7}{2}k$ before the molecule dissociates. The electronic contribution is quenched at all temperatures at which the molecule exists. This general behaviour is typical of diatomic molecules. (Exceptions to this remark include H_2, where the moment of inertia is so small that the rotational heat capacity is substantially reduced below 50 K, and the halogens Cl_2, Br_2 and I_2 which are massive and soft molecules with low vibrational frequencies, in which the vibrational heat capacity is appreciable at room temperature.)

The sum of the contributions

Problem 9.3

9.1.3 *Pressure of electromagnetic radiation*

We saw in §8.3.4, when we considered the energy density of equilibrium radiation, that electromagnetic radiation exerts a pressure on its container equal to one-third of its energy density. We obtained this result by treating the photons as a semi-classical gas. Here we shall derive the same result by considering the quantum states of the photons, as an illustration of the use of the formula $P = -\sum p_i(\mathrm{d}E_i/\mathrm{d}V)$.

In this formula, E_i stands for the quantised energy of the *whole* of the radiation in the container. This is the sum of the energies of a large number of photons, each of which will be in some sort of standing electromagnetic wave mode: for a rectangular container the possible photon modes will look very much like the modes for the gas atom shown in Fig. 5.1. Imagine now that we increase the volume of the container *while keeping its shape unchanged* – for a rectangular box this means that we expand the length, breadth and height in the same proportion. Because the standing wave modes have to fit inside the box, *all the photon wavelengths for every mode have to change in the same*

p. 48

Fig. 9.3. Thermal predictions for rotational properties of a rigid, asymmetric diatomic molecule of moment of inertia I. Upper curve: mean rotational energy $\bar{\varepsilon}_K$. Lower curve: rotational heat capacity.

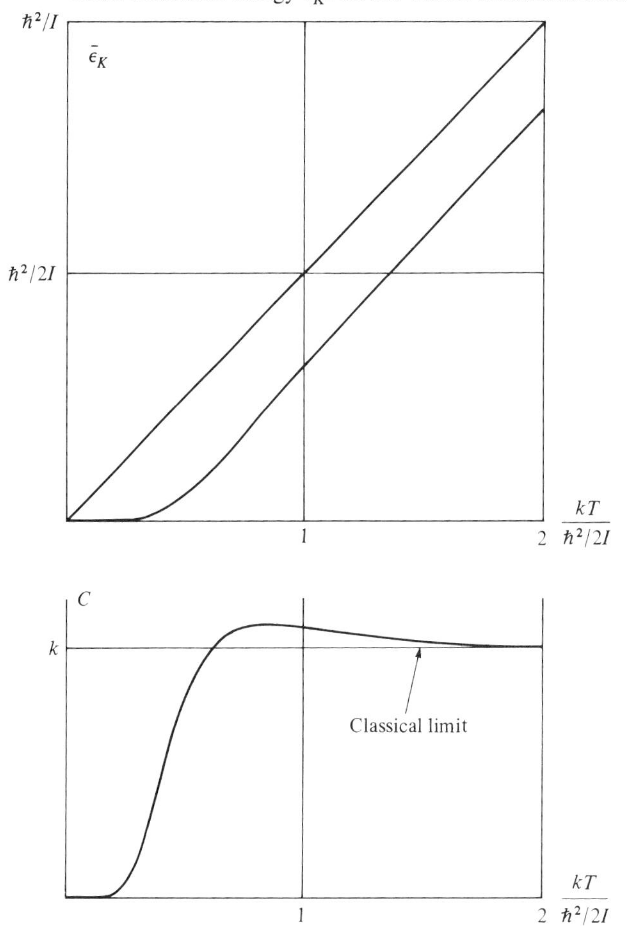

proportion, so that $\lambda \propto L$, where L is the length of the box, for every photon. But for photons the energy ε is $h\nu$, and $\lambda\nu = c$, so we find that $\varepsilon_\nu \propto \nu \propto 1/\lambda \propto 1/L \propto V^{-\frac{1}{3}}$. Since this is true for every photon, it must also be true for E_i, the total energy of all the photons in the box. Thus $E_i \propto V^{-\frac{1}{3}}$, and consequently $\mathrm{d}E_i/\mathrm{d}V = -\frac{1}{3}E_i/V$. If we substitute this expression into our formula for P we find that

$$P = -\sum p_i(\mathrm{d}E_i/\mathrm{d}V)$$

$$= \tfrac{1}{3}\sum p_i E_i/V, \quad \text{or}$$

▶ $$P = \tfrac{1}{3}E/V, \tag{9.7}$$

the radiation pressure is one-third of the energy density, in agreement with our earlier calculation.

9.2 Second method: calculation of the free energy from the partition function

The second method of calculation is more formal than the first, but usually leads quickly to the same mathematical expressions. It is worth learning thoroughly, because it is more general – the same calculation can be made to yield the value of almost any quantity – and because it forms the basis of more advanced methods, which are often referred to in the theoretical literature. The method is based on the following calculation. Using the Boltzmann distribution we may write the canonical entropy as

Compare with §4.4

$$S = -k\sum p_i \ln p_i$$
$$= -k\sum p_i(-E_i/kT - \ln Z)$$
$$= E/T + k\ln Z, \tag{9.8}$$

since $\sum p_i = 1$. Remembering that the Helmholtz free energy F is defined as $E - TS$, we deduce that

§7.1

▶ $$F = -kT\ln Z, \tag{9.9}$$

which may also be written in the alternative form $\mathrm{e}^{-F/kT} = \sum \mathrm{e}^{-E_i/kT}$. To calculate functions of state using this result we first write down the partition function Z; if the state energies of the system are known this is straightforward. Equation (9.9) then gives F as a function of T and the constraints, since the state energies depend on the constraints. This is particularly convenient, for these variables are in fact the natural variables for F: we know that $\mathrm{d}F = -S\,\mathrm{d}T - P\,\mathrm{d}V +$ similar reversible work terms, and we may therefore deduce at once that $S = -(\partial F/\partial T)_V$, while $P = -(\partial F/\partial V)_T$, and the responses to other constraints may be found in the same way. The internal energy E may also be found quickly from Z, for

$$E = \sum E_i\,\mathrm{e}^{-E_i/kT}/\sum \mathrm{e}^{-E_i/kT}$$
$$= -(\partial Z/\partial\beta)_V/Z$$
$$= -(\partial \ln Z/\partial\beta)_V, \tag{9.10}$$

where $\beta = 1/kT$. Let us summarise these results:

$$\text{Energy} \quad E = -(\partial/\partial\beta)_V(\ln Z),$$

▶ $$\text{Entropy} \quad S = (\partial/\partial T)_V(kT \ln Z), \tag{9.11}$$

$$\text{Pressure} \quad P = (\partial/\partial V)_T(kT \ln Z).$$

This method is often the best to use when more than one thermodynamic parameter is required for a small system at fixed temperature.

9.2.1 *Magnetisation and entropy of a paramagnetic salt with $S=\frac{1}{2}$*

We shall have more to say of magnetic systems and paramagnetic salts in particular in Chapter 15. Here we shall consider a very simple salt as an example of calculation using the partition function, and all that we need note is that, if the spin quantum number S is $\frac{1}{2}$, each paramagnetic ion in the salt has a permanent magnetic moment m whose magnitude m_0 is the *Bohr magneton* $\mu_B = e\hbar/2m_e$, equal to 9.3×10^{-24} J T^{-1}. In an external magnetic field of fixed direction $\mathscr{B}_E$ the magnetic moment has two possible states, the parallel state with energy $-m_0\mathscr{B}_E$, and the anti-parallel state with energy $+m_0\mathscr{B}_E$. When the magnitude $\mathscr{B}_E$ changes, the reversible work done on the system by the field is $-m\,d\mathscr{B}_E$, where m is the mean magnetisation parallel to the external field.

§15.2

In this simple two-level system the partition function for a single ion contains only two terms:

The partition function

$$Z = \sum e^{-E_i/kT} = e^{m_0\mathscr{B}_E/kT} + e^{-m_0\mathscr{B}_E/kT} = 2\cosh(m_0\mathscr{B}_E/kT). \tag{9.12}$$

To illustrate the uses of the partition function, we shall derive from it the mean magnetisation m, and the magnetic energy, heat capacity and entropy, as functions of $\mathscr{B}_E$ and T. To start with m, we note that dF contains the reversible work term $-m\,d\mathscr{B}_E$, and so deduce that

$$m = -(\partial F/\partial\mathscr{B}_E)_T = (\partial/\partial\mathscr{B}_E)_T(kT \ln Z). \tag{9.13}$$

(This formula is the analogue of the formula (9.11) which gives the pressure.) Using our expression for Z we deduce after a little algebra that

▶ $$m = m_0 \tanh(m_0\mathscr{B}_E/kT). \tag{9.14}$$

This relationship is plotted in Fig. 9.4. Note that for large fields (of either sign) or low temperatures the magnetic moment *saturates* at the value $\pm m_0$. This is because when $|m_0\mathscr{B}_E| \gg kT$ the Boltzmann probability for the anti-parallel state is negligibly small, and virtually all the spins lie parallel to the applied field. The largest field which we can produce easily in the laboratory is a few tesla, which makes the energy $m_0\mathscr{B}_E$ of order 10^{-22} J. Since kT at room temperature is about 4×10^{-21} J we cannot readily approach saturation at room temperature, but we may do so at temperatures of order 1 K, and the above expression agrees well with observations. At room temperature $m_0\mathscr{B}_E/kT$ will normally be small, and, since $\tanh x \simeq x$ for small x, we see that m will be approximately proportional to the external field,

The saturation field

§15.3

▶ $$m \simeq (m_0^2/kT)\mathscr{B}_E. \tag{9.15}$$

Curie magnetisability

The ratio $m/\mathscr{B}_E$ is the *magnetisability* α_m of the ion, and we see that at normal temperatures and fields α_m is proportional to $1/T$, a result known as *Curie's law*.

If we calculate the mean magnetic energy of the ion E_m using the formula $E_m = -(\partial/\partial\beta)_{\mathscr{B}}(\ln Z)$ we find after a little algebra that $E_m = -m_0\mathscr{B}_E \tanh(m_0\mathscr{B}_E/kT)$ which implies that $E_m = -m\mathscr{B}_E$, where m is the

Fig. 9.4. Predicted properties of a paramagnetic ion with $S=\frac{1}{2}$ in an external magnetic field $\mathscr{B}_E$. Top curve: magnetisation as a function of field at fixed temperature. Lower curves: magnetic energy E_m, heat capacity $C_{\mathscr{B}}$ and entropy S_m as a function of temperature in a given field.

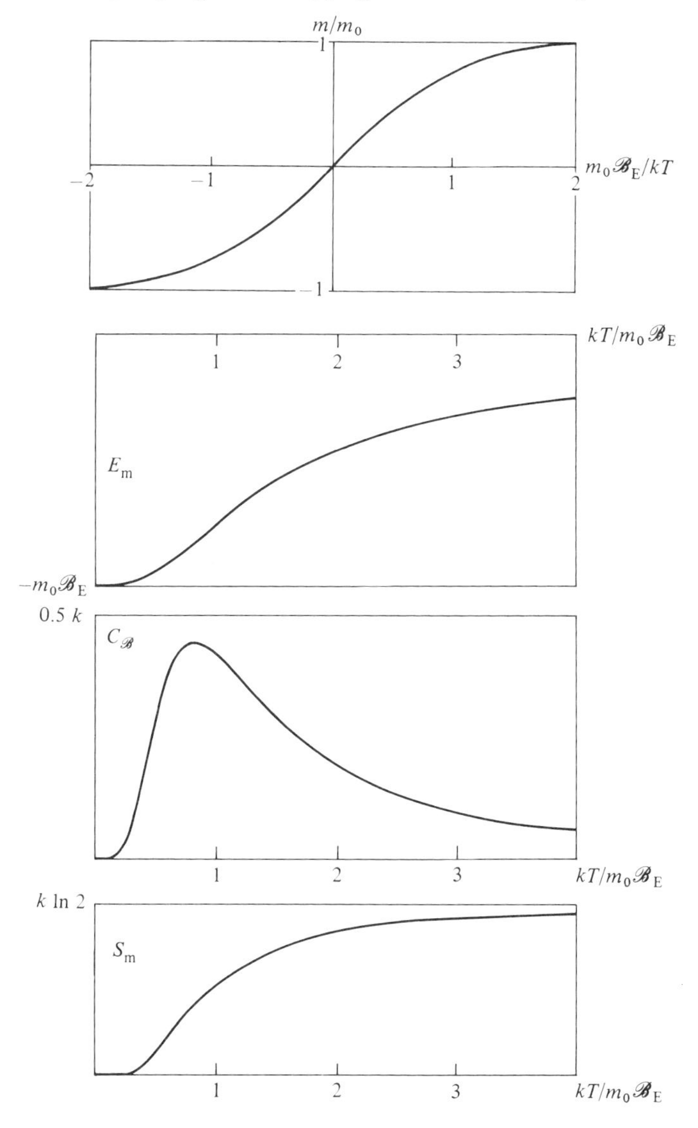

mean magnetisation. This, of course, is to be expected in this case because $E_m = -m_{\pm}\mathscr{B}_E$ for the two states involved. The mean energy is negative, and is shown as a function of $kT/m_0\mathscr{B}_E$ in Fig. 9.4; and the corresponding heat capacity, given by $C_{\mathscr{B}} = (\partial E/\partial T)_{\mathscr{B}} = k(m_0\mathscr{B}_E/kT)^2 \operatorname{sech}^2(m_0\mathscr{B}_E/kT)$ is shown below it. This heat capacity as a function of T is typical of the heat capacities for *systems having a small number of states all lying within a narrow range of energy*. At low temperatures in such systems all the energy is in the ground state, and the heat capacity is quenched because the level spacing is much greater than kT. At very high temperatures the states become equally probable; the mean energy then again becomes constant, and the heat capacity is again negligible. But when kT is of the same order as the level spacing the energy rises rapidly with temperature and we have a large heat capacity. It is not uncommon in solids to find a group of states spread over a narrow range of energies in this fashion. (Such a situation is often produced in paramagnetic solids, in zero applied magnetic field, by the internal *crystal fields* which they contain.) When this occurs it is always associated with a characteristic peak in the heat capacity, known as a *Schottky anomaly*, similar to the curve shown in Fig. 9.4.

Schottky anomalies

Problem 9.2

The *magnetic entropy* of the salt can be found from the formula $S_m = (\partial/\partial T)_{\mathscr{B}_0}(kT \ln Z)$, which leads after a little algebra to the result

$$S_m = k[\ln\{2\cosh(m_0\mathscr{B}_E/kT)\} - (m_0\mathscr{B}_E/kT)\tanh(m_0\mathscr{B}_E/kT)]. \quad (9.16)$$

This relation is plotted as a function of $kT/m_0\mathscr{B}_E$ in Fig. 9.4. Notice that at low temperatures $S \rightarrow 0$ in accordance with the third law, while at high temperatures, where the probability is spread more and more uniformly over the two states, the entropy approaches $k \ln 2$, as one would expect. The heat capacity, which is shown above the entropy and was obtained from E_m as $(\partial E_m/\partial T)_{\mathscr{B}}$ could, of course, have equally well been obtained from S_m as $T(\partial S_m/\partial T)_{\mathscr{B}}$. The most important thing to notice about the result is that the *entropy is a function of* $\mathscr{B}_E/T$. It follows that, if other forms of entropy such as lattice entropy are negligible at low temperatures, then when the applied field $\mathscr{B}_E$ is reduced reversibly and adiabatically, S_m and hence $\mathscr{B}_E/T$ must be constant. Thus $T \propto \mathscr{B}_E$, and *the temperature falls as the field is lowered*, approaching absolute zero for small fields. This is the technique of *cooling by adiabatic demagnetisation*, used to attain the lowest temperatures so far reached, of about 10^{-6} K. We shall have more to say about adiabatic demagnetisation later.

Adiabatic demagnetisation

§15.4

9.3 Third method: calculation of the entropy as $k \ln g$ or $k \ln W$ for large systems

When we wish to evaluate theoretically a function of state for a *large* system, it is best to try first to break the system down conceptually into *independent microsystems*, each with its own set of energy states. Where this can be done we may calculate Boltzmann averages or find the free energy from the partition function for the separate microsystems, and this is usually the simplest way to proceed. (We have done this already in this chapter, when we

Split into microsystems if possible

considered as independent subsystems the molecules of the diatomic gas, the different photon modes in the box of radiation, and the individual ions of the paramagnetic salt.)

Sometimes, however, we are forced to consider the quantum states of a macroscopic system *as a whole*, because its parts interact strongly and do not have well-defined, independent states of definite energy. If the overall states of the macroscopic system are known we may still evaluate its thermodynamic parameters from them, and the most efficient way of doing so is often to start by writing $S = k \ln g$, using the *large system approximation*, where g is the density of states for the large system. This method gives S as a function of its natural variables, the energy and the constraints, and other quantities may be found from S, using the relations

§4.3
§6.7

§6.4
§6.7

$$\blacktriangleright \quad \begin{aligned} 1/T &= (\partial S/\partial E)_V, \\ P/T &= (\partial S/\partial V)_E, \end{aligned} \qquad (9.17)$$

with similar relations for other responses to the constraints.

This third method of calculation gives the thermodynamic variables as functions of E and the constraints, the natural variables for S. Since we usually want them as functions of T and the constraints, this method is at first sight slightly less convenient than the two methods described earlier. However, it is *not* usually helpful to apply these two earlier methods to large systems, because *they usually only lead, after a good deal of effort, to results which could have been written down at sight, using $S = k \ln g$*. Why this happens is easily understood by examining the mathematics involved. The partition function $\sum \exp(-E_i/kT)$, for instance, may be written in a large system as $\int g(E) \exp(-E/kT)\, \mathrm{d}E$ where $g(E)$ is the density of states. Now the factor $g(E) \exp(-E/kT)$ is proportional to the energy distribution function $f(E)$, and *is always very sharply peaked at the mean energy for large systems*. Thus the integral for Z always reduces to an expression of the form $g(E) \exp(-E/kT)\, \Delta E$, where E is the mean energy, and ΔE is an energy of the same order as the range of energy fluctuations. Since we can ignore $\ln \Delta E$ in comparison with $\ln g$ we deduce that F can be written as $E - kT \ln g(E)$. But since $F = E - TS$, the result of all these manipulations is just to discover that $S = k \ln g$, *which we knew already*. Similar integrals involving the same sharply peaked factor appear when we try to calculate Boltzmann averages, and they reduce in the same way to expressions involving g. So for large systems it is usually best to *start* with $S = k \ln g$, and accept the slight inconvenience that the variables appear as functions of energy rather than temperature.

Methods 1 and 2 are inefficient for large systems

§3.4

§4.3

It is not uncommon to find for large systems that g can be expressed as Wg_0, where W is a *number of possible arrangements* for states of the same energy. In such systems there is, as we noted in Chapter 4, a corresponding contribution to the entropy equal to $k \ln W$, known as the *configurational entropy*. Calculation of the configurational entropy is often an important step in determining the entropy of large systems.

§4.3

9.3.1 *The equilibrium concentration of vacancies in a solid*

Consider a simple solid such as a metal or a solidified noble gas which has a close-packed crystal structure in which each atom is surrounded by 12 nearest neighbours. At $T=0$ all the atomic sites in the crystal lattice are filled. At a finite temperature, however, there will be an equilibrium concentration of *vacancies*, lattice sites containing no atom. Let us calculate this equilibrium concentration of vacancies. We are dealing here with a set of strongly interacting particles, and we shall have to work with the possible states of the crystal *as a whole*; this provides a good example of a calculation involving configurational entropy.

We start by observing that, in equilibrium, the free energy of the crystal F must be at a minimum with respect to changes in the number of vacancies present. Thus if we move a single atom from a bulk site to a surface site so as to create one extra vacancy the corresponding change in free energy δF should be zero. Now F is equal to $E-TS$. Let us split S into three parts,

§7.2

$$S=S_c+S_b+S_s, \tag{9.18}$$

corresponding to the fact that the density of states is the *product* of three terms, $g(E)=g_c(E)W_bW_s$ (and $S=k\ln g$ for a large system). Here $g_c(E)$ is the density of states (of lattice vibrations, etc.) for any single configuration of atoms in the lattice, W_b is the number of bulk configurations corresponding to different arrangements of the vacancies in the bulk of the crystal, and W_s is the number of different possible surface configurations. $S_b=k\ln W_b$ and $S_s=k\ln W_s$ are the corresponding configurational entropies, while $S_c=k\ln g_c(E)$. We may therefore write δF as

$$\delta F=\delta F_c-T\delta S_b-T\delta S_s=0, \tag{9.19}$$

where $\delta F_c=\delta E-T\delta S_c$ is the change in free energy which occurs when we move an atom from a particular bulk site to a particular surface site without permitting lattice rearrangements to occur. The quantity δF_c is positive, and the main contribution to it is the increase of potential energy which occurs when the number of nearest-neighbour bonds of the transferred atom is reduced. (There is also a small contribution to δF_c due to changes in the free energies of the lattice vibrations which occur when the vacancy is created.)

The meaning of δF_c

To understand the value of δF_c we need to know the type of bulk site which the atom is leaving and the type of surface site to which it is transferred. If the density of vacancies in the bulk is low, we may safely assume that the atom is leaving a normal bulk site in which it is surrounded by 12 close-packed neighbours. The free surface, on the other hand, is not an ideal close-packed plane. In thermal equilibrium it will have on it islands of new crystal plane of various sizes and shapes which themselves contain holes of various sizes (Fig. 9.5). The new atom may therefore be absorbed on the surface in many different types of site, having between zero and six nearest neighbours in the plane (as well as three in the plane below). But in calculating δF_c we are interested not in where the atom goes *first* but in where it goes *on average*, after thermal

Growth points and the value of δF_c

equilibrium is reestablished on the surface. Now in a *complete* plane there are three nearest-neighbour bonds per atom lying in the plane (each atom has six bonds, but each bond involves two atoms). Thus *on average* each atom added to the surface must add three new bonds lying in the plane. On average, then, the new atoms may be regarded as having been added at points such as D in the diagram, where they have *half* as many neighbours as they do in the bulk (in the case of the close-packed lattice, three in the plane and three in the plane below, making six in all). Such points are known as *growth points.* (Note that adding an atom at a growth point does not destroy the growth point but simply moves it sideways one step. This leaves the number of growth points unchanged, and therefore leaves the surface in thermal equilibrium.) Evidently δF_c must be calculated for transfer of an atom from a bulk site with 12 neighbours to a surface growth point with six. Thus δF_c is approximately six nearest-neighbour bond energies, which is, in fact, the binding energy per atom for atoms in the bulk of the solid.

The change δS_b in the bulk configurational entropy may be found as follows. When we have N atoms and N_v vacancies the total number of lattice sites is $N+N_v=N_s$. Thus the number of vacancy configurations W_b is just the number of ways of distributing N atoms and N_v vacancies amongst the $N+N_v$ sites. In the language of the statistics of combinations this is $\binom{N_s}{N}$ and thus

$$W_b=(N+N_v)!/N!\,N_v! \tag{9.20}$$

On increasing the number of vacancies by one, W_b is *multiplied* by $(N+N_v+1)/(N_v+1)$ which in a large crystal we may write approximately as $(N+N_v)/N_v$. Thus the change in the bulk configurational entropy is

$$\delta S_b=k\ln[(N+N_v)/N_v]. \tag{9.21}$$

Fig. 9.5. Pattern of atoms on the close-packed surface of a growing crystal, showing islands of new growth superimposed on the plane below. The sites A to G have decreasing numbers of neighbours in the plane. A site such as D which has three neighbours in the plane is known as a 'growth point'. New atoms may be assumed to go to a growth point on average.

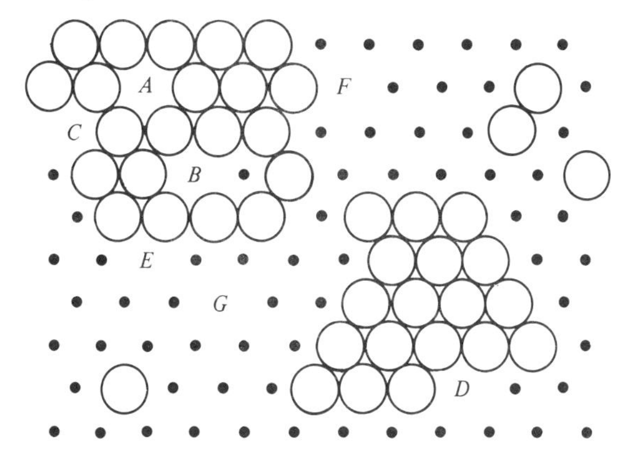

(Strictly speaking, this formula requires correction because a vacancy in a *surface* site does not count as a vacancy, but in a large crystal this correction is negligible.) To discuss the change in the surface configurational entropy δS_s, we note that the surface has a definite configurational entropy per unit area, at a given temperature. If we write the volume as $N_s a^3$ (where a is of the order of the interatomic spacing) then the surface area will be of order $N_s^{\frac{2}{3}} a^2$, and the *change* in surface area when one vacancy is created and N_s increases by one is of order $N_s^{-\frac{1}{3}} a^2$. In a large crystal we observe that this is very much less than the surface area corresponding to one atom – not surprisingly, because the extra atom added to the surface covers up, to first order, as much surface as it creates. Now the surface configurational entropy cannot be more than a few k per atom in the surface. It follows that $\delta S_s \ll k$, and is therefore negligible compared with δS_b. Thus

$$\delta F = \delta F_c - T\delta S_b - T\delta S_s$$
$$\simeq \delta F_c - kT \ln [(N + N_v)/N_v] = 0.$$

On rearranging this result we find that the proportion of vacant sites in the bulk of the crystal is given by

$$\blacktriangleright \quad N_v/N_s = e^{-\delta F_c/kT}. \tag{9.22}$$

This proportion is low at most temperatures. For many solids, however, δF_c is about $3kT$ at the melting temperature, which implies a vacancy concentration of a few percent. Experiment confirms this prediction. Near the melting point the equilibrium concentration of vacancies has appreciable effects on the heat capacity, diffusion coefficient and, in the case of metals, the electrical resistivity.

Problems

9.1 In §9.1.1 we found $E(T)$ and $S(T)$ for the harmonic oscillator by method 1. Obtain the same results by method 2, first calculating $Z(T)$ for the oscillator and then using (9.11); and by method 3, first calculating $S(E) = k \ln W$ for a large system of N oscillators with the help of (2.9), and then using (9.17).

9.2 A small system has only two energy levels. The lower level has energy 0 and degeneracy g_1 (i.e. it consists of g_1 different states all having the same energy), and the upper level has energy ε and degeneracy g_2. Use method 1 to show that the mean energy in the system is equal to $\varepsilon[(g_1/g_2) \exp(\varepsilon/kT) + 1]^{-1}$. Sketch the energy and the corresponding heat capacity as functions of temperature, and explain why one expects to see behaviour of this general type in any system having a small number of energy levels spread over a finite range of energy (Schottky anomaly).

The Cr^{3+} ions in chromium methylamine alum have two energy levels, both doubly degenerate. Between 0.4 and 1 K the corresponding heat capacity per ion has the form $C = k(T_0/T)^2$ where $T_0 = 0.14$ K. Deduce the energy separation between the energy levels.

9.3 The potential energy ϕ between the two atoms in a hydrogen molecule, calculated using the Born–Oppenheimer approximation, may be fitted by the function $\phi = \phi_1 \exp(-2\alpha r) - \phi_2 \exp(-\alpha r)$, where $\phi_1 = 1.72 \times 10^{-17}$ J, $\phi_2 = 6.93 \times 10^{-18}$ J and $\alpha = 2.0 \times 10^{10}$ m^{-1}. Find $\hbar^2/2I$, where I is the moment of inertia of the molecule, and $h\nu$, where ν is its vibration frequency. Hence, with the help of Figs. 9.1 and 9.3, estimate roughly the temperatures below which the rotation and vibration heat capacities would be fully quenched if hydrogen were an asymmetric molecule. For what *really* happens in hydrogen, see problem 12.9.

9.4 Show that the partition function Z_r for the rotation states of an asymmetrical diatomic molecule such as CO is $\sum (2K+1) \exp[-\hbar^2 K(K+1)/2IkT]$, and use this result to confirm (9.6). Show that at sufficiently high temperatures the sum for Z_r may be converted to an integral with $\varepsilon_K = \hbar^2 K(K+1)/2I$ as variable. Hence find expressions for the corresponding rotational free energy, entropy, heat capacity and energy, valid at high temperatures. How high must T be for these expressions to be valid?

Occupation of interstitial sites

9.5 A crystalline solid contains N identical atoms, N lattice sites, and N interstitial sites to which atoms may be transferred at the cost of some energy. If n atoms are on interstitial sites, show that the configurational entropy is $2k \ln [N!/n!(N-n)!]$. If the transfer of an atom from a particular lattice site to a particular interstitial site increases the free energy by δF_c, show by minimising the total free energy that the equilibrium proportion of atoms on interstitial sites n/N is $(1 + \exp \delta F_c/2kT)^{-1}$.

[Assume that n/N is small, and that vacancies are very rare.]

Model for rubber elasticity

9.6 The elasticity of rubber, like the elasticity of a gas, is due principally to its entropy rather than its potential energy. To explore this idea, consider a simplified model in which the polymer chains consist of a large number N of identical links, each of which can only lie parallel or anti-parallel to a particular direction, all configurations of the chain having the same energy. If each link has an internal free energy equal to $10kT \ln(T/T_0)$, show that the total free energy per link takes the form $F = 10kT \ln(T/T_0) - kT[\ln 2 - \frac{1}{2}(1+x)\ln(1+x) - \frac{1}{2}(1-x)\ln(1-x)]$, where $x = L/L_{max}$ is the ratio of the length of the chain to its maximum length. Sketch the variation of tension with length for a chain at room temperature. If the chain is stretched reversibly and adiabatically from $x=0$ to $x=\frac{1}{2}$, find the change in temperature.

10

Waves in a box

When we have a uniform medium through which waves can travel enclosed by boundaries which are held at temperature T, we must expect to find some density of *thermally excited waves* in the medium. In this chapter we shall consider the question of how much thermal energy such waves possess. This type of analysis is widely used in statistical thermodynamics. We shall illustrate it by finding the spectral energy density of equilibrium electromagnetic radiation in an empty cavity, and the lattice heat capacity of a block of crystalline solid.

10.1 The density of modes in k space

The first step in calculating the energy density for thermally excited waves is to find how many wave modes there are. If we ignore the boundaries we always find that a uniform medium has *plane-wave solutions*. Using the complex representation, the amplitude ξ of a plane wave may be written as

$$\xi = \xi_0 \exp\left[i(\mathbf{k}\cdot\mathbf{r} - \omega t)\right]. \qquad (10.1)$$

Plane waves

This represents a sinusoidal wave with plane wave fronts normal to the vector $\mathbf{k}$. This vector is known as the *k vector*. Its magnitude k is equal to $2\pi/\lambda$, where λ is the wavelength. The quantity ω is the *angular frequency* of the wave. The wave fronts travel at velocity c, the *phase velocity*, equal to ω/k, in the direction of $\mathbf{k}$. For some waves, such as electromagnetic waves travelling in empty space, c is constant, and $\omega = ck$. Such waves are said to be *non-dispersive*. In other cases we find on solving the wave equation that ω is some less simple function of k, $\omega(k)$. In such cases the wave velocity ω/k is a function of k, and the wave is said to be *dispersive*. The relation between ω and k is called the *dispersion relation*.

Dispersion

At the boundaries, the wave has to satisfy *boundary conditions*, and the plane waves are reflected. A *single* plane wave is not a good solution, but combinations of incident and reflected waves at the same frequency can always be found which do satisfy the boundary conditions. Such solutions are known as *standing waves*. Let us, for instance, consider a medium whose

Standing waves in a rectangular box

boundaries form a rectangular box of dimensions L_x, L_y and L_z in the three co-ordinate directions, with one corner at the origin, and with boundary condition $\xi=0$. In this case the standing waves take the simple form

$$\xi=\xi_0 \sin(k_x x)\sin(k_y y)\sin(k_z z)\exp(-i\omega t), \tag{10.2}$$

where the boundary conditions limit k_x, k_y and k_z to the discrete values

$$k_x=\pi n_x/L_x, \quad k_y=\pi n_y/L_y, \quad k_z=\pi n_z/L_z. \tag{10.3}$$

p. 48

Here n_x, n_y and n_z are the *wave indices* which are positive integers. (We met waves of this form before, and they are illustrated in Fig. 5.1.) These particular standing waves consist of eight super-imposed running waves, with wave vectors $(\pm k_x, \pm k_y, \pm k_z)$. For a standing wave of given wave indices we may find the angular frequency ω from the dispersion relation.

k space

To help us assess the number of allowed wave modes it is useful to introduce *k space*, a three-dimensional space in which we plot points k_x, k_y, k_z for the discrete positive values of k which correspond to the different sets of wave indices. These points lie on a rectangular lattice in k space (Fig. 10.1), whose unit cell has the volume $\pi^3/(L_x L_y L_z)$. It turns out that we are usually interested in values of k much larger than $1/L_x$. On the scale of interest, then, the points which represent the allowed modes are very closely but uniformly packed. Since $L_x L_y L_z$ is just V, the volume of the wave medium in *real* space, we observe that the

▶ density of standing wave modes in k space $=V/\pi^3$. (10.4)

We have, of course, only demonstrated this result for the special case of rectangular boundaries on which $\xi=0$. For boundaries of arbitrary shape and other boundary conditions the standing waves involve many more running waves and may become extremely difficult to analyse. However, in 1911 Weyl solved the considerable problem of showing that a result analogous to the one just obtained holds for sufficiently large boxes of *any* shape and *any* reasonable boundary condition.

For any given type of wave, knowing the density of standing wave modes in k space, we can use the dispersion relation to deduce the *frequency density of modes*. From the frequency density the energy density can be found, as we shall see.

10.2 The spectral distribution for temperature radiation

Temperature radiation or *black body radiation* is the electromagnetic radiation which exists in equilibrium inside a cavity whose walls are held at temperature T. If we place a small flat plate inside such a cavity, it will necessarily come to equilibrium with the radiation at the same temperature, and the rate of spontaneous emission of radiation by the plate at temperature T must therefore balance its rate of absorption from the cavity at temperature T. Since this must always happen, whatever the details of the cavity and the orientation of the plate, it is clear that *the equilibrium radiation in the cavity must be a function of temperature only, isotropic, and independent of the size or shape of the cavity, or the nature of its walls.*

This means that we may safely make use of the special case considered in the previous section, of a rectangular cavity and almost perfectly reflecting walls (corresponding to setting $\xi=0$ on the boundaries). For electromagnetic

Fig. 10.1. Calculation of the energy of thermally excited electromagnetic waves. The upper diagram shows the allowed k values for standing waves in a box of volume V. The allowed values fill one octant of k space at density V/π^3. Waves in a given range of frequencies lie within a spherical shell. The lower diagram shows the corresponding spectral energy density ε_ω as a function of frequency. The Rayleigh–Jeans formula uses the classical energy of kT per mode. It fits at the lowest frequencies, but gives infinite energy at large frequencies: *the ultraviolet catastrophe.* Planck's quantum formula fits the observations exactly.

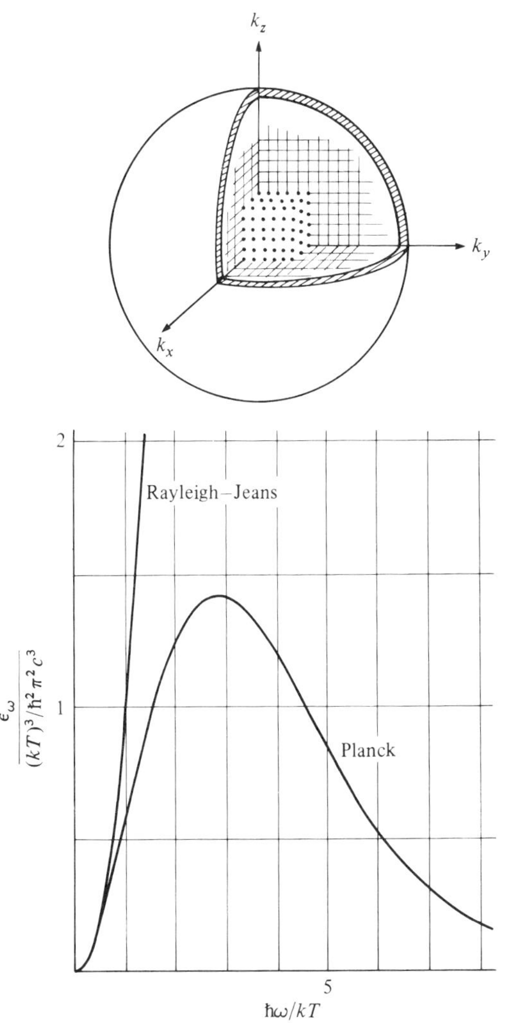

waves the dispersion relation is just $\omega = ck$, where c is the velocity of light and $k = (k_x^2 + k_y^2 + k_z^2)^{\frac{1}{2}}$ is the distance of the representative point from the origin in k space. Thus if we choose modes which lie within a spherical shell in k space of radius k and thickness $\mathrm{d}k$, we are choosing modes whose angular frequencies lie between ck and $c(k + \mathrm{d}k)$ (Fig. 10.1). The number of such modes is given by

The frequency density of cavity modes

$$\mathrm{d}n = \left(\frac{V}{\pi^3}\right)(4\pi k^2\, \mathrm{d}k)(\tfrac{1}{8})(2).$$

Here, V/π^3 is the density of standing wave modes in k space, $4\pi k^2\, \mathrm{d}k$ is the volume of the spherical shell, the factor of $\frac{1}{8}$ allows for the fact that only positive wave indices are to be counted, corresponding to positive values of k_x, k_y and k_z, and the factor of 2 allows for the two possible independent polarisations of electromagnetic waves. If we rewrite this result in terms of angular frequency we find that

$$\blacktriangleright \quad g_\omega = V\omega^2/(\pi^2 c^3), \tag{10.5}$$

where g_ω is the *frequency density of modes* allowing for polarisation, equal to $\mathrm{d}n/\mathrm{d}\omega$.

Now it can be shown that the energies in the different modes are independent – if different modes are superimposed their field energies can be added together. According to *classical* statistical mechanics it is possible to show that the mean energy in each mode should be equal to kT at temperature T. Using (10.5), therefore, it was originally thought that the *spectral energy density* ε_ω, the energy per unit volume and per unit angular frequency, should be given by

§13.5

The Rayleigh–Jeans formula and the ultra-violet catastrophe

$$\varepsilon_\omega = kT\omega^2/(\pi^2 c^3). \tag{10.6}$$

This classical result was derived by Rayleigh before the advent of quantum theory, and his theory was extended and very fully checked by Jeans. This *Rayleigh–Jeans formula* is shown in Fig. 10.1. It agrees with observation at the lowest frequencies, but *is obviously wrong at high frequencies, where it predicts an infinite total energy* when integrated over all frequencies, because there is no limit to the number of modes available at high frequencies, a result which became known as the 'ultra-violet catastrophe'. It was this intractable difficulty with classical theory which first led Planck to the idea of the quantum. The energy in a radiation mode of angular frequency ω is, in fact, quantised in units of $\hbar\omega$, like an oscillator, and Planck's oscillator formula shows that, if we omit the zero-point energy, the energy per mode should be given by $\hbar\omega/(\mathrm{e}^{\hbar\omega/kT} - 1)$ rather than kT. Thus, although we do indeed have an infinite number of modes available at high frequencies, the quantisation ensures that *the modes beyond a certain frequency are quenched, and the total energy is finite.* Using the oscillator formula with (10.5) the spectral energy density becomes

§9.1.1

Planck's radiation formula

$$\blacktriangleright \quad \varepsilon_\omega = \frac{\hbar\omega^3}{\pi^2 c^3 (\mathrm{e}^{\hbar\omega/kT} - 1)}. \tag{10.7}$$

This is the celebrated *Planck radiation formula*, first proposed in 1900. It fitted the experimental data on cavity radiation so well that physicists were forced to take seriously the, at that time, bizarre notion of quantisation.

The Planck formula is plotted as a function of frequency in Fig. 10.1. Note that the spectral energy density has a fairly sharp maximum where $\hbar\omega \simeq 3kT$. At the surface temperature of the Sun, about 6000 K, this peak is near the centre of the visible spectrum. For a body at room temperature it lies in the infra-red, and such bodies emit negligible visible radiation. The area under the curve is the total energy density $\varepsilon = E/V$, given by

The total energy density

$$\varepsilon = \int_0^\infty \varepsilon_\omega \, d\omega = \frac{(kT)^4}{\hbar^3 \pi^2 c^3} \int_0^\infty \frac{x^3 \, dx}{e^x - 1},$$

where $x = \hbar\omega/kT$, so, since the value of the integral is $\pi^4/15$, we have

▶ $$\varepsilon = (\pi^2/15\hbar^3 c^3)(kT)^4. \qquad (10.8)$$

This confirms our earlier conclusion, based on thermodynamic reasoning, that the total energy density is proportional to T^4.

§8.3.4

10.3 Absorption and emission

When radiation falls on matter it may be reflected or transmitted, or it may be *absorbed*, increasing the temperature of the matter on which it falls. Conversely, all matter spontaneously *emits* radiation, at a level depending on its temperature. As we have already noted, a body placed inside a cavity at temperature T must itself come to equilibrium with the cavity when it reaches temperature T. We can use this fact to make some useful deductions about the rates of absorption and emission.

The black body emitter

Consider first an *ideal black body*, a body which completely absorbs all radiation falling on it. When such a body is in thermal equilibrium with the temperature radiation inside a cavity it must be precisely replacing the radiation which it absorbs by a corresponding emission. Thus *a black body must emit pure temperature radiation*, corresponding to its own temperature.

The energy of the radiation emitted by unit area of a black body in unit time is therefore equal to the energy of the equivalent radiation which falls on the area. For radiation at normal incidence this energy flux would be εc. For radiation incident at θ to the normal it would be $\varepsilon c \cos\theta$. Thus, for isotropic radiation the total flux arriving may be written as $\int \varepsilon c \cos\theta \, df$, where df is the fraction of the radiation flowing at θ to the normal. By considering areas on the surface of a sphere whose centre lies on the area of interest it is easy to see that $df = 2\pi r^2 \sin\theta \, d\theta / 4\pi r^2 = \frac{1}{2}\sin\theta \, d\theta$. Thus the total flux is equal to $\int_0^{\pi/2} \frac{1}{2}\varepsilon c \cos\theta \sin\theta \, d\theta = \frac{1}{4}\varepsilon c$, or

▶ Flux leaving unit area of a black body in unit time is σT^4, (10.9)

Stefan's law

where the constant σ is the *Stefan–Boltzmann* constant $\pi^2 k^4/(60\hbar^3 c^2)$, equal to 5.67×10^{-8} W m^{-2} K^{-4}. This is *Stefan's law*.

The emissivity is equal to the absorption coefficient

If we have a body which at angular frequency ω absorbs a fraction e_ω of the radiation falling on it then, by considering its equilibrium with the temperature radiation in a cavity, we see that at frequency ω it must emit a fraction e_ω of the full black body radiation, for otherwise the distribution of cavity radiation would be changing with time. In this context e_ω is known as the *emissivity*. We see that *the emissivity is always equal to the absorption coefficient at any given frequency*. This is in accordance with the principle of detailed balance which requires that the rate of absorption of photons by the body must always be equal to the corresponding rate of emission, in thermal equilibrium. We shall have more to say later about the implications of detailed balance in emission and absorption.

§ 16.7

The frequency dependence of the emissivity

It is important to remember that the emissivity may vary with frequency. For instance an ionised gas has an *absorption spectrum*, meaning that the gas only has a large absorption coefficient at certain special frequencies, corresponding to certain electronic transitions. Correspondingly the gas has a large emissivity at the same frequencies, so that if it is sufficiently hot it will emit strongly at these frequencies, giving a corresponding *emission spectrum*. Again, a body placed in sunlight will usually warm up. If we can arrange that the body has a high emissivity in the visible (where the Sun's radiation is concentrated) and a low emissivity in the infra-red (where its own radiation is concentrated) it will warm up more than usual, a phenomenon known as the *greenhouse effect*. It may be used, for instance, to regulate the temperature of artificial satellites.

10.4 Lattice heat capacity of a solid

An isotropic solid can support two types of elastic wave, a transverse *shear wave* with velocity c_T and two possible polarisations, and a longitudinal *compression wave* with velocity c_L and one polarisation only. The velocities c_T and c_L can be found if the elastic moduli and density of the material are known. It therefore seems at first sight that the heat capacity of an elastic solid could be calculated by a straightforward adaptation of Planck's radiation formula to the case of elastic waves.

The real modes of the lattice

However, a crystalline solid is not an elastic continuum: it contains discrete atoms with forces acting between them, arranged in a lattice structure. To illustrate the difference let us compare a simple system of three equal masses held between two fixed points by four equal light elastic strings with a corresponding continuous heavy elastic string, in which the same mass is evenly distributed (Fig. 10.2). For both systems it turns out that there are standing waves of transverse oscillation of the form $\xi = \xi_0 \sin(kx) \exp(-i\omega t)$, where $k = n\pi/L$. For the continuous string the corresponding transverse wave velocity c_T is constant and we have an infinite set of modes of transverse oscillation equally spaced in frequency (Fig. 10.2, left). The three masses, on the other hand, have only three modes of transverse oscillation in the y direction corresponding to their three spatial degrees of freedom, with $k = \pi/L$, $2\pi/L$ and

$3\pi/L$. These modes are sketched on the right of Fig. 10.2; it is easy to check that they satisfy the equations of motion and to find their frequencies, which are plotted below.

On comparing these modes with the waves on the elastic string, we notice two important differences. First, the modes for the discrete masses are *dispersive*: when the wavelength becomes comparable with the spacing between the masses the phase velocity ω/k becomes less than the velocity for transverse waves on the equivalent string and decreases with increasing k. Secondly, *there is a limit to the number of distinct modes.* If we ask what becomes of the modes for $n>3$, that is, for wavelengths less than two mass spacings, we find for $n=4$ a non-mode – no displacements at all – for $n=5$ a mode which turns out to be identical with $n=3$, for $n=6$ a mode identical with $n=2$, and so on. Thus although, in a formal sense, we may write down short wavelength modes with $n>3$, we find that they are not new modes, but old modes in an alternative description.

This brief discussion of masses and elastic strings makes it clear that the *exact* calculation of lattice heat capacities for a solid containing many

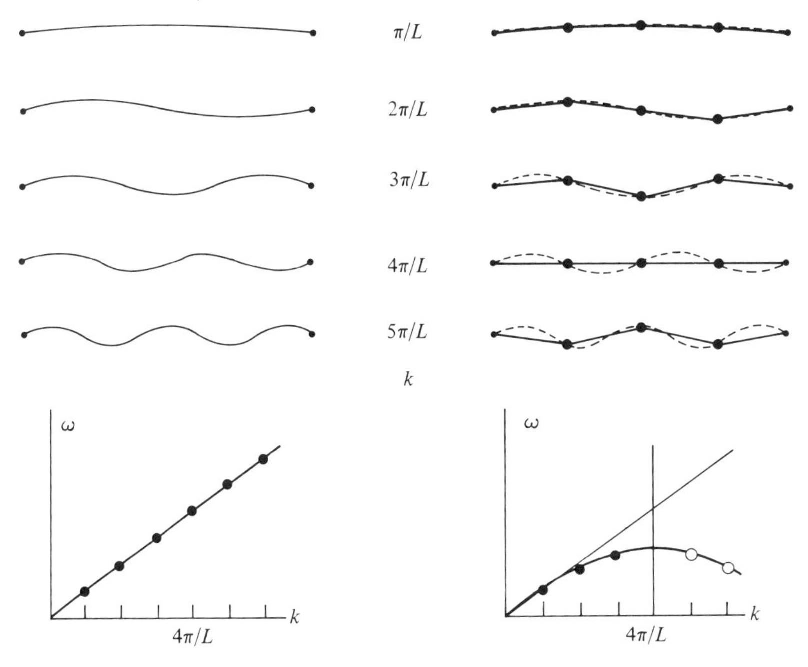

Fig. 10.2. Transverse standing waves on a continuous heavy elastic string of length L (left) compared with transverse waves on an equivalent light elastic string with three discrete masses attached to it (right). The tension and mass per unit length are the same for the two cases. The continuous string has a linear dispersion plot of ω against k, and an infinite series of standing wave modes. The three masses have a non-linear dispersion plot, and only three modes: modes with $k>4\pi/L$ are repeat descriptions of modes with $k<4\pi/L$.

discrete atoms will be a complex affair. What we have to do is to calculate from our knowledge of the interatomic forces (the analogues of the elastic strings) all the true vibrational modes of the solid and their frequencies. If the lattice contains N atoms, it has $3N$ spatial degrees of freedom, and we ought therefore to find $3N$ modes. Techniques for finding such modes are explained in textbooks of solid state physics, which also explain how the frequencies of such modes may be *measured* using slow neutron spectroscopy. A typical frequency density of phonon modes g_ω for a real solid, aluminium, is shown in Fig. 10.3. When we apply quantum mechanics to such a system we find that the energies of the modes are independent and quantised like the energies of an oscillator of the corresponding frequency. The quanta of oscillation are known as *phonons* (by analogy with photons). Thus, once g_ω has been calculated or measured, we may write down the energy in the system at temperature T as

The general result for the lattice energy

$$E = \int_0^\infty g_\omega(\omega)\hbar\omega/(e^{\hbar\omega/kT} - 1)\, \mathrm{d}\omega, \tag{10.10}$$

using Planck's formula for the energy in an oscillator (again omitting the constant zero-point energy), and so deduce the heat capacity.

Such calculations were first carried out by Born and von Kármán in 1912, and have been greatly elaborated since. Provided the values of g_ω are correct, they give excellent agreement with measured lattice heat capacities. They are, however, difficult to carry through, and have to be done afresh for each solid of interest. It is therefore useful to consider two approximations, each involving a deliberate simplification of the phonon spectrum so that only one parameter is needed to describe the solid. The first and crudest simplification is the *Einstein model*, in which we treat all the $3N$ modes as though they had the same frequency ω_E, which might, for instance, be the mean mode frequency. At high temperatures the Einstein heat capacity approaches $3Nk$, as, of course, it must for any model having the right number of modes. This result agrees with experiment, and is known as *Dulong and Petit's law*. Many solids are already in the Dulong–Petit limit at room temperature. At lower temperatures the Einstein model shows the heat capacity falling off as it would for a single oscillator (compare Figs. 10.3 and 9.1), and is quite successful at medium temperatures.

The Einstein approximate spectrum

pp. 88, 110

At the lowest temperatures the Einstein model fails. It does so because it then treats all the modes as quenched and therefore predicts an exponentially small heat capacity. In fact, however low the temperature, there are always some unquenched modes of low frequency, and it is in those modes that most of the thermal energy resides. It was pointed out by Debye that at the lowest temperatures, these modes have wavelengths much greater than the interatomic spacing, and for them it would be valid to treat the solid as a continuous elastic medium, with constant values of c_T and c_L. For such an elastic continuum we may, by analogy with Equation (10.5) for the electromagnetic case, at once write down the frequency density of modes allowing for the

The low temperature limit

§ 10.2

number of polarisations as

$$g_\omega = (V/2\pi^2)(2/c_T^3 + 1/c_L^3)\omega^2 = \alpha\omega^2. \tag{10.11}$$

This frequency density is shown as the dashed curve in Fig. 10.3; note that it does indeed agree with the exact calculation of g_ω for aluminium at low frequencies. Since at low temperatures only the low-frequency modes matter, we may in that limit adapt the analysis of temperature radiation and apply it as it stands to the elastic waves. The analogue of (10.8) shows that the energy density should be equal to $(\pi^2/30\hbar^3)(2/c_T^3 + 1/c_L^3)(kT)^4$, corresponding to a heat capacity given by

$$C_V = (2\pi^2k^4/15\hbar^3)(2/c_T^3 + 1/c_L^3)T^3. \tag{10.12}$$

This prediction fits the experimental lattice heat capacity accurately in all solids at the lowest temperatures. C_V as given by (10.12) is known as the *Debye T^3 low temperature heat capacity.*

It was clear from the success of the Einstein model at intermediate temperatures that, at such temperatures, the details of the frequency spectrum are not very important, so long as the total number of modes is correct and their mean frequency is approximately correct. Debye therefore made the further suggestion that a model which used the continuum spectrum (10.11) for *all* frequencies up to a cut-off ω_D, chosen to make the total number of modes equal to $3N$, might provide a useful approximation. This *Debye spectrum* is shown in Fig. 10.3 and it is clearly very different from the true spectrum. It will, however, automatically give the correct heat capacity at low temperatures (where only the low-frequency modes are important) and at high temperatures

The Debye approximate spectrum

Fig. 10.3. Lattice heat capacity. The solid curve on the left shows the measured frequency density of lattice modes for aluminium (after C. B. Walker, 1956), compared with the Einstein model, in which all $3N$ modes have the same frequency ω_E, and with the Debye model, which uses the frequency density for an elastic continuum, cut off at a frequency ω_D which gives a total of $3N$ modes. The corresponding predictions for the lattice heat capacity are shown on the right. The Einstein frequency ω_E is here chosen as $0.75\omega_D$, the mean frequency of the Debye spectrum. The real heat capacity for aluminium is close to Debye's prediction.

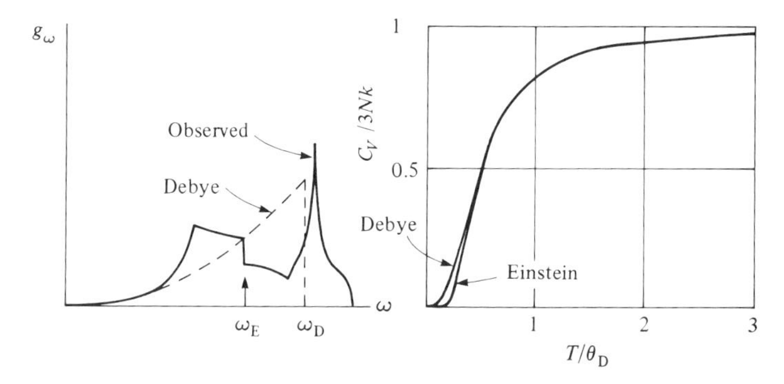

(where only the number of modes is important). We may hope that it will give a reasonably good interpolation between them.

To determine the heat capacity for the Debye spectrum we first find ω_D. Integrating (10.11) shows that $3N = \int_0^{\omega_D} g_\omega \, d\omega = \frac{1}{3}\alpha\omega_D^3$. This fixes ω_D as $(9N/\alpha)^{\frac{1}{3}}$. Using this cut-off, the Debye expression for the total energy is

$$E = \int_0^{\omega_D} \alpha\omega^2 \, d\omega \, \hbar\omega/(e^{\hbar\omega/kT} - 1). \tag{10.13}$$

We change the variable of integration to x, equal to $\hbar\omega/kT$, and introduce the *Debye temperature* θ_D as a convenient measure of ω_D by writing $k\theta_D = \hbar\omega_D$. In these terms the energy expression becomes

The Debye interpolation formula for the lattice thermal energy

$$E = 3NkT\left[3\left(\frac{T}{\theta_D}\right)^3 \int_0^{\theta_D/T} \frac{x^3 \, dx}{e^x - 1}\right]. \tag{10.14}$$

This is the *Debye interpolation formula* for the lattice energy at temperature T. The heat capacity derived from this result is plotted in Fig. 10.3, and it takes the form of a universal function of T/θ_D. Thus θ_D is the only parameter of the solid needed to evaluate the Debye heat capacity. It may be written in terms of macroscopic measurable quantities, as follows:

$$k\theta_D = \hbar\omega_D = \hbar(9N/\alpha)^{\frac{1}{3}} = \hbar(18\pi^2 N/V)^{\frac{1}{3}}(2/c_T^3 + 1/c_L^3)^{-\frac{1}{3}}. \tag{10.15}$$

For most solids the Debye heat capacity agrees with experiment within a few percent at intermediate temperatures, and provides a useful practical approximation to the truth.

Problems

10.1 The tungsten filament of a 240 V, 100 W light bulb emits most of its power as black body radiation with a peak in the frequency distribution at $\lambda = 1.7 \times 10^{-6}$ m. The resistivity of tungsten is approximately equal to $\alpha(T - 100\text{ K})$ where $\alpha = 2.8 \times 10^{-10}\,\Omega$ m K^{-1}, and its emissivity when hot is 0.5. Estimate the temperature, length and radius of the filament. With the help of Fig. 10.1 estimate the energy radiated by the filament in the region where the eye is efficient (5×10^{-7} m $< \lambda < 6 \times 10^{-7}$ m).

10.2 A box with insulated, reflecting walls containing temperature radiation is slowly increased in size, its shape remaining unchanged. Show that the energy distribution remains a black body distribution, and that $PV^{\frac{4}{3}} = \text{const.}$ According to a simplified cosmology the radius of the Universe has been increasing steadily since the 'big bang', about 1.3×10^{10} years ago. Interstellar space is now filled with 'cosmic background' black body radiation at 3 K, which is thought not to have interacted appreciably with the rest of the Universe since very soon after the big bang. Estimate the effect of the cosmic background on the temperature of the Earth now. Would it ever have been large enough to affect the weather? [Age of the Earth $= 4.7 \times 10^9$ years.]

The cosmic background radiation

10.3 Explain why the reflection coefficient of a metal surface should be equal to $1 - e$, where e is the emissivity. Hence show that the net radiation flux density between two closely spaced surfaces of emissivity e and temperatures T_1 and

T_2 is equal to $e\sigma(T_1^4 - T_2^4)/(2-e)$. Liquid helium at 4.2 K is contained in a spherical tank of radius 40 cm, which is insulated by a vacuum jacket containing 'superinsulation', a loose sandwich of 60 layers of aluminised plastic film. If radiation is the only important source of heat gain, how long does it take a full tank to boil away? [Latent heat of evaporation of helium = 2.73×10^6 J m^{-3}. Emissivity of bright aluminium = 0.03.]

10.4 The Debye temperature of copper is 310 K. With the help of Fig. 10.3 estimate what fraction of the total vibrational energy of copper at room temperature is represented by zero-point energy of vibration.

10.5 The dispersion relation for surface tension ripples on a liquid is $\omega^2 = \Gamma k^3/\rho$, where Γ is the surface tension and ρ is the density. By finding the two-dimensional equivalent of (10.4), and the surface tension equivalents of (10.5) and (10.8), show that the energy density of ripples in the low temperature limit takes the form $\beta T^{\frac{7}{3}}$. Deduce that if Γ is independent of T the corresponding surface free energy density is equal to $-\frac{3}{4}\beta T^{\frac{7}{3}}$. Find the value of this quantity for liquid helium at 2 K, and comment on your result. [For He at 2 K, ρ = 150 kg m^{-3} and $\Gamma = 3.1 \times 10^{-4}$ N m^{-1}. $\int_0^\infty x^{\frac{4}{3}}/(e^x - 1)\, dx = 1.68$.]

10.6 When the volume of a solid is changed by thermal expansion or the application of pressure the interatomic forces change, with consequent changes in the phonon frequencies. It is often a good approximation to assume that all the frequencies change in the same proportion, and that $\nu_{ph} \propto V^{-\gamma}$, where γ is the *Grüneisen constant*. Show that when this is true C_V may be expressed as $f(TV^\gamma)$. Deduce that

The Grüneisen expansion relation

$$(\partial C_V/\partial T)_V = -(V/\gamma)(\partial/\partial T)_V[(\partial P/\partial V)_T(\partial V/\partial T)_P],$$

and hence obtain the *Grüneisen relation*

$$\alpha = \gamma C_V/K_T V.$$

where α is the volume expansion coefficient $(\partial V/\partial T)_P/V$ and K_T is the isothermal bulk modulus $-V(\partial P/\partial V)_T$. [Since K_T varies little with T, this relation shows that at constant P the expansion of the solid is proportional to the thermal vibration energy which it contains.]

11

Systems with variable contents

So far we have assumed that a thermodynamic system contains a fixed number of particles. There are many situations where this is an inconvenient limitation. We may, for instance, wish to discuss the diffusion of molecules through some porous solid which separates two containers of gas, A and B. In such a situation it is convenient to speak of 'the gas in container A' and 'the gas in container B' as thermodynamic systems, but if we wish to do this we must recognise that the systems so defined do not contain fixed numbers of particles. Other systems having variable contents include a lump of ice surrounded by water (if we think of the ice and the water as separate systems), a metal conductor into which electrons can flow from an external circuit, a plasma in which the proportions of ionised and unionised atoms may change, and a reacting mixture of chemical reagents. Any such system will contain one or more *species* of particle, and we can specify its *contents* by quoting the numbers of particles N_s of each of the species s which it contains. In this chapter we shall develop a general formalism suitable for handling such situations. The following chapter is concerned with the special case of non-interacting identical particles, and other applications will appear later in the book.

11.1 Semi-permeable membrane and particle reservoir

It is convenient to introduce at the outset two notional devices. They are both extremely hard to construct in practice, but may exist in principle, and it is useful to imagine that they exist for the purposes of thermodynamic argument. The first is the *ideal semi-permeable membrane*, a membrane which is supposed to allow one species of particle to pass through it, but blocks all others. Practical semi-permeable membranes do exist; they have pores of uniform diameter which allow small molecules through and block larger ones. One can invent other 'in principle' devices which could separate particles of different species by using, for instance, strong gravitational, electric or magnetic fields to exploit the differences in mass and electric and magnetic polarisability of different species. We shall assume that ideal membranes can exist in principle. The second notional device is the *ideal particle reservoir*. This is an imaginary system which contains a large number N_0 of particles of the same species, *all having the same energy* ε, so that the total energy of the reservoir is $N_0\varepsilon$. The reservoir is to be so constructed that *its entropy does not change when particles are added or removed.* We shall see in the next chapter how such reservoirs could, in principle, be set up.

§§ 12.4, 12.5

11.2 Chemical potential

Suppose that we have a thermodynamic system which can exchange particles of species s with an ideal particle reservoir whose particle energy is ε_s. If the system and reservoir are not in equilibrium particles will flow in such a direction that the total entropy increases. But by definition the entropy of the reservoir does not change in a particle flow, so, if the system is otherwise isolated, we have

$$\mathrm{d}S_{\text{tot}} = \mathrm{d}S = (\partial S/\partial E)_{N_s}\,\mathrm{d}E + (\partial S/\partial N_s)_E\,\mathrm{d}N_s \geqslant 0, \tag{11.1}$$

where N_s is the number of particles of species s in the system. Now, since energy is conserved, $\mathrm{d}E = -\varepsilon_s\,\mathrm{d}N_0 = \varepsilon_s\,\mathrm{d}N_s$, and $(\partial S/\partial E)_{N_s} = 1/T$, as usual. Our condition may therefore be rewritten as

§ 6.4

$$\varepsilon_s\,\mathrm{d}N_s/T + (\partial S/\partial N_s)_E\,\mathrm{d}N_s \geqslant 0.$$

At this point we introduce a new quantity μ_s, the *chemical potential* inside the system for the sth species, by writing

The definition of chemical potential

$$\blacktriangleright \quad \mu_s/T = -(\partial S/\partial N_s)_{E,V}. \tag{11.2}$$

In terms of μ_s our condition may be written as

$$(\varepsilon_s - \mu_s)\,\mathrm{d}N_s \geqslant 0. \tag{11.3}$$

Thus the direction of particle flow depends on the sign of $\varepsilon_s - \mu_s$: particles may flow into the system from the reservoir if $\varepsilon_s > \mu_s$, and in the opposite direction if $\varepsilon_s < \mu_s$. This gives us a first picture of the meaning of μ_s: it is a characteristic energy for particles of a particular species inside a system, which determines the direction of flow between the system and an ideal particle reservoir, according to the sign of $\varepsilon_s - \mu_s$. It is measured in joules per particle.

We may extend this picture at once to the question of particle exchange between two ordinary systems A and B. In general two systems which can exchange particles can also exchange energy, so we have

$$\begin{aligned}\mathrm{d}S_{\text{tot}} &= \mathrm{d}S_A + \mathrm{d}S_B = (\partial S_A/\partial E_A)\,\mathrm{d}E_A + (\partial S_B/\partial E_B)\,\mathrm{d}E_B + \sum_s (\partial S_A/\partial N_{As})\,\mathrm{d}N_{As} \\ &\quad + \sum_s (\partial S_B/\partial N_{Bs})\,\mathrm{d}N_{Bs} \\ &= (1/T_A - 1/T_B)\,\mathrm{d}E_A + \sum_s (-\mu_{As}/T_A + \mu_{Bs}/T_B)\,\mathrm{d}N_{As} \geqslant 0, \end{aligned} \tag{11.4}$$

For particle exchange the chemical potentials are equalised in equilibrium

since both energy and particles are conserved. When the two systems have reached equilibrium, S_{tot} must be maximised with respect to *all* possible changes of both E_A and N_{As}. We deduce from (11.4) that in equilibrium T_A must be equal to T_B as usual and, in addition, μ_{As} must be equal to μ_{Bs} for each species. Thus, *when systems can exchange particles not only the temperature but also the chemical potential for each species concerned must be equalised in equilibrium.*

Chemical potential is also involved in the problem of *internal* equilibrium when some chemical or particle reaction can occur between the species present inside the system. If the N_s values inside an isolated system change, then

$$dS_{tot} = dS = \sum_s (\partial S/\partial N_s)_E \, dN_s = -\sum_s \mu_s \, dN_s/T \geqslant 0. \quad (11.5)$$

The chemical potentials fix the directions of chemical reactions

Now if the values of N_s change through a chemical reaction the various dN_s values are not independent. When a reaction occurs, such as

$$2H_2 + O_2 \rightarrow 2H_2O$$

we define a *reagent number* R_r, which is the number of molecules of the rth species disappearing, and a *product number* P_p which is the number of molecules of pth species appearing, when the reaction occurs once. In this case, $R_{H_2} = 2$, $R_{O_2} = 1$ and $P_{H_2O} = 2$. The dN_s values are related to the corresponding R_r and P_p values, and the condition may be rewritten as

$$\blacktriangleright \quad \sum R_r \mu_r \geqslant \sum P_p \mu_p, \quad (11.6)$$

or, in words, the total chemical potential of the reagents must be greater than the total chemical potential of the products. *It is the chemical potential which determines the direction in which chemical reactions occur*: this is the reason for its name.

11.3 Energy relations involving the chemical potential

If we have a system with variable contents, we must recognise that the entropy S is a function of N_s as well as E and V. We must therefore write the general change in S as

$$dS = (\partial S/\partial E)\, dE + \sum_s (\partial S/\partial N_s)\, dN_s + (\partial S/\partial V)\, dV + \text{similar constraint terms}$$

$$= dE/T - \sum_s \mu_s \, dN_s/T + P\, dV/T \quad + \text{similar work terms.}$$

§6.8

On rearranging we find that the expression for the change in energy which we derived earlier is now extended, and reads

$$\blacktriangleright \quad dE = T dS - P\, dV + \text{other work terms} + \sum_s \mu_s \, dN_s. \quad (11.7)$$

Particle heats

We shall refer to the new terms of the form $\mu_s \, dN_s$ as *particle heats*; they represent energy which, in a reversible change, arrives with the particles entering the system. If we suppose that the system exchanges *heat* with an ordinary heat reservoir, but exchanges *particles* only with ideal particle reservoirs, then, in a reversible change, the terms $\mu_s\,dN_s$ represent particle heat which has come from the ideal particle reservoirs and the term $T dS$ represents ordinary heat which has come from the heat reservoir. In real life, as we noted above, systems which can exchange particles can always in practice exchange heat, too, by the same mechanism, and we cannot draw any clear operational distinction between heat and particle heat. We simply have to say that the total heat energy which has moved reversibly from one system into the other is equal to $T dS + \sum \mu_s\,dN_s$.

When we allow the N_s values to vary, we still define the various thermodynamic potentials in the same way (so that F remains equal to $E - TS$, for instance), but the formulae for their differentials have to be extended (Table 7.1). We have, for instance, in place of (7.2),

§7.1

$$dF = -S\,dT - P\,dV + \sum_s \mu_s\,dN_s,$$

▶ $$dH = T dS + V dP + \sum_s \mu_s\,dN_s, \qquad (11.8)$$

$$dG = -S\,dT + V dP + \sum_s \mu_s\,dN_s.$$

We may derive from such expressions a number of new Maxwell relations involving μ_s. From the last relation, for instance, since $\partial^2 G/\partial T\,\partial N_s = \partial^2 G/\partial N_s \partial T$, we discover that $(\partial \mu_s/\partial T)_{P,N_s} = -(\partial S/\partial N_s)_{P,T}$. Such relations may be constructed from (11.8) when they are required.

The connection between G and μ

An important property of the chemical potential is apparent from the last relation of (11.8). If we are interested in a homogeneous system, we may imagine ourselves building the system up, starting from an empty system, by gradually adding more material, but keeping the *intensive* variables T, P and μ_s constant. In such a process dG is just $\sum \mu_s\,dN_s$, and since we started with an empty system we find on summing these terms that

▶ $$G = \sum \mu_s N_s \text{ in a homogeneous system.} \qquad (11.9)$$

This provides a link between (7.8) and (11.6), two apparently different criteria fixing the direction in which chemical reactions occur. It also makes it clear that in a homogeneous system containing only one species – a pure substance – the chemical potential is just the Gibbs function per particle.

§7.4

We may extend our range of thermodynamic potentials by introducing

Thermodynamic potentials involving μ

$$\begin{aligned}
\Phi_E &= E - \sum \mu_s N_s & d\Phi_E &= T dS - P\,dV - \sum N_s\,d\mu_s \\
\Phi = \Phi_F &= F - \sum \mu_s N_s & d\Phi_F &= -S\,dT - P\,dV - \sum N_s\,d\mu_s \\
\Phi_H &= H - \sum \mu_s N_s & d\Phi_H &= T dS + V dP - \sum N_s\,d\mu_s \\
\Phi_G &= G - \sum \mu_s N_s & d\Phi_G &= -S\,dT + V dP - \sum N_s\,d\mu_s
\end{aligned} \qquad (11.10)$$

which generate still further Maxwell relations. Of these potentials the most important is Φ_F, usually written Φ and named by Gibbs the *grand potential*, for

Problem 11.3

reasons which will appear in the next section. The grand potential is similar in rôle to the free energy: it is the effective mechanical potential energy for a system which is in permanent contact with a heat and particle reservoir. In a homogeneous system we may use (11.9) and deduce that

$$\begin{aligned} \Phi_E &= TS - PV \quad & \Phi_F &= -PV, \\ \Phi_H &= TS \quad & \Phi_G &= 0. \end{aligned} \tag{11.11}$$

The last relation combined with the expression for $d\Phi_G$ shows that

The Gibbs–Duhem relationship

▶ $$-S\,dT + V\,dP - \sum N_s\,d\mu_s = 0 \quad \text{in a homogeneous system,}$$

the *Gibbs–Duhem relationship*. This connection between the differentials of the intensive variables is often useful in finding derivatives of the chemical potentials: for instance, we see at once that in a pure substance $(\partial\mu/\partial T)_P = -S/N$.

11.4 The Gibbs distribution

We must now extend the question which led originally to the Boltzmann distribution and ask: if a system can exchange *both heat and particles* of a single species with a much larger reservoir, what is the probability in equilibrium of finding it in any one of its possible quantum states? The argument is very similar to the earlier one and should be read in conjunction with it. The system now has access to a much wider range of possible quantum states, corresponding to the complete range of possible contents, and much greater than the range of possible states for *fixed* contents. But it remains true that each of the accessible joint states of the system plus the reservoir is equally probable in equilibrium, and that therefore the probability p_i of finding the system in one particular state i is still proportional to the number of states N_i which the *reservoir* then has access to. As before, this is equal to $g_R\delta E$, where g_R is the density of states of the reservoir and δE is the energy range of the accessibility convention. The only difference is that g_R is now a function of two variables: it depends not only on the energy E_R in the reservoir, but also on the number of particles N_R in it.

§ 3.3

We shall see shortly that the large system approximation $S = k \ln g$ still holds when systems can exchange particles, so, using the definitions of T and μ, we may write for the reservoir

$$(\partial \ln g_R/\partial E_R)_{N_R} = 1/kT, \quad (\partial \ln g_R/\partial N_R)_{E_R} = -\mu/kT. \tag{11.12}$$

Because the reservoir is large, we must treat both T and μ as constants for small changes in E_R and N_R (as for the Boltzmann distribution), and we deduce that

$$g_R \propto e^{(E_R - \mu N_R)/kT},$$

for the ranges of E_R and N_R of interest. Since $E_R + E$ and $N_R + N$ are conserved we deduce that $p_i \propto g_R \propto \exp[-(E_i - \mu N_i)/kT]$, or, when properly normalised,

The Gibbs distribution

▶ $$p_i = \frac{e^{-(E_i - \mu N_i)/kT}}{\Xi}, \quad \text{where } \Xi = \sum_G e^{-(E_i - \mu N_i)/kT}. \tag{11.13}$$

This is the *Gibbs distribution*. The sum over states is called a *grand sum*, because it is taken not only over states of all energies, but also *over states having all possible contents, from zero particles upwards*. The quantity Ξ is the *grand partition function*.

We naturally define, now, a *grand canonical entropy* S_G, equal to $-k\sum_G p_i \ln p_i$. Since the system can exchange particles as well as heat with the reservoir, its probability is spread over a much larger number of states, of order $g\,\Delta E\,\Delta N$ where ΔN is the range of fluctuation of N, and the grand entropy is clearly greater than the canonical entropy. However, as with $\ln \Delta E$ in Chapter 4, we find when we take the logarithm that $\ln \Delta N$ is completely negligible compared with $\ln g$, so that we may safely write S_G as $k \ln g$, which means that the difference between S and S_G may be ignored in practice.

The grand canonical entropy

§4.3 The large system approximation

We find, however, that $kT\ln \Xi$ is significantly different from $kT\ln Z$ because Ξ contains the extra factor $\exp(-\mu N_i/kT)$. Using the expression for the grand entropy we have

$$TS = -kT\sum_G p_i \ln p_i$$

$$= -kT\sum_G p_i[-(E_i - \mu N_i)/kT - \ln \Xi] \quad \text{using (11.13)}$$

$$= E - \mu N + kT\ln \Xi.$$

Now $E - TS - \mu N$ is the grand potential $\Phi = \Phi_F$, so we have

$$\Phi = -kT\ln \Xi, \tag{11.14}$$

The grand potential in terms of Ξ

which may also be written in the alternative form

$$e^{-\Phi/kT} = \sum_G e^{-(E_i - \mu N_i)/kT}.$$

This result provides a method for calculating thermodynamic variables analogous to method 2 of Chapter 9. The grand partition function Ξ may be written down as a function of T, μ and the constraints, and S, P and N may be found as derivatives of Φ. As we shall see, for systems of identical particles this calculation is often easier than proceeding by way of Z and F for fixed N.

§9.2

§12.6

We have limited outselves in this section to systems containing a single species, but the extension to many species is straightforward.

Problems

11.1 Using (5.14) and the fact that S is an extensive variable, show that $S = Nk \ln(\beta V T^{\frac{3}{2}}/N)$ for an ideal monatomic gas, where β is a constant for the gas. Write down F for the gas, and hence show that μ is equal to $-kT\ln(\beta V T^{\frac{3}{2}}/Ne^{\frac{5}{2}})$. Two mixtures of gases are separated by a semi-permeable membrane. Assuming that the free energies for dilute mixtures of gases are additive, show that for any gas which can pass through the membrane the *partial pressure* (the part of the pressure exerted by that gas) must be the same in equilibrium on the two sides of the membrane.

Partial pressures are equalised in diffusive equilibrium

11.2 Show by considering the mechanical equilibrium that the number density n for an ideal monatomic gas at height h in a gravitational field g is proportional to $\exp(-mgh/kT)$, where m is the atomic mass. Modify the argument of problem 11.1 to find the chemical potential of a sample of N atoms of gas in volume V at height h, and show that this expression is independent of height for a gas in equilibrium under gravity.

§7.2

11.3 Prove the analogue of (7.4) for the grand potential Φ. A homogeneous system is held at fixed T and μ and slowly expanded from zero volume. What happens to P, $\Phi=\Phi_F$, G and Φ_G?

§12.3

11.4 A site in a semiconductor either is empty, or contains one electron with energy ε_1 (which may be spin up or spin down), or contains a pair of electrons of opposite spin, which because of their Coulomb repulsion have a considerably higher energy, ε_2. Treating the site as the system and the electrons in the rest of the material as a reservoir of temperature T and chemical potential μ, use the Gibbs distribution to show that if $\varepsilon_2-2\mu \gg kT$ the probability of finding an electron on the site is approximately $2/[\exp(\varepsilon_1-\mu)/kT+2]$. [Contrast this result with the Fermi distribution of Chapter 12: here the energies of the individual electrons are not independent.]

12

Indistinguishable particles

The particles of a given species are *indistinguishable*. In this chapter we shall explore the meaning of this concept. We shall find that the individual particles cannot normally be regarded as *systems* in the usual sense, and do not obey Boltzmann statistics. This has some important effects on the thermodynamics of systems containing indistinguishable particles.

12.1 Identifiable system and indistinguishable particle

We have frequently used the idea of a *thermodynamic system* in this book. In this section I aim to establish the point that a particular *particle* of a given species – a given benzene molecule, for instance – should *not* normally be thought of as a system in the usual sense. The distinction between a system and a particle is not a matter of size – some very small objects may properly be regarded as systems, and one can imagine circumstances in which quite large objects should properly be regarded as particles. A thermodynamic system such as a lump of aluminium is an object which *contains* particles, but *is not identified with a particular set of labelled particles.* We identify it rather by its occupied states. If I have two lumps of aluminium before me I mean by 'the left-hand lump' certain occupied lattice sites on my left, which are distinct from the occupied sites of 'the right-hand lump', and I am *not* concerned with the question of which atom is in which lump. Each of the lumps has its own excited states, which are distinct from those of the other. We may describe the joint states by quoting the state of the left-hand lump and the state of the right-hand lump, and every such combination represents a distinct state of the joint system. We used this picture when we established the notion of temperature, and when we derived the Boltzmann and Gibbs distributions. *Systems* when in contact with a reservoir definitely obey Boltzmann statistics, or, if they can exchange particles with the reservoir, Gibbs statistics.

'Systems' are identified by distinct sets of states

§§ 3.1, 3.3, 11.4

The particles themselves, however, may not usually be regarded as systems. Consider the *photons* in a cavity containing radiation. The various cavity modes *are* systems: each of them contains both energy and particles (photons), and each of them has its own set of states, distinct from those of other modes. We describe the state of the whole cavity by quoting the *occupations*, the number of photons in each mode. But the photons themselves are *not* systems in the same sense. We may think of transitions, if we like, in terms of a photon jumping from mode 1 to another mode of the same frequency, mode 2. But we *cannot* enumerate the states of the modes of the same frequency by labelling the photons $a, b, c, \ldots$ and considering all possible distributions of the labelled photons amongst the modes, for this would treat as distinct certain states which are really identical: a state described by saying that photon a is in mode 1 and photon b is in mode 2 is the *same* as the state described by saying that a is in 2 and b is in 1, because all that really matters is the number of photons in each mode.

It may seem misleading to be thinking of photons as particles in this sense, because they are easily created and destroyed, unlike electrons or protons. It turns out, however, that the same principles apply to *all* identical particles. Let us distinguish between *identical* particles and *similar* particles. By similar particles I mean hypothetical particles which are almost identical but can be distinguished in principle, by some very small difference in mass, perhaps. For any pair of particles a and b we may write down a Schrödinger wave function of the general form $\psi(a, b)$ where a and b represent all the space and spin co-ordinates of the two particles. Suppose now that the two particles change places, so that the new wave function takes the form $\psi(b, a)$. For *similar* particles (if such objects existed) the new state, though it would indeed be hard to distinguish from the old one, would normally be different from it in principle. When we are dealing with *identical* particles of the same species, however, experiment shows that this situation never arises. Instead, *the only wave functions which exist have the property that $\psi(b, a)$ actually represents the same state as $\psi(a, b)$*. This is the *exchange principle* of quantum theory.

Exchange of identical particles leaves the state unchanged

To illustrate the distinction between a particle and a system consider the following wave function, which is meant to represent the states of two adjacent paramagnetic ions in a crystalline salt:

$$\psi=(1/\sqrt{2})[\phi_{\uparrow L}(a)\phi_{\downarrow R}(b)+\phi_{\uparrow L}(b)\phi_{\downarrow R}(a)]. \tag{12.1}$$

In this wave function a term such as $\phi_{\uparrow L}(a)\phi_{\downarrow R}(b)$ represents a state in which ion a occupies the left-hand site with spin up and ion b occupies the right-hand site with spin down. The ions a and b are here identical particles, and you will see that the wave function does obey the exchange principle: exchanging the ions a and b leaves this particular wave function unchanged (the first term changes into the second and the second changes into the first). Indeed, we cannot say which ion is on which site: a and b both have probability of $\frac{1}{2}$ of being found on either the left or the right. But, although the ions themselves are indistinguishable, the left- and right-hand *sites* are distinguishable systems. Each site has two states (spin up and spin down) and every possible combination of site states is distinct. For instance, the wave function quoted has the left-hand site spin up and the right-hand site spin down. This is distinct from the wave function in which the spins on the sites are reversed (left down and right up). As identifiable systems, the *sites* obey Boltzmann statistics and, indeed, we assumed this when we calculated the magnetisation of a paramagnet.

§9.2.1

To summarise, a *system* is an object which *contains* energy and particles and is identified not as a particular set of particles, but as a particular and distinct set of states which can be occupied independently of the states of any other system. *Identical particles*, on the other hand, are objects which all have access to the *same* states, and their wave functions obey the exchange principle. This limits the range of joint states available to identical particles, and they do not obey Boltzmann statistics, as we shall see.

12.2 Bosons and fermions

The exchange principle requires that for identical particles $\psi(b, a)$ should represent the same state as $\psi(a, b)$. In quantum theory two wave functions representing the same state can differ, at most, by a phase factor, so $\psi(b, a)$ must be equal to $e^{i\alpha}\psi(a, b)$. Exchanging the particles a second time brings us back to the initial state, so the *square* of the phase factor must be unity. Thus we have two possible cases: either

$$\psi(b, a) = +\psi(a, b) \quad \text{symmetric,}$$

or (12.2) Symmetry conditions

$$\psi(b, a) = -\psi(a, b) \quad \text{antisymmetric.}$$

All species of particle fall into one of these two classes. Species with integral spin such as α particles have wave functions which are symmetric under exchange of any pair of particles: they are known as Bose–Einstein particles or *bosons*. Species with half-integral spin such as electrons, protons and neutrons have wave functions which are anti-symmetric under particle exchange: they are known as Fermi–Dirac particles or *fermions*. Note that, because the Hamiltonian is itself symmetric, it follows from the time-dependent Schrödinger equation that all wave functions retain their symmetry character as time proceeds. Thus symmetric boson wave functions cannot be scattered into non-symmetric states – and indeed no non-symmetric boson states exist. In the same way, fermions are only found in anti-symmetric states.

The Bose or Fermi character of more complex particles can be discovered by exchanging their contents one pair at a time. A hydrogen molecule, for instance, contains two protons and two electrons. Two such molecules may be interchanged completely by exchanging *four* pairs of fermions. This leaves the wave function unchanged, so H_2 must be a boson. By such arguments we see that complex particles containing an even number of fermions are bosons, but those containing an odd number of fermions are themselves fermions.

Larger particles

The symmetry conditions restrict the number of quantum states available very greatly. Let us consider the special case of particles which are non-interacting or weakly interacting. In this situation we may think of the individual particles as occupying well-defined *particle states* ϕ_j. Thus for a hypothetical assembly of similar, but not identical, particles, free of symmetry conditions, we would be involved with assembly wave functions of the product type

$$\psi(a, b, \ldots) = \phi_1(a)\phi_1(b)\phi_2(c)\phi_4(d) \ldots$$

$$n_1 = 2,\ n_2 = 1,\ n_3 = 0,\ n_4 = 1, \ldots, \quad (12.3)$$

in which particular particles occupy definite particle states. Such wave functions may be characterised by quoting the *occupation numbers* n_j, the numbers of particles occupying the various particle states ϕ_j. However, for these similar but not identical particles the occupation numbers would not specify the wave function fully, for there would be many different ways of distributing the N distinct particles amongst the particle states: there would be in fact $N!/(n_1!\,n_2!\ldots)$ such permutations. For real *identical* particles, on the other hand, the symmetry conditions restrict our freedom severely. When particles b and c are interchanged the above wave function changes into $\psi' = \phi_1(a)\phi_1(c)\phi_2(b)\phi_4(d)\ldots$. Thus if the wave function contains a term of the form ψ the symmetry condition for bosons, for instance, requires that it must *also* contain a term of the form ψ' having the same amplitude and phase. By exchanging all possible pairs of particles in this way we discover that a boson wave function containing ψ must also contain terms corresponding to *all* the possible permutations of the particles amongst the occupied particle states, each having the same amplitude and phase. A state so constructed is called a *symmetrised state*. For a pair a, b of identical bosons in states 1, 2, for instance, the symmetrised wave function is $(1/\sqrt{2})[\phi_1(a)\phi_2(b) + \phi_1(b)\phi_2(a)]$. Note that, whatever permutation we start from, the symmetry condition always generates the *same* symmetrised state: *for a given set of occupation numbers there is only one boson state.* The n_j values specify the state completely. There is only one way of occupying the required states because the particles are indistinguishable. For each state j the occupation number n_j can take the values 0, 1, 2, . . . , for bosons.

Occupation numbers

The symmetrised boson state

Problem 12.2

When we try to apply the fermion symmetry conditions in the same way a problem arises. For fermions, if the wave function contains state ψ it must also contain state ψ' with the same amplitude but *opposite* sign. This creates no difficulty if the particles exchanged are in different particle states. But consider exchanging particles a and b in state ψ above. Since a and b are both in particle state ϕ_1, for this exchange it happens that the new state is the *same* as the old state. The fermion symmetry condition thus requires that if the wave function contains state ψ it must also contain $-\psi$. In other words, it requires that such a state can only appear with zero amplitude: evidently *we cannot have more than one identical fermion in the same particle state.* This is Pauli's *exclusion principle*. For fermions, therefore, the occupation numbers n_j may only take the values 0 and 1. Provided this requirement is satisfied we *can* construct an *anti-symmetrised state function* for fermions: it is a combination

The exclusion principle

of terms corresponding to all the possible permutations of the particles amongst the occupied states each with the same amplitude, even permutations having positive sign and odd permutations negative sign. For a pair a, b of identical fermions in states 1, 2, for instance, the anti-symmetric state function takes the form $(1/\sqrt{2})[\phi_1(a)\phi_2(b) - \phi_1(b)\phi_2(a)]$. Once again, *for a given set of occupation numbers there is only one fermion state*: the occupation numbers specify the state completely.

The anti-symmetrised fermion state

12.3 The Bose–Einstein and Fermi–Dirac distributions

When the particles in a system have strong forces acting between them the individual particle states are not well defined. We know that, in equilibrium with a reservoir, the states of the system *as a whole* obey the Boltzmann or Gibbs distributions but we can say nothing about the occupation of the *particle* states. If the particles are *weakly interacting*, on the other hand, we *can* describe the states of the assembly by giving the occupation numbers n_j for the various particle states ϕ_j. For identical particles we must not speak of the probability that any *particular* particle is in some particle state ϕ_j, but we *may* speak of the *mean occupation number* $\bar{n}_j$, the average number of particles in state ϕ_j. Let us find these mean occupation numbers for weakly interacting boson and fermion states in contact with a reservoir.

If the system can exchange both energy and particles with the reservoir, we observe that, although the particles themselves are not systems, *the individual particle states* ϕ_j *are 'systems'*, in much the same way that the photon modes of a cavity are systems. The individual boson state j has a set of 'states' corresponding to its various possible occupation numbers n_j. Their 'state energies' E_{n_j} are $0, \varepsilon_j, 2\varepsilon_j, 3\varepsilon_j, \ldots$, and their 'contents' N_{n_j} are $0, 1, 2, 3, \ldots$. Each such 'system' may be populated from the reservoir independently of the other 'systems' (or particle states), and the states of the assembly as a whole may be enumerated by listing every possible combination of 'states' (or occupations) of the individual 'systems'. It follows that *we may apply the Gibbs distribution to the separate particle states*, treated as systems, and we find that

The particle states behave as 'systems'

§11.4

$$p_{n_j} = \mathrm{e}^{-(E_{n_j} - \mu N_{n_j})/kT} / \Xi_j$$

$$= \mathrm{e}^{-n_j(\varepsilon_j - \mu)/kT} / \Xi_j, \quad \text{where } \Xi_j = \sum_{n_j} \mathrm{e}^{-n_j(\varepsilon_j - \mu)/kT}. \tag{12.4}$$

We can now evaluate $\bar{n}_j$ very simply as $\sum_{n_j=0}^{\infty} n_j p_{n_j}$. The algebra is similar to the algebra which we used in deriving Planck's oscillator formula and I shall not repeat it here. The conventional symbol for the mean occupation number $\bar{n}_j$ is f_j. We find that

§9.1.1

$$▶ \quad f_j = \frac{1}{\mathrm{e}^{(\varepsilon_j - \mu)/kT} - 1} \quad \text{for bosons.} \tag{12.5}$$

This is the *Bose–Einstein distribution* (or Bose distribution, for short). Since f_j is a mean occupation number and not a probability, it may be greater than unity. It is worth noting that this calculation is almost identical to the calculation of the mean number of photons in a given photon mode. The only difference is that we do not have to get the photons from anywhere – they can be *created* if the energy is available. Thus we use the Boltzmann distribution in finding the photon occupation and μ does not appear in the result. Here we are dealing with *conserved* particles, which have to be obtained from the reservoir; thus we have to use the Gibbs distribution and consequently μ does appear in the result.

A parallel derivation applies to fermions. A fermion state j is a 'system' with just two states, having occupations 0 and 1, and energies 0 and ε_j. The mean occupation $\bar{n}_j$ in this case is $\sum_{n_j=0}^{1} n_j p_{n_j}$. This is the same as the probability of being in the second, occupied state. Using the Gibbs distribution we find that

$$f_j = \bar{n}_j = p_1 = e^{-(\varepsilon_j - \mu)/kT}/(1 + e^{-(\varepsilon_j - \mu)/kT}),$$

or

$$\blacktriangleright \quad f_j = \frac{1}{e^{(\varepsilon_j - \mu)/kT} + 1} \quad \text{for fermions.} \tag{12.6}$$

This is the *Fermi–Dirac distribution* (or Fermi distribution, for short). Its form is shown in Fig. 12.1. In this case f_j lies between 0 and 1, since n_j cannot be greater than 1 for fermions.

In obtaining these distributions we assumed that the particle state j was interacting with a reservoir of energy and particles having temperature T and chemical potential μ. This need not be an external reservoir, however. In a large assembly the very large number of partly occupied particle states other than j constitute an effective reservoir which is much larger than the 'system', state j itself. Thus the Fermi and Bose distributions hold inside large isolated assemblies in equilibrium, as well as for assemblies in contact with an external reservoir. It should be noted carefully, however, that the Fermi and Bose

Fig. 12.1. The Fermi occupation number f at $T=0$ and at a non-zero temperature, for energies near the chemical potential μ, which is assumed to remain constant and equal to the Fermi energy ε_F. The arrow shows how electrons are transferred from below ε_F to above ε_F as the system is warmed.

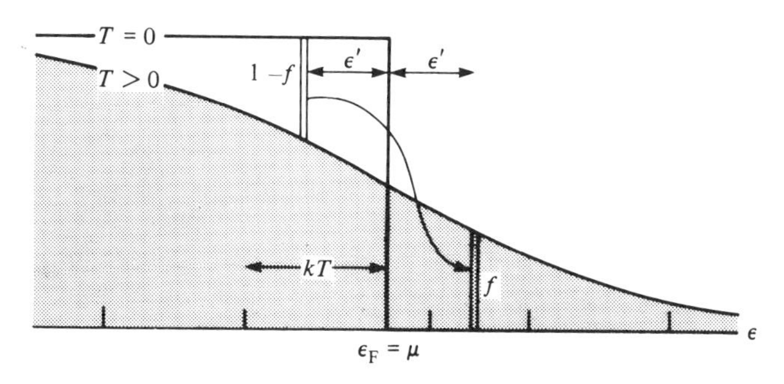

distributions only hold if the particle states are *independent*: the energy or occupation of one particle state must not be affected when the occupation of another particle state is altered.

Problem 11.4

12.4 Properties of the Fermi–Dirac distribution

It turns out that there are few real examples of weakly interacting bosons, so I shall not devote much space to the Bose–Einstein distribution. The Fermi distribution, on the other hand, is widely used, especially in its application to electrons in metals and semi-conductors, and we shall therefore examine its properties carefully.

We start with the low temperature limit. Inspection of (12.6) shows that, as $T \to 0$, the Fermi distribution approaches the limit

$$\begin{aligned} f_j &= 1 \quad \text{for } \varepsilon_j < \mu \\ &= 0 \quad \text{for } \varepsilon_j > \mu. \end{aligned} \tag{12.7}$$

This distribution is shown in Fig. 12.1: all states lying below the chemical potential in energy are occupied, and all states above it are empty. This is easy to understand. At absolute zero the assembly is in its ground state. We can only place one fermion in each particle state, so we may, in imagination, build up the state of lowest energy by adding fermions one at a time, putting each one into the lowest available particle state as we do so, until we have accommodated all the fermions. At the end of this process the energy of the last particle state to be occupied is the *Fermi energy*, ε_F (Fig. 12.2, left). Evidently, according to (12.7), μ must be equal to ε_F at $T=0$.

The Fermi energy

To find the value of ε_F we need to know the *density of particle states*, $g(\varepsilon)$. Taking the particle ground state as the origin of energy, we can write the

Fig. 12.2. The density of electron states $g(\varepsilon)$ and density of occupation $g(\varepsilon)f(\varepsilon)$ for a free-electron gas. At $T=0$ all states up to ε_F are filled, and the shaded area is equal to the number of electrons in the system. For $kT \ll \varepsilon_F$, the usual situation in a metal, the changes in occupation are limited to energies near ε_F. The chemical potential μ is fixed by the requirement that the shaded area remains equal to N. In practice this means that μ remains very close to ε_F.

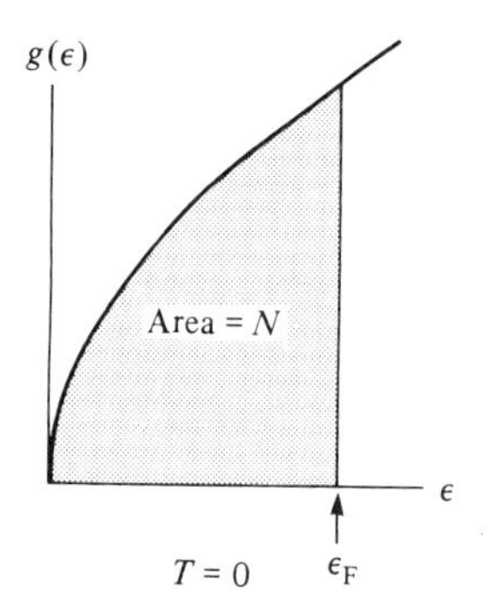

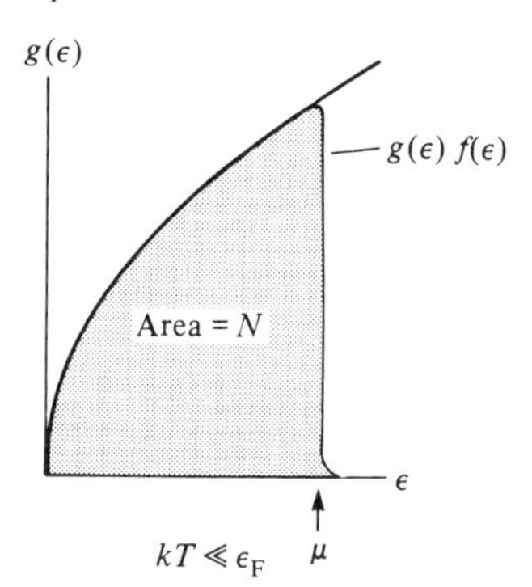

number of fermions in the system at $T=0$ as

$$N=\int_0^{\varepsilon_F} g(\varepsilon)\,\mathrm{d}\varepsilon. \tag{12.8}$$

If we know the form of the density of particle states $g(\varepsilon)$ we can fix ε_F by finding what upper limit is required in this integral to give the right number of particles. For instance, according to the *free electron model* the conduction electrons of a metal can be treated as a gas of non-interacting particles. For a gas of non-interacting electrons one can show that

Problem 12.4

$$g(\varepsilon)=(V/2\pi^2)(2m_e/\hbar^2)^{\frac{3}{2}}\varepsilon^{\frac{1}{2}}. \tag{12.9}$$

If we substitute this density of states in (12.8) we find that

$$\varepsilon_F=(\hbar^2/2m_e)(3\pi^2 n)^{\frac{2}{3}}, \tag{12.10}$$

where n is the number density of electrons. For the number density corresponding to a typical metal we find that ε_F is a few eV – about 10^{-18} J – which is much greater than kT for the temperatures at which metals exist. In *real* metals the interactions between each electron and the lattice ions modify the free electron density of states substantially, but it remains true that the Fermi energy is normally several eV.

At a finite temperature, the Fermi distribution takes the form shown in Fig. 12.1. We notice that there is now some occupation in particle states just above the Fermi energy, and some depletion of states just below it. But this disturbance is confined to a range of energies within about kT of the Fermi energy; at higher energies the states remain virtually empty, and at lower energies they remain completely full, as shown in Fig. 12.2. If $kT \ll \varepsilon_F$, as is always the case in metals, we may safely treat $g(\varepsilon)$ as a constant over the region where the occupation has changed, equal to its value at the Fermi level, g_F. In general, the chemical potential μ varies with temperature, as we shall see. But if $kT \ll \varepsilon_F$ it is good approximation to treat μ as constant, equal to the Fermi energy, and Fig. 12.1 is plotted on this assumption. We may see that the assumption is self-consistent in the following way. For the Fermi distribution it is easy to show that

μ is independent of T if $kT \ll \varepsilon_F$

$$1-f=1/(e^{-(\varepsilon-\mu)/kT}+1). \tag{12.11}$$

If we compare this with the expression for f, we see that $1-f$ measured at an energy $\varepsilon=\mu-\varepsilon'$ lying *below* the chemical potential is equal to f measured at $\varepsilon=\mu+\varepsilon'$ *above* the chemical potential. Thus, as is shown in Fig. 12.1, if we keep μ fixed and equal to ε_F, and treat the density of states as a constant in the region of interest, the mean number of electrons *lost* from the occupied states below the Fermi energy just balances the mean number *gained* by the empty states above the Fermi level, so we still have the correct number of electrons without any need to modify the value of μ.

We may use the same picture in calculating the *electronic heat capacity* C_{el}. In raising the temperature from zero to T, we may imagine that we shift a number of electrons equal to $f(\varepsilon')g_F\,\mathrm{d}\varepsilon'$ from a point at energy $\varepsilon_F-\varepsilon'$ to a point at energy $\varepsilon_F + \varepsilon'$ (Fig. 12.1), where $f(\varepsilon')$ is the *Fermi function*, $1/(e^{\varepsilon'/kT}+1)$. In this process the energy added for each electron is equal to $2\varepsilon'$, so the total heat added is given by

The electronic heat capacity of metals

$$E(T)-E(0)=\int_0^\infty 2\varepsilon' f(\varepsilon')g_F\,d\varepsilon'$$
$$=2g_F(kT)^2\int_0^\infty \frac{x\,dx}{e^x+1}$$
$$=\tfrac{1}{6}\pi^2 g_F(kT)^2, \qquad (12.12)$$

since the value of the integral is $\pi^2/12$. By differentiating with respect to T, we find that the electronic heat capacity is given by

▶ $$C_{el}=\tfrac{1}{3}\pi^2 g_F k^2 T. \qquad (12.13)$$

For the free electron model $g_F = 3N/2\varepsilon_F$, so $C_{el} = (\frac{1}{2}\pi^2 Nk)(kT/\varepsilon_F)$, which is *much smaller* than the heat capacity of the equivalent classical monatomic gas, $3Nk$. It is not hard to see why: in the classical gas *all* the particles gain energy as it warms up, but in the electron gas, as shown in Fig. 12.2, *only the electrons within about kT of the Fermi energy gain thermal energy*. The observed electronic heat capacity of metals is in good agreement with (12.13). It may be measured readily at temperatures of order 1 K, where the lattice heat capacity is small enough not to swamp it.

§5.2

Problem 12.4

The *entropy* of an assembly of independent fermions is easily written down if we remember that each fermion state j is a 'system' with two 'states', occupied and unoccupied, having Gibbs probabilities f_j and $1-f_j$. It therefore has a grand entropy $-k\sum_G p \ln p$ equal to $-k[f_j \ln f_j + (1-f_j)\ln(1-f_j)]$, and the entropy of the complete assembly is given by

The entropy of independent fermions

▶ $$S=-k\sum_j [f_j \ln f_j + (1-f_j)\ln(1-f_j)]. \qquad (12.14)$$

You may like to confirm that our expression for C_{el} (12.13) may be derived from this result. It is also worth noting that the entropy of a fermion state is maximised when $f_j = \frac{1}{2}$. It follows that we could in principle construct a *particle reservoir* for fermions by setting up a very large number of degenerate fermion states, all of energy ε, and mean occupation equal to $\frac{1}{2}$. In such a system all the fermions have the same energy ε, and the entropy does not vary with occupation, to first order.

A fermion particle reservoir: see §11.1

Finally, we recall that it is the gradient of μ which determines the direction of particle flow at constant temperature. For a charged particle such as an electron the state energies ε_j include a contribution $-e\phi$ from the electrostatic potential ϕ. It follows that μ includes the same contribution, and for this reason μ for an electron is often referred to as the *electrochemical potential*. Note, then, that the driving force for electrons is the gradient of electrochemical potential and *not* the gradient of electric potential; that is, *the electric field is not the driving force for electrons*. We shall follow up this idea in more detail later.

§11.2

§15.6

12.5 The variation of μ with T at fixed density

We frequently have to deal with Bose and Fermi systems in which the volume and the number of particles N are fixed. If the particles are weakly interacting we know that a Bose or Fermi distribution will be established inside the system, having some definite value of chemical potential μ, but we do not know what this value is. However, inspection of the Bose and Fermi distribution functions shows that the occupation must rise if μ increases. Thus increasing μ is equivalent to increasing N at a given temperature; clearly, the correct choice of μ is the one which gives the correct value of N when substituted into the formula

$$N=\sum_j f_j=\int_0^\infty f(\varepsilon)g(\varepsilon)\,\mathrm{d}\varepsilon$$

$$=\int_0^\infty \frac{g(\varepsilon)\,\mathrm{d}\varepsilon}{\mathrm{e}^{(\varepsilon-\mu)/kT}\pm 1}, \tag{12.15}$$

where we have taken the particle ground state as the origin of energy. What this means is shown in pictorial terms for fermions in Fig. 12.2. The quantity $f(\varepsilon)g(\varepsilon)$ is the *density of occupation* in energy. If this is plotted as a function of ε the area under the curve represents the total number of particles in the system. *We have to vary our choice of μ so that this area remains constant as the temperature is raised.*

We shall be content with a broad picture of how μ actually varies with T. The details depend on the density of particle states $g(\varepsilon)$ for the system concerned. As an illustration I show in Fig. 12.3 how μ varies with T for a gas of non-interacting particles, where $g(\varepsilon)\propto\varepsilon^{\frac{1}{2}}$. For fermions, μ is approximately constant if $kT\ll\varepsilon_F$, as we noted earlier. As T rises, however, μ starts to fall. The reason is easy to understand. If we had kept μ constant, it is clear that N would have increased with T, because the fact that g increases with energy would then require that the number of particles *gained* by empty states above the Fermi level would have been greater than the number *lost* by filled states below the Fermi level. Since N is in fact constant it is clear that μ must fall to compensate for this effect. The equivalent density of occupation curves for a range of temperatures are shown in Fig. 12.4(*a*). Notice how the chemical potential, marked by the arrows, has to move to the left to keep the area under each curve the same. When $kT>\varepsilon_F$ the chemical potential becomes negative and decreases rapidly with rising temperature.

§ 12.4

For bosons the situation is rather more complicated. The chemical potential cannot be positive for bosons, for otherwise the lowest-lying particle states would have a non-physical *negative* occupation. At absolute zero *all* the bosons are in the particle ground state, since this gives the assembly the smallest possible energy. In this limit $\mu=0$. As we raise the temperature we at first only remove a small proportion of the bosons from the particle ground state; this state is said to have *macroscopic occupation*: a substantial fraction of all the bosons remains in it. The chemical potential is now very slightly

The Bose condensation

negative, but the shift is not measurable, and in fact the separation between the chemical potential and the ground state energy is much less than the spacing between the ground state and the next particle state. For the states other than the ground state the occupation is described by the occupation density $f(\varepsilon)g(\varepsilon)$ as usual. The curves of fg at low temperatures for bosons are shown in

Fig. 12.3. Variation of μ with T for a gas of non-interacting particles at fixed density. T_c is the Bose–Einstein condensation temperature. The dots correspond to the curves of Fig. 12.4.

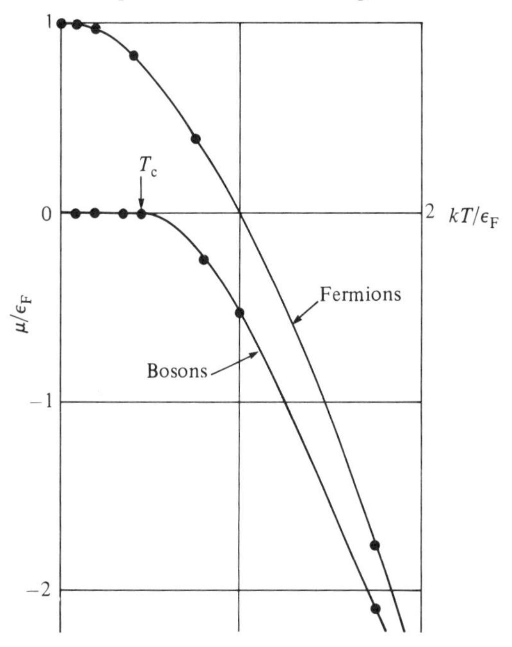

Fig. 12.4. The occupation density $f(\varepsilon)g(\varepsilon)$ at various temperatures for a gas of non-interacting particles at fixed density. (*a*) Fermions. (*b*) Bosons for $T \leqslant T_c$. (*c*) Bosons for $T \geqslant T_c$. The arrows show the position of the chemical potential, and the thin line represents the density of states $g(\varepsilon)$. The temperatures chosen are shown in Fig. 12.3. In (*a*) and (*c*) the areas under the curves are constant and represent the total number of particles. In (*b*) the areas represent the particles other than those condensed in the particle ground state at $\varepsilon = 0$.

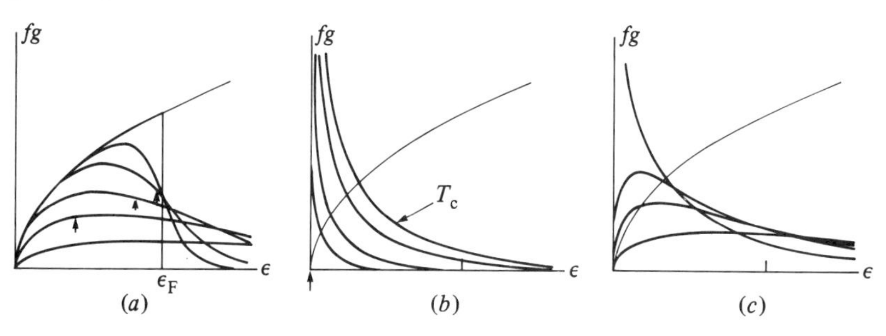

Fig. 12.4(*b*). Note that, although they diverge at $\varepsilon = 0$, the areas under the curves do not. At these low temperatures the area represents only a fraction of all the particles, the other bosons being in the particle ground state. Eventually we reach the *Bose–Einstein condensation temperature* T_c at which the fraction of particles in the ground state has fallen to zero. At this point the chemical potential starts to fall (and the system has an anomaly in its heat capacity). The occupation density $f(\varepsilon)g(\varepsilon)$ is shown for higher temperatures in Fig. 12.4(*c*). From now on the area under the curves is constant. At the highest temperatures μ becomes large and negative and the occupation densities for the Fermi and Bose cases become almost identical.

I must make clear that the phenomenon of Bose–Einstein condensation just described, in which a finite fraction of particles occupy the particle ground state, has not been observed in practice, because no examples of weakly interacting bosons at the required density are known. It is sometimes suggested that the Bose–Einstein condensation is a model for the appearance of *superfluidity* in liquid ^{4}He, and that the superfluid particles may be identified with the atoms in the particle ground state; and, indeed, if liquid helium is treated as a weakly interacting gas, the calculated value of T_c is close to the observed critical temperature for superfluidity, which is 2.17 K. The link is, however, a tenuous one. The theory of superfluidity does have something in common with the concept of macroscopic occupation of a single particle state, but it is *not* a theory of weakly interacting particles, and liquid ^{4}He is *not* a gas of weakly interacting bosons.

Problem 12.5

This is a convenient point to observe that it is easy to construct in principle a particle reservoir for bosons. A system consisting of a *single* boson state with macroscopic occupation has the required properties: its entropy is zero, and all the particles in it have the same energy.

A boson particle reservoir: see § 11.1

12.6 Non-condensed gases

In Chapter 5 we examined an elementary treatment of ordinary monatomic gases which ignored altogether the fact that the atoms of the gas are indistinguishable particles. In this section we shall reexamine ordinary gases, taking the indistinguishability into account more clearly. A gas of weakly interacting particles in which the mean state occupations f_j are substantial ($f \sim 1$) is said to be *condensed*. In a condensed gas the distinction between fermions and bosons is important. However, we have just seen that at high temperatures, for both fermions and bosons, μ becomes large and negative, and the two distributions become more alike. It is not hard to see why this happens. The two distributions both have an energy width of order kT. If $kT \gg \varepsilon_F$ the width of the distribution is much greater than it was at low temperatures and it follows that the occupation numbers for the individual states must be very small, for otherwise we should have far too many particles. Thus $\exp(-\mu/kT)$ must be large, and μ must be negative and much greater than ε_F in magnitude. It follows that both Bose and Fermi distributions may be approximated as

$$f = \frac{1}{e^{(\varepsilon-\mu)/kT} \mp 1} \simeq e^{-\varepsilon/kT}\, e^{\mu/kT}. \tag{12.16}$$

This result for *uncondensed gases* is very important. It shows that for them the occupation is proportional to the Boltzmann factor for the state; in other words, it shows that the *occupation numbers are the same for uncondensed gases as they would have been if the particles had been identifiable systems.* This is a very helpful result. It can be used, for instance, to justify the treatment of the ideal monatomic gas which I gave in Chapter 5. In that chapter we first considered a *single* atom in a box as a system obeying the Boltzmann distribution: that was a correct argument. We then said that if we add further atoms they exert their pressure on the walls independently of each other. That is also correct provided one atom does not affect the occupation of states by another. Our present result shows that this assumption is valid *so long as the gas is uncondensed.* (Note carefully that *particles* obey the Boltzmann distribution *as an approximation* when uncondensed; but *identifiable systems* obey the Boltzmann distribution *exactly.*)

The condition for non-condensation is that $kT \gg \varepsilon_F$, or, using the expression for ε_F,

§ 12.4

▶ $$kT \gg (\hbar^2/2m)(3\pi^2 n)^{\frac{2}{3}}. \tag{12.17}$$

The condition for non-condensation

This expression holds for bosons as well as fermions. Note that the factors making for non-condensation are *high* temperature, *low* density and *high* mass. You may like to confirm that air at room temperature and pressure satisfies this condition easily, but that electrons in a metal do not. The condition may be expressed in a different way by writing kT as a kinetic energy $h^2/(2m\lambda_T^2)$, where λ_T is the *de Broglie wavelength for a particle of energy kT.* Using λ_T we may rewrite the condition as

Problem 12.5

$$\lambda_T \ll s, \tag{12.18}$$

where $s = n^{-\frac{1}{3}}$ is the typical spacing between the particles. When this condition holds we may describe the individual particles by wave packets which do not often overlap. This confirms the general picture: *the gas is uncondensed when the occupation is low for all states and two particles very rarely try to occupy the same state.*

It will be apparent by now that the treatment of the monatomic gas given in Chapter 5 was really a treatment of an uncondensed gas. Let us now calculate the thermodynamic parameters of such a gas in a more formal way, using the ideas set up in the last two chapters. Let us first find the grand partition function for state j, which may be written for both fermions and bosons as

$$\Xi_j = \sum_{n_j} e^{-n_j(\varepsilon_j-\mu)/kT}. \tag{12.19}$$

If the gas is uncondensed μ will be large and negative, so we ignore the small boson terms with $n_j > 1$ and write for both bosons and fermions

$$\Xi_j \simeq 1 + e^{-(\varepsilon_j - \mu)/kT}. \tag{12.20}$$

The corresponding grand potential Φ_j is given by

$$\begin{aligned}\Phi_j &= -kT \ln \Xi_j \\ &\simeq -kT \ln(1 + e^{-(\varepsilon_j - \mu)/kT}) \\ &\simeq -kT e^{-(\varepsilon_j - \mu)/kT},\end{aligned} \tag{12.21}$$

using the fact that the exponential term is always small in an uncondensed gas. Summing the contributions of all particle states (since they behave like independent systems) we have for the whole uncondensed gas

$$\Phi \simeq -kT \sum_j e^{-(\varepsilon_j - \mu)/kT}. \tag{12.22}$$

§§ 5.1, 10.1

As we noted in Chapter 5, the quantum states of the gas atoms must fit into a rectangular container as the modes of the photons did in Chapter 10, and thus the density of particle states in k space is equal to V/π^3, in the first octant of k space, as shown in Fig. 10.1, and the number of states with k in the range dk will be equal to $Vk^2\,dk/2\pi^2$. Remembering that the state energy ε_k is equal to $\hbar^2k^2/2m$ we may therefore write Φ as

$$\begin{aligned}\Phi &\simeq -k_B T e^{\mu/k_B T} \int_0^\infty (Vk^2\,dk/2\pi^2)\, e^{-\hbar^2k^2/2mk_B T} \\ &= -(k_B T/2\pi^2) e^{\mu/k_B T} V(2mk_B T/\hbar^2)^{\frac{3}{2}} \int_0^\infty x^2 e^{-x^2}\,dx \\ &= -k_B T e^{\mu/k_B T} V(mk_B T/2\pi\hbar^2)^{\frac{3}{2}},\end{aligned} \tag{12.23}$$

since the value of the integral is $\sqrt{\pi}/4$. (We temporarily write Boltzmann's constant as k_B, to distinguish it from the magnitude of the wave vector.) By the usual methods we deduce that

§ 11.4

$$N = -(\partial\Phi/\partial\mu)_{V,T} = V e^{\mu/kT}(mkT/2\pi\hbar^2)^{\frac{3}{2}},$$

$$P = -(\partial\Phi/\partial V)_{\mu,T} = kT e^{\mu/kT}(mkT/2\pi\hbar^2)^{\frac{3}{2}} = NkT/V,$$

and

$$S = -(\partial\Phi/\partial T)_{\mu,V} = Vk\, e^{\mu/kT}(mkT/2\pi\hbar^2)^{\frac{3}{2}}(\tfrac{5}{2} - \mu/kT). \tag{12.24}$$

From the second result we recover the usual *equation of state*. From the first we find the *expression for the chemical potential* of an atom in an uncondensed monatomic gas,

Microscopic expressions for μ and S in an uncondensed monatomic gas

▶ $$\mu = -kT \ln\left[(mkT/2\pi\hbar^2)^{\frac{3}{2}} V/N\right]. \tag{12.25}$$

From the third result combined with the expressions for N and μ we obtain the *Sackur–Tetrode expression* for the entropy of an uncondensed monatomic gas,

▶ $$S = Nk \ln\left[(mkT/2\pi\hbar^2)^{\frac{3}{2}} e^{\frac{5}{2}} V/N\right]. \tag{12.26}$$

§ 5.6

We thus at last discover the value of the constant in (5.14).

12.7 The law of mass action

Let us generalise the expression for μ just obtained. For any system of uncondensed bosons or fermions we may write, using (12.22),

$$\Phi = -kT e^{\mu/kT} \sum_j e^{-\varepsilon_j/kT}$$

$$= -kT e^{\mu/kT} \zeta(T, V), \tag{12.27}$$

where ζ is the *one-particle partition function*, the partition function for a single particle alone in the system. We deduce that

$$N = -(\partial\Phi/\partial\mu)_{T,V} = e^{\mu/kT}\zeta,$$

so

$$\blacktriangleright \quad \mu = kT \ln (N/\zeta). \tag{12.28}$$

The expression for μ obtained in the previous section is a special example of this result for the case of a free atom.

We know that the condition for chemical equilibrium may be written as $\sum R_r\mu_r = \sum P_p\mu_p$. If the chemical reaction takes place between uncondensed bosons or fermions we may rewrite this condition using (12.28) as

§ 11.2

$$\sum R_r \ln (N_r/\zeta_r) = \sum P_p \ln (N_p/\zeta_p),$$

or

$$\blacktriangleright \quad \frac{\prod N_r^{R_r}}{\prod N_p^{P_p}} = K_N(T, V), \quad \text{where } K_N = \frac{\prod \zeta_r^{R_r}}{\prod \zeta_p^{P_p}}. \tag{12.29}$$

This is one of several ways of expressing the *law of mass action*. To explore its meaning, consider the reaction $2H_2 + O_2 \rightarrow 2H_2O$. When this reaction is in chemical equilibrium we have

$$\frac{N_{H_2}^2 N_{O_2}}{N_{H_2O}^2} = K_N(T, V).$$

This result should apply to water vapour at low density. The constant K_N, known as the *equilibrium constant*, though small is not zero, and thus if we have a large number of water molecules present, the equation shows that there must be small numbers of hydrogen and oxygen molecules present as well. If we remove this hydrogen and oxygen, water molecules will dissociate until the equation is again satisfied. If we add hydrogen, hydrogen will combine with oxygen until the equation is again satisfied. The kinetic basis of the result is easy to understand. For the reaction to proceed *forwards*, *two* hydrogen molecules and *one* oxygen molecule must meet in a small region. The probability that this will occur in a given time is proportional to $N_{H_2}^2 N_{O_2}$, and the rate at which the reaction proceeds forwards must be proportional to this factor. In the same way, for the reaction to proceed *backwards two* water molecules must meet, and the reverse rate must be proportional to $N_{H_2O}^2$. In chemical equilibrium the forward and reverse rates must be equal (an example of detailed balance) and so the ratio $N_{H_2}^2 N_{O_2}/N_{H_2O}^2$ must be constant.

The kinetic model of mass action

The calculation given above confirms this kinetic idea and also predicts the value of the equilibrium constant as a function of T and V. To get some idea of what this means in practice, notice that in a gas we may write ζ as

The value of the equilibrium constant

$$\begin{aligned}\zeta &= \sum_j e^{-\varepsilon_j/k_B T} \\ &= \sum_{k,i} e^{-(\hbar^2 k^2/2m+\varepsilon_i)/k_B T} \\ &= \sum_k e^{-\hbar^2 k^2/2mk_B T} \sum_i e^{-\varepsilon_i/k_B T} \\ &= \zeta_t \zeta_i, \end{aligned} \tag{12.30}$$

where $\mathbf{k}$ is the wave vector for the translational motion of the molecule having kinetic energy $\hbar^2 k^2/2m$, i is a general label for the various internal molecular states of vibration, rotation, etc., and we have again temporarily written Boltzmann's constant as k_B. Here, ζ_t is the *translational partition function*, which, as we saw in § 12.6, is equal to $(mkT/2\pi\hbar^2)^{\frac{3}{2}}V$, and ζ_i is the *internal molecular partition function*, which I shall write as $\exp(-F_i/kT)$, where F_i is the contribution of the internal modes of the molecule to its free energy. Using these quantities we may write the equilibrium constant as

▶ $$K_N = \left(\prod \zeta_{tr}^{R_r} / \prod \zeta_{tp}^{P_p}\right) e^{\Delta F_i kT}, \tag{12.31}$$

where ΔF_i is the total change in internal free energy which occurs in the reaction. The final factor shows that the reaction tends to proceed in the direction of smaller F_i, as one would expect: in the reaction $2H_2 + O_2 \rightarrow 2H_2O$, for instance, there is a substantial fall in internal free energy, and this is why under normal conditions there are very few hydrogen and oxygen molecules present in equilibrium. But we must note that if, when the reaction occurs once, the total number of molecules increases by M, the *first* term will contain a factor V^{-M}. *Thus an increase of volume will drive the reaction in the direction of increasing the number of particles present.* If we increase the volume enough, almost all the water molecules will eventually dissociate. This principle has important effects in astronomical chemistry. The gas clouds which are sometimes found in galaxies have a much higher proportion of small radicals than is usual in the laboratory, and therefore exhibit a number of unfamiliar line spectra.

Low density favours smaller molecules

We must remember that the law of mass action applies only to reactions between weakly interacting and uncondensed particles. But it is not limited to *chemical* reactions: it may be applied also, for instance, to fundamental particle reactions. It may be applied to cases in which particles are created or destroyed. For instance, it is shown in texts of solid state physics that, in a semi-conductor,

$$N_e N_h = \left[\left(\frac{m_e kT}{2\pi\hbar^2}\right)^{\frac{3}{2}} V\right]\left[\left(\frac{m_h kT}{2\pi\hbar^2}\right)^{\frac{3}{2}} V\right] e^{-(E_g - kT\ln 4)/kT}, \tag{12.32}$$

where N_e and N_h are the numbers of electrons and holes present, m_e and m_h are their masses, and E_g is the energy gap of the semi-conductor. This is an example of (12.31) applied to the reaction of *pair annihilation* $e+h\rightarrow 0$. (An electron–hole pair has internal entropy $k \ln 4$ because each particle has spin $\frac{1}{2}$.)

12.8 Entropy of mixing

When two different substances are allowed to mix an irreversible process occurs and there is a rise in the entropy. We may find the magnitude of this rise for non-degenerate gases by using the Sackur–Tetrode expression (12.26). Suppose we have N_A molecules of gas A and N_B molecules of gas B held separately at the same temperature and pressure (so that $V_A/V_B = N_A/N_B$). If they are allowed to mix by opening a connecting valve they both eventually occupy the same volume $V = V_A + V_B$. In spite of this fact, they remain distinct systems, each having access to its states as though the other were absent. We may thus apply the Sackur–Tetrode expression to each gas separately, and we find that the change in entropy on mixing is equal to

§ 12.6

$$\Delta S = k[N_A \ln (V/V_A) + N_B \ln (V/V_B)],$$

as though each gas had undergone a Joule expansion into an empty container independently of the other. Writing N for $N_A + N_B$ we see that the *entropy of mixing* is given by

$$\blacktriangleright \quad \Delta S = k[N_A \ln (N/N_A) + N_B \ln (N/N_B)]. \tag{12.33}$$

Note that if the molecules of A and B had been *identical*, the situation would have been different, because, after mixing they would constitute a *single* system of N molecules and not two distinct systems. The entropy change would thus have been given by

$$\Delta S = Nk \ln (V/N) - N_A k \ln (V_A/N_A) - N_B k \ln (V_B/N_B) = 0,$$

corresponding to the fact that mixing *identical* gases at the same temperature and pressure is a reversible process.

It is instructive to deduce the entropy of mixing by a purely classical argument. As shown in Fig. 12.5, a mixture of N_A molecules of gas A and N_B molecules of gas B occupying volume V may be separated reversibly into separate samples of A and B each of the same volume by forcing it slowly through a pair of semi-permeable membranes. If the pistons in cylinders (a) and (b) are withdrawn at the same rate as the piston is pushed into (c) the net work done will be zero, for the partial pressure of each gas is equalised on the two sides of the membrane through which it passes. If no heat enters, we see that the internal energy and hence the temperature is unchanged in this process. The entropy is also unchanged, since the process is reversible. We may now return to our starting point by first compressing each gas isothermally and independently to the original total pressure and then allowing them to mix. In the isothermal compressions the entropy of the two gases falls, ΔS being equal to $N_A k \ln (V_A/V) + N_B k \ln (V_B/V)$. On mixing, the entropy must return to its original value: it must rise on mixing by an amount equal to the

Problem 11.1

§ 5.6

Fig. 12.5. A process for separating a mixture of gases isothermally and reversibly. The three cylinders have the same volume V. The membrane α allows gas A to pass, but blocks gas B. Membrane β passes B but blocks A.

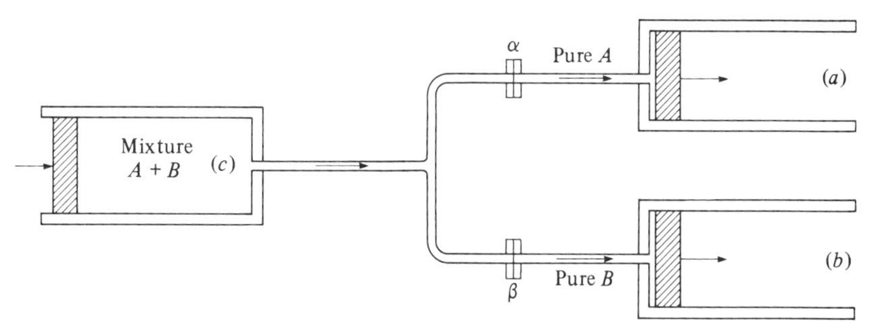

fall during the compressions. Inspection shows that this argument confirms the result (12.33) quoted above. (The argument of course fails if A and B are identical, because the gases cannot be separated in the same way using semi-permeable membranes.)

Problems

12.1 A system containing N identical fermions has $2N$ distinct particle states, $2N/3$ with energy zero, $2N/3$ with energy ε and $2N/3$ with energy 2ε. Show that μ is independent of T. Find and sketch an expression for the heat capacity as a function of T.

12.2 Two spinless identical non-interacting bosons oscillate at frequency ν in the same one-dimensional harmonic potential. Taking the ground state as the zero of energy, find Z for the system. Show that at high temperatures $Z \simeq \frac{1}{2}\sum_{n=0}^{\infty}\sum_{m=0}^{\infty}\exp[-(n+m)h\nu/kT]$, and that at low temperatures $Z \simeq 1 + \exp(-h\nu/kT)$. Deduce that at high T the heat capacity is the same as that for *two* separate oscillators, whereas at low T it is the same as that of a *single* oscillator, and explain how this difference arises.

12.3 When a magnetic field $\mathscr{B}_E$ is applied to a free-electron metal, the electron energies are shifted by $\mp\mu_B\mathscr{B}_E$, depending on whether the spins are up or down, where $\mu_B\mathscr{B}_E \ll \varepsilon_F$. Show that $\int_{-\infty}^{\infty}[f(\varepsilon-\varepsilon')-f(\varepsilon)]\,d\varepsilon=\varepsilon'$, and use this result to demonstrate that the corresponding *Pauli susceptibility* is approximately independent of T, and equal to $3kT/\varepsilon_F$ times the Curie susceptibility for similar spins arranged at the same density on a crystal lattice.

Pauli paramagnetism

§9.2.1

12.4 Remembering that an electron of wave vector $\mathbf{k}$ has energy $\hbar^2k^2/2m_e$ and two possible spin states, use (10.4) to prove (12.9) and (12.10). Predict ε_F and g_F for 1 mole of potassium, which has $A_r = 39$ and density 860 kg m^{-3}. Below 4 K a plot of C_m/T against T^2, where C_m is the molar heat capacity of potassium, is a straight line with intercept 2.1×10^{-3} J K^{-2} mol^{-1} at $T^2=0$, and slope 2.6×10^{-3} J K^{-4} mol^{-1}. Deduce experimental values for g_F and θ_D for potassium. Why does the experimental value of g_F not agree perfectly with the theoretical value?

12.5 Estimate the thermal de Broglie wavelength λ_T for N_2 at room temperature and for ^{4}He at 4 K. Use Fig. 12.3 to find also the room temperature value of the chemical potential for electrons in heavily doped n-type germanium, and the critical temperature T_c for Bose condensation of liquid ^{4}He, treated (incorrectly) as a gas of non-interacting bosons. (The conduction electrons in the Ge have energy $\hbar^2k^2/2m^*$, where $m^* = 0.2m_e$, and number density $9 \times 10^{23}\ \mathrm{m}^{-3}$. Density of liquid $^4\mathrm{He} = 125\ \mathrm{kg\ m}^{-3}$.)

12.6 Find an expression for μ analogous to (12.25) for a two-dimensional gas occupying an area A at potential energy $-\phi$. If helium atoms may be adsorbed from a gas at pressure P onto a solid surface where they move freely and without interaction, deduce that the surface density of adsorbed atoms is equal to $(P/kT)(2\pi\hbar^2/mkT)^{\frac{1}{2}} \exp(\phi/kT)$ where m is the mass of a helium atom and $-\phi$ its potential energy when adsorbed. When would you expect this formula to be valid?

12.7 If air is regarded as 1 volume of O_2 mixed with 4 volumes of N_2, find the minimum work required to extract 1 mole of pure O_2 from air at the same temperature and pressure. Where does the work go? Find a way of recovering it, using a semi-permeable membrane.

12.8 Show that K_N for the ionisation reaction $\mathrm{He} \rightarrow \mathrm{He}^+ + \mathrm{e}$ is to a good approximation $\exp(\mathrm{e}\phi/kT)(2\pi\hbar^2/m_e kT)^{\frac{3}{2}}/V$, where ϕ is the ionisation potential of He, equal to 24.6 V. Find the proportion of He which is ionised at 10^4 K (i) at atmospheric pressure, and (ii) at $10^{-2}\ \mathrm{N\ m}^{-2}$.

12.9 The two protons in a H_2 molecule are described by a product wave function $\psi(1,2) = \phi_K(\mathbf{r}_1, \mathbf{r}_2)\chi(1,2)$, where ϕ_K describes the rotation, with rotational quantum number K as in §9.1.2, and χ describes the spin state. ϕ_K is symmetric under particle exchange for even K and anti-symmetric otherwise. In o-H_2 the proton spins are parallel, and there are three possible symmetric spin states χ. In p-H_2 the spins are anti-parallel, and there is one anti-symmetric spin state. The energy is $K(K+1)\hbar^2/2I$, independent of spin state, where $I = 5.4 \times 10^{-48}\ \mathrm{kg\ m}^2$. The proton spin state is not easily changed, and it may take some hours for o-H_2 and p-H_2 to come into mutual equilibrium. Show that in equilibrium at room temperature there should be three times as much o-H_2 as p-H_2. Estimate the molar heat capacity of hydrogen (i) after being kept at 30 K for some days, (ii) after fast cooling to 30 K, and (iii) if warmed extremely slowly from 30 K.

Ortho- and para-hydrogen: see problem 9.3

13

Classical statistical mechanics

It is not always necessary to use quantum mechanics in describing the behaviour of a thermodynamic system. For many systems ordinary Newtonian mechanics is an excellent approximation, and quantum statistics would involve unnecessary complications. In such cases it is better to use the classical form of statistical thermodynamics, developed by Boltzmann and Gibbs before the advent of quantum theory. In this chapter I shall present an outline development of this classical approach, parallel to the quantum development of Chapters 2, 3 and 4, but less detailed. Some of the ideas mentioned here are taken further in the final chapter.

13.1 The classical limit of quantum theory

So far, we have described the thermodynamic system in terms of its quantum states. Our line of approach has been that, if we know the quantum state of the system at time zero, we may use Schrödinger's time-dependent equation to determine the state at later times. Knowing this, we may find how the thermodynamic state, represented by the probabilities $\tilde{p}_i$ of the approximate energy states, changes with time. In classical mechanics, on the other hand, we specify the mechanical state of the system at time zero by quoting the *positions and momenta of the particles* inside the system, and we must use Newton's laws of motion to find how the mechanical state changes with time. The classical analogue of $\tilde{p}_i$ is the probability of finding the system in a given mechanical state, and we have to discover how this probability changes with time. This question will occupy us in the following three sections. Here we must first discuss the question: when is the classical limit valid?

The classical description

To approach this problem it is helpful to have a simple illustration in mind. Consider, for instance, the case of a particle moving in one dimension in the triangular potential shown in Fig. 13.1. This potential corresponds to a restoring force of fixed magnitude which changes sign at $x=0$, so that the particle is always driven back to the origin. In quantum theory the states of definite energy, or *stationary states*, are the standing wave solutions of the Schrödinger time-independent equation. Finding the stationary states for this triangular potential is a well-known problem in quantum theory. The solutions are known as Airy integrals and one of them is shown in Fig. 13.1(*c*). They correspond to a ladder of bound state energy levels, which get more closely spaced as the energy rises. In classical mechanics, on the other hand, the particle can have *any* energy ε, and we think of it as bouncing backwards and forwards between its *turning points*, the points where the potential energy $\phi(x)$ is equal to ε. Its mechanical state at any instant may be given by quoting the position co-ordinate x and the corresponding momentum co-ordinate p_x. We may also alternatively describe this mechanical state by giving the position of a *representative point* having co-ordinates x and p_x in an imaginary plane, which we refer to as the *phase space* for the particle. The phase space for this particular system is shown in Fig. 13.2. As time proceeds the representative point moves along a definite *trajectory* in phase space, at a definite rate. In fact, at any given point in phase space we may obtain $\mathrm{d}x/\mathrm{d}t$ from p_x, and obtain $\mathrm{d}p_x/\mathrm{d}t$ from x, by finding the force acting at x and using Newton's second law. In our simple example the trajectories are closed, and you may like to check, by solving the equation of motion, that they consist in this case of two pieces of parabola, back-to-back, as shown in the diagram. The trajectories further from the origin correspond to greater energies. In classical mechanics any energy is possible.

The phase space description of the mechanical state

To discuss the validity of this classical description we must compare it with the corresponding quantum description. However, the quantum description closest to the classical one is not the stationary state of definite energy, but a state known as a *wave packet*, shown in Fig. 13.1(*b*). A wave packet is a localised group of waves which moves to and fro like the particle. It is not a stationary state, and its energy is slightly blurred, or uncertain: it may in fact be constructed by superimposing with suitable amplitudes and phases contributions from several stationary states having slightly different energies. The wave packet is spread out over a finite distance or position uncertainty Δx, and its momentum in the x direction is also uncertain by an amount Δp_x. According to Heisenberg's *uncertainty principle*, these quantities obey the relation

Wave packets and the uncertainty principle

$$\Delta p_x \, \Delta x \geqslant h/4\pi, \tag{13.1}$$

Fig. 13.1. Motion of a particle in one dimension in potential $\phi(x)$, to illustrate the classical limit. (*a*) Energies as a function of position, and classical motion. (*b*) An equivalent wave packet. (*c*) The equivalent stationary state. The corresponding phase space is shown in Fig. 13.2.

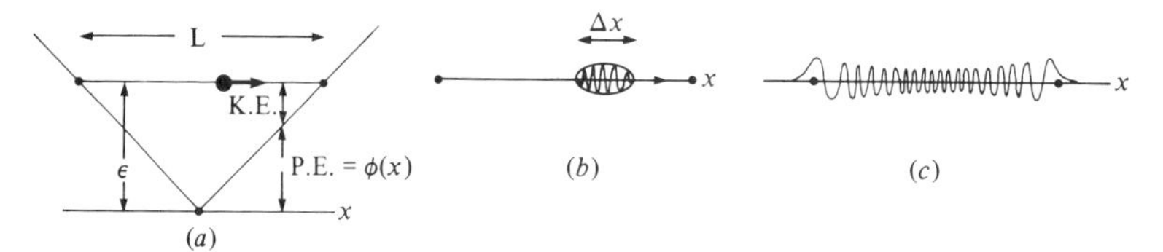

where h is Planck's constant. In terms of phase space we have to say that, for a wave packet, the representative point is *blurred*: the point is evidently smeared out over a phase space area of order h or more.

With the nature of the wave packet in mind we may understand the conditions which must be satisfied if the classical description is to be valid. We require first that *the wave packet should move in the same way as the classical particle*, and secondly that *the blurring of the wave packet or representative point should be on a scale small compared with the scales important in the problem.*

The conditions for the classical limit

One can show that the packet will move in the classical manner provided that Δp_x is small compared with p_x itself (which is equivalent to saying that the packet must be large enough to contain many de Broglie wavelengths), and provided that Δx is small compared with the distance over which p_x changes appreciably, which in this problem we may take as the distance L between the turning points. It is evident that, since the uncertainty

The condition for classical motion

Fig. 13.2. Phase space trajectories for the one-dimensional motion shown in Fig. 13.1. Representative points initially in the box at A continue to occupy constant area at later times as they move, but the area is drawn out gradually into a long filament. After many circuits the points originally in A are spread uniformly along the strip between the two trajectories. If the two trajectories shown correspond to neighbouring quantum states of definite energy, the area between them is equal to h.

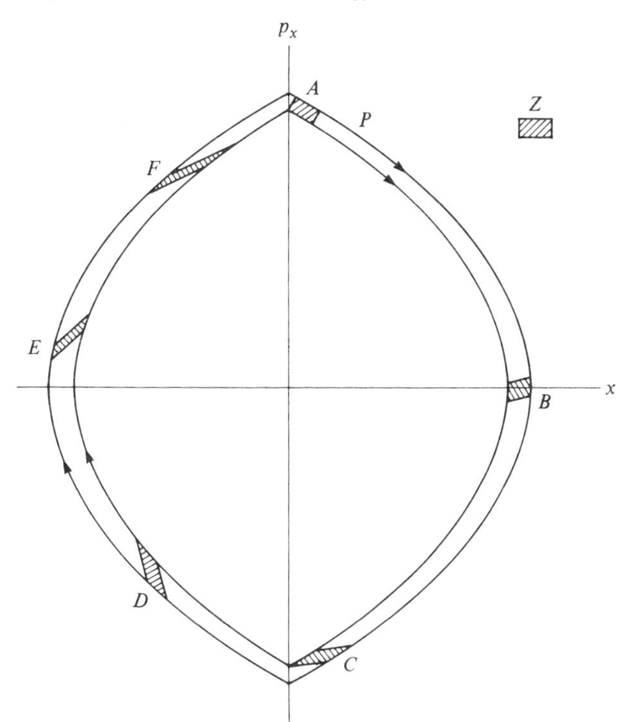

principle requires that $\Delta p_x \, \Delta x \gg h/4\pi$, we can only construct a wave packet which satisfies these conditions if

$$|p_x| L \gg h/4\pi. \tag{13.2}$$

If this condition is satisfied we may assert that a suitably constructed wave packet will move in the classical manner.

We also need to be able to ignore the blurring of the representative point in phase space if the classical description is to be valid. We shall see later that the representative point obeys a Boltzmann distribution: the probability of finding it at a particular point in phase space is proportional to $\exp(-E/kT)$, where E is the corresponding energy. This probability will change appreciably if E changes by an amount of order kT. Clearly, we may only use the classical description if the energy blurring ΔE associated with the finite size of the representative point is much smaller than this. Now we shall also see later that in classical statistics the total energy of a single particle is itself of order kT. Thus we may express our requirement that $\Delta E \ll kT$ by specifying that $\Delta E/E$ should be small. If we meet the requirements for classical motion discussed above ($\Delta p_x \ll p_x$ and $\Delta x \ll L$) we observe that $\Delta E/E$ will automatically be small, for if $\Delta p_x \ll p_x$ the fractional uncertainty in the kinetic energy will be small, and if $\Delta x \ll L$ the fractional uncertainty in the potential energy will also be small. Thus the condition for classical motion of the wave packet *also* ensures that we may safely apply the classical Boltzmann distribution to the wave packet. It is convenient to write this condition in terms of the *thermal de Broglie wavelength* λ_T, the de Broglie wavelength for a particle of kinetic energy kT, given by the relation $kT = h^2/2m\lambda_T^2$. Since $|p_x| \approx h/\lambda_T$ we may rewrite the joint condition for classical motion and for application of classical statistics in terms of λ_T as

§ 13.4

The condition for application of the classical distribution

▶ $$\lambda_T \ll 4\pi L. \tag{13.3}$$

If the system contains a number of identical particles another condition has to be satisfied: we have to be sure that the wave packets for different particles do not often overlap, for otherwise, as we have seen, the quantum effects of the exchange principle would become important. In other words, the system must be *uncondensed*. We derived the condition for this earlier. The condition was that

The condition for non-condensation

§ 12.6

▶ $$\lambda_T \ll s, \tag{13.4}$$

where s is the interparticle spacing.

When these two conditions are satisfied, classical statistics gives a good description of the system. Using the expression for λ_T we note that the conditions which favour the classical limit are *high particle mass*, *high temperature*, and *large interparticle spacing*. For a particle of relative molecular mass 30 at room temperature λ_T is about 3×10^{-11} m. Thus ordinary gases at room temperature are clearly in the classical limit so far as translation is concerned (but the classical limit does not necessarily apply to

Examples of classical systems

their internal motions). In ordinary liquids, too, where the interparticle distance is of order 3×10^{-10} m or more the classical limit is an acceptable approximation (but this is *not* true of liquid hydrogen or helium at temperatures of a few K). The classical limit also applies in most areas of plasma physics, and in discussing fluctuations in macroscopic quantities, such as thermal voltage noise in electrical circuits. We shall use it in a number of applications later in this book.

Problem 12.5

Finally, let us reverse our approach and ask how the quantum states of definite energy should be regarded from the classical point of view. We could imagine that we take a blurred representative point in phase space, corresponding to a wave packet, and draw it out parallel to the trajectory. Since it is getting longer, the uncertainty principle allows it also to get narrower, and in the limit we have a trajectory which has become completely sharp (corresponding to a definite energy), but for which *the position of the representative point on the trajectory has become completely uncertain.* This is the classical picture of a quantum state of definite energy. In real space the wave packet has spread out until it has joined up on its own tail (Fig. 13.1). When it does so, it must join up with the correct phase θ, and this is the reason for the quantisation of energy. In fact, the phase change in the wave function when we pass once around the trajectory must be a multiple of 2π. In the classical limit, where the potential varies little in one de Broglie wavelength, we may identify $\partial\theta/\partial x$ with $p_x/\hbar$, where p_x is the classical momentum. Thus

The classical picture of quantum states of definite energy

$$\oint d\theta = \oint (\partial\theta/\partial x)\, dx = \oint p_x\, dx/\hbar = 2\pi n,$$

or

$$\oint p_x\, dx = nh. \tag{13.5}$$

This is the *Bohr–Sommerfeld quantum condition*, and it applies wherever the particles are in the classical limit. (Strictly speaking, in our example the particle is non-classical near its turning points, where $|p_x| \to 0$. It can be shown that when this is taken into account the above rule takes the very similar form $\oint p_x\, dx = (n + \frac{1}{4})h$.) Now the integral on the left of this expression is just the area of the closed trajectory in phase space. Thus successive allowed trajectories are separated by strips of area h, and *in the classical limit the area of phase space per quantum state of definite energy is equal to* h. We shall later make use of this fact in relating classical and quantum statistics. We also note that in the classical limit, since $L \gg \lambda_T$, n will be large, and *the energy spacing* δE *between successive states will be small compared with the typical particle energy* kT. This is always true in the classical limit. Note, however, that we cannot use it as a *criterion* for the classical limit: in *large* systems, as we have seen, δE is always extremely small, but there are many large systems which are not in the classical limit. The proper criteria for the classical limit are equations (13.3) and (13.4).

§ 2.3

13.2 The phase space picture of the approach to equilibrium

When dealing with quantum statistics we wished to know how the probability $\tilde{p}_i$ of finding the system in quantum state i changed with time. In discussing the classical picture of the approach to equilibrium we are involved with the classical analogue of $\tilde{p}_i$ which, as we shall see, is the *probability density in phase space*, $\tilde{\rho}$. For a one-dimensional system like the one whose trajectories are illustrated in Fig. 13.2, $\tilde{\rho}$ means simply the probability per unit area of finding the representative point in a given neighbourhood. We mean here the instantaneous probability at a fixed time. As we noted earlier, we may think of such probabilities as belonging to a large statistical population or *ensemble* of possible experimental situations, corresponding to our limited knowledge of the microscopic state of the system. We may imagine a corresponding large number of representative points, one for each state in the ensemble, each moving along its own trajectory. From this point of view $\tilde{\rho}$ is the normalised density of ensemble representative points per unit area of phase space.

The probability density $\tilde{\rho}$

p. 141

§2.1

As defined so far $\tilde{\rho}$ has dimensions of reciprocal phase space area, while, of course, $\tilde{p}_i$ was dimensionless. These are, indeed, the usual conventions. In this book, however, I propose to make $\tilde{\rho}$ more closely analogous to $\tilde{p}_i$, by noting, as we saw in the previous section, that the area of phase space per quantum state is equal to h. If I then choose to measure phase space area in units of h, the probability per unit area $\tilde{\rho}$ becomes exactly analogous to $\tilde{p}_i$, and this is a convenience in comparing classical and quantum calculations. I therefore choose to write the *dimensionless area* dΩ corresponding to a small rectangle in the x–p_x plane as

Dimensionless units of phase space

$$d\Omega = dx\, dp_x/h. \tag{13.6}$$

I then take $\tilde{\rho}$ to be the classical probability per unit dimensionless area in phase space. This quantity is itself dimensionless, and corresponds as closely as possible to $\tilde{p}_i$.

As time goes on, in general, $\tilde{\rho}$ at a given point in phase space will change as the representative points move along their trajectories. In this motion, some neighbourhoods are *mutually accessible* and others are not. For instance, in Fig. 13.2, the shaded area Z is not accessible from A, because no trajectory reaches Z from A. The shaded areas A and B, on the other hand, are mutually accessible in the sense that there are trajectories which pass from A to B and from B to A. In this particular illustration it is obvious that all neighbourhoods at the same energy are mutually accessible.

Accessibility

Suppose we start at time zero knowing that the system is definitely in neighbourhood A, and represent the corresponding thermal state by some ensemble of representative points, all inside the shaded area A. As time proceeds, each point moves along its own trajectory. Although the points move in a broadly similar way, the speed at which they move depends slightly on exactly where they are within the shaded area, and as a consequence the occupied area is gradually distorted as it moves. At position F, after almost a

The approach to equilibrium

complete cycle, the occupied area has been drawn out into a narrow strip. After several thousand cycles, it will be drawn out into a very long filament, which will be wound many times around the area between the two marked trajectories. Initially, $\tilde{\rho}$ measured at a fixed position such as P will be changing with time, and will show a strong pulse once per cycle as the shaded area passes P. But after many cycles the stretching of the strip means that these pulses become blurred, until eventually $\tilde{\rho}$ will become uniform and independent of time in all of the mutually accessible neighbourhoods. The system has now reached statistical or thermal *equilibrium.*

The probability distribution in this equilibrium state is easily determined. It is easy to show for our illustration (I leave this as an exercise) that an occupied region such as the shaded area at A *retains the same area as it moves.* We can deduce from this that *in the equilibrium state* $\tilde{\rho}$ *must take the same value in all of the accessible neighbourhoods.* The argument is that, since the occupied region retains the same area as it moves, $\tilde{\rho}$ measured within the moving region must be independent of time. It follows that in the equilibrium state $\tilde{\rho}$ must be the same in all mutually accessible regions, for if it were not then $\tilde{\rho}$ measured at a fixed point would fluctuate as the different occupied regions passed the point. This result is the classical equivalent of saying that in equilibrium $\tilde{p}$ is the same for all mutually accessible quantum states: it is the classical form of the *principle of equal equilibrium probability* in this simple system.

Problem 13.1

The principle of equal equilibrium probability

§§2.6, 13.4

13.3 Liouville's theorem

We must now generalise these one-particle, one-dimensional ideas to the case of N particles, which may be strongly interacting, moving in three dimensions. We now have $3N$ cartesian co-ordinates x_i and $3N$ corresponding momentum variables p_i. (Here i is a label which runs over particles as well as dimensions: x_1, x_2 and x_3 are the co-ordinates x, y, and z for the *first* particle, x_4, x_5 and x_6 are the equivalent co-ordinates of the *second* particle, and so on.) Thus phase space now has $6N$ dimensions. But it remains true that a *single* representative point in this space represents the state of the complete system of N particles. If the particle interactions are appreciable the representative point may follow an extremely tortuous path in phase space. Once again, we define the statistical state of the system in terms of a dimensionless probability density $\tilde{\rho}$. In place of a phase space area we now have a phase space hypervolume. We shall continue to define the extent of phase space in a dimensionless way by writing the elementary hypervolume $d\Omega$ as

Generalised phase space

$$d\Omega = \prod_i (dp_i\, dx_i)/h^{3N}. \tag{13.7}$$

How does a representative point move in phase space? If the system is isolated and its particles interact through simple potential forces, it is convenient to introduce the *Hamiltonian*, a function of position in phase space, which in this case is just the total energy,

$$\mathscr{H}(p_i, x_i) = \sum_i p_i^2/2m_i + \phi(x_1, x_2, \ldots). \tag{13.8}$$

Here the terms in p_i^2 represent kinetic energy and ϕ is the total potential energy of the system. Now $\partial\mathscr{H}/\partial p_i = p_i/m_i = u_i = \mathrm{d}x_i/\mathrm{d}t$. Also, if we write $f_i = -\partial\phi/\partial x_i$, we observe that f_i represents a particular component of force acting on a particular particle. According to Newton's second law, this force is equal to $\mathrm{d}p_i/\mathrm{d}t$. Using these results we see that

$$\mathrm{d}x_i/\mathrm{d}t = \partial\mathscr{H}/\partial p_i, \quad \mathrm{d}p_i/\mathrm{d}t = -\partial\mathscr{H}/\partial x_i. \tag{13.9}$$

Hamilton's equations

These are *Hamilton's equations*. Hamilton's equations show how the representative point of an isolated system moves in phase space and determine the details of its trajectory. (If you are familiar with Hamilton's formulation of classical mechanics, you will know that these equations may be greatly generalised. In the first place, the x_i in these equations may be replaced by *any* convenient set of position variables q_i which describe the spatial configuration of the system, provided that p_i is simultaneously replaced by the appropriate *canonical momentum*. For instance, one of the position variables q_i might refer to the position of the centre of mass of a molecule, or to the angle of rotation of a macroscopic object such as a lightly suspended mirror, in which cases the corresponding p_i would refer to the momentum of the molecule as a whole, or the *angular* momentum of the mirror. In addition, with a suitably modified Hamiltonian, Hamilton's equations may be used to describe motion under more general types of force which cannot be described by a simple potential, and to describe the behaviour of classical fields. Thus the formulation which we are using here is a very general one.)

With the help of Hamilton's equations, let us explore the nature of the equilibrium state in the phase space of $6N$ dimensions. Suppose that at time zero we have a *uniform distribution*, $\tilde{\rho} = \text{constant}$, for all parts of phase space. It is easy to show that this uniform distribution is also an *equilibrium* distribution. Consider a small hyperbox in phase space whose widths in the various dimensions are written as Δx_i and Δp_i, and whose dimensionless hypervolume $\Delta\Omega$ is equal to $\prod_i \Delta x_i \, \Delta p_i/h^{3N}$. Motion of representative points in the x_j direction brings points into this box from the negative x_j direction, while other points leave the opposite face of the box in the positive x_j direction (Fig. 13.3). Consider a short time $\mathrm{d}t$. In this time the points move a distance $\dot{x}_j \, \mathrm{d}t$ in the x_j direction, and the points which enter the box are those which originally occupied a dimensionless hypervolume $\Delta\Omega(\dot{x}_j \, \mathrm{d}t/\Delta x_j)$. The corresponding flux of points into the box will be equal to $\tilde{\rho} \, \Delta\Omega(\dot{x}_j/\Delta x_j)$. If $\dot{x}_j$ does not vary in the x_j direction there will be a similar flux of points leaving the opposite face of the box, and the net inflow of points will be zero. But if $\dot{x}_j$ increases in the x_j direction there will be a net inward flux equal to

$$-(\tilde{\rho} \, \Delta\Omega/\Delta x_j)(\partial\dot{x}_j/\partial x_j) \, \Delta x_j. \tag{13.10}$$

If we sum over similar terms for flow in each of the $6N$ directions, and remember that the total inward flux must be equal to $\Delta\Omega(\partial\tilde{\rho}/\partial t)$ we find that

$$\partial\tilde{\rho}/\partial t = -\tilde{\rho}\sum_i (\partial\dot{x}_i/\partial x_i + \partial\dot{p}_i/\partial p_i). \tag{13.11}$$

At this point we may use Hamilton's equations to evaluate the term in brackets. When we do this we find that the right-hand side of this expression is *zero*. Thus $\partial\tilde{\rho}/\partial t = 0$ everywhere, and therefore *the uniform distribution is an equilibrium distribution.*

It follows that, if we follow a small group of points as they move through the uniform distribution, they must stay at constant density – which implies that *they continue to occupy a constant hypervolume.* Thus

> In any isolated system a group of representative points occupying a small hypervolume of phase space will continue to occupy a constant hypervolume as it moves through phase space.

This important result is *Liouville's theorem.* We have proved it by considering the uniform distribution. However, since the distortion of phase space volumes as they move depends only on Hamilton's equations, and has nothing to do with the distribution of points inside them, the result must be true for *any* distribution.

13.4 The fundamental results in the classical formulation

Liouville's theorem shows that any small group of points in the statistical population remains at constant density as it moves through phase space. It follows that, if the system is to be in equilibrium there can be no

Fig. 13.3. Flow of representative points into and out of a hypervolume $\Delta\Omega$ of phase space due to motion in the x_j direction: only three dimensions can be shown in the diagram. The heavy lines represent the hypervolume $\Delta\Omega$. In time dt representative points originally in the shaded hypervolume A enter it, and representative points originally in the shaded hypervolume B leave it.

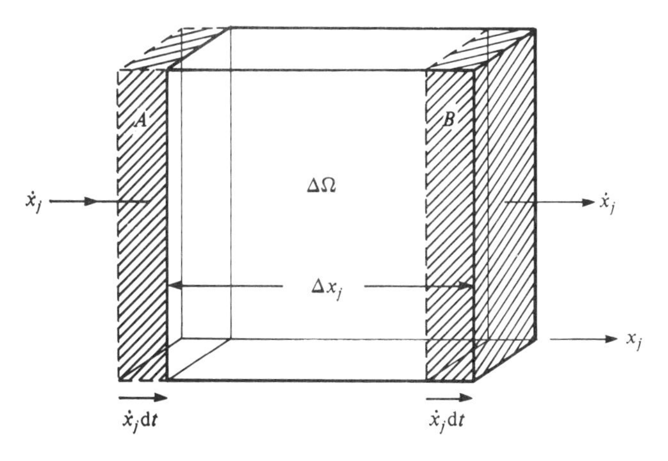

gradient of $\tilde{\rho}$ measured along any trajectory, for if there were the value of $\tilde{\rho}$ measured at a fixed point on the trajectory would be changing with time. Moreover, if $\tilde{\rho}$ is constant along all trajectories $\partial\tilde{\rho}/\partial t$ will be zero everywhere, and the system will be in equilibrium. It follows that *the isolated system will be in equilibrium if, and only if, $\tilde{\rho}$ takes the same value in all mutually accessible localities.* (In the classical formulation two localities are said to be *linked* if a trajectory passes through both of them, and all localities which are linked, directly or indirectly, are said to be *mutually accessible.*) This result is the classical form of the principle of equal equilibrium probability. In an isolated system, localities can only be mutually accessible if they correspond to the same energy. As in quantum statistics, we shall assume that there are no further restrictions on accessibility, so that all localities of the same energy *are* mutually accessible. This is the classical version of the *ergodic assumption.*

The principle of equal equilibrium probability

§2.6

§§2.5, 19.1

We know that the classical analogue of 'number of states' is the 'dimensionless hypervolume' in phase space, Ω. We may introduce a quantity γ which is the analogue of the 'density of states' in quantum statistics by observing that the classical states of fixed energy lie on an *energy hypersurface* in phase space. For a small increase in energy $\mathrm{d}E$ we move to a neighbouring hypersurface. The hypervolume $\mathrm{d}\Omega$ lying between these surfaces is the classical analogue of the number of quantum states between E and $E+\mathrm{d}E$. Thus the classical analogue γ for the density of states is given by

The classical equivalent of the density of states

$$\gamma = \mathrm{d}\Omega/\mathrm{d}E. \tag{13.12}$$

If we make use of this quantity, we may take over into classical mechanics the quantum analysis of temperature and the Boltzmann distribution given in Chapter 3, essentially unchanged. Consider Fig. 3.1, for instance. The classical analogue of the density of states g_A for system A is $\gamma_A = \mathrm{d}\Omega_A/\mathrm{d}E_A$, where Ω_A is the dimensionless hypervolume of phase space for system A. For system B, γ_B is similarly defined and is the analogue of g_B. If E_A lies in a narrow range of energy $\mathrm{d}E_A$, and E_B lies in a narrow range $\mathrm{d}E_B$, the corresponding hypervolumes of their phase spaces are $\gamma_A\,\mathrm{d}E_A$ and $\gamma_B\,\mathrm{d}E_B$. If we consider the *joint* system we have to describe it by a *joint phase space*, having more dimensions than either of the separate phase spaces. The hypervolume of this joint phase space corresponding to the state just described is the *product* $\gamma_A \mathrm{d}E_A \gamma_B \mathrm{d}E_B$ of the separate hypervolumes. In the quantum case it was the dots in the plane, the accessible joint states, which carried the same probability in equilibrium. In classical statistics we have to say that in equilibrium the probability density is evenly spread over the accessible hypervolume of the joint phase space. Thus in classical theory we expect that the final distribution will correspond to the variation of $\gamma_A\gamma_B$ along the sloping lines in the diagram. Just as $g_A g_B$ was very sharply peaked for large systems, so is $\gamma_A\gamma_B$, and we can safely say that the joint system will move in such a direction that $\gamma_A\gamma_B$ increases. Thus in classical statistics we may define the temperature of a large system by analogy with (3.3), writing

p. 26

The classical definition of T

$$1/kT = \partial \ln \gamma / \partial E. \qquad (13.13)$$

In the same way, we may take over the derivation of the Boltzmann distribution given in §3.3. Instead of finding the probability that the system of interest is in a particular state, we wish to find the probability density ρ per unit dimensionless phase space hypervolume. The modified argument runs that, for a small range $\mathrm{d}E$ in the energy of the system, the available *joint* hyperspace volume is $\gamma_R \delta E_{tot} \gamma \, \mathrm{d}E$, where $\gamma_R = \partial \Omega_R / \partial E_R$ for the reservoir, $\gamma = \partial \Omega / \partial E$ for the system of interest, and δE_{tot} is the total energy width used in the accessibility convention. The probability of finding the joint system within the specified energy ranges is proportional to this product. The corresponding phase space hypervolume for the system of interest is $\gamma \, \mathrm{d}E$, so we deduce that the probability ρ per unit hypervolume is proportional to γ_R. From this point onwards the argument is exactly analogous to the quantum argument, and we find that the canonical probability density is given by

The classical Boltzmann distribution

$$\blacktriangleright \quad \rho = \mathrm{e}^{-E/kT} / Z_{cl}, \qquad (13.14)$$

where Z_{cl} is the *classical partition function*, the integral over phase space

$$Z_{cl} = \int \mathrm{e}^{-E/kT} \, \mathrm{d}\Omega, \qquad (13.15)$$

which is the classical analogue of the quantum sum over states.

§4.1

By analogy with the argument of Chapter 4 we naturally define the *classical entropy* as

The classical definition of entropy

$$\tilde{S} = -k \int \tilde{\rho} \ln \tilde{\rho} \, \mathrm{d}\Omega. \qquad (13.16)$$

Its properties are similar to those of the quantum entropy. It increases with time through the process which we examined for a special case in §13.2: the representative points originally in a compact region in phase space gradually become widely separated as the region is drawn out into a more and more elongated filament. In this process representative points originally in quite separate regions are mixed together more and more closely within the mutually accessible parts of phase space, in rather the same way that separate volumes of red and yellow paint in a tin become mixed as they are stirred, until eventually $\tilde{\rho}$ becomes uniform within each accessible region. To complete the parallel with our quantum treatment I should at this point give a formal proof that $\tilde{S}$ cannot decrease in an isolated system. However, this proof, the classical H-theorem, raises some fundamental questions of principle which I prefer to leave until the final chapter. For the time being we shall assume the result without proof, by analogy with the quantum result.

§19.3

13.5 The classical equipartition theorem

In classical theory the energy is a function of the q_i and p_i variables, $E(q_i, p_i)$. It sometimes happens that there is a *separable squared term* in the energy associated with some particular position variable q_j (or, equivalently, some particular momentum variable). We mean by this that the energy can be written as

$$E=\alpha q_j^2+E'[q_{i\neq j},p_i], \tag{13.17}$$

where E' is independent of q_j, and depends only on the remaining $6N-1$ phase space co-ordinates. This situation is more common than might be supposed. It normally arises whenever q_j is a position co-ordinate describing *departure from mechanical equilibrium*, for small q_j at least: for a harmonic oscillator, for instance, the potential energy takes the form αx^2, where x is the departure from equilibrium. It also arises for the momentum variables in the expression for kinetic energy: the classical kinetic energy of a rotating molecule, for instance, may be written as $L^2/2I$ where L is the angular momentum and I is the moment of inertia.

In such a situation, let us calculate the mean value of the energy αq_j^2 associated with q_j. Using the classical Boltzmann distribution we find that

$$\begin{aligned}\overline{\alpha q_j^2}&=\int \alpha q_j^2\rho\,\mathrm{d}\Omega\\ &=\int \alpha q_j^2\,\mathrm{e}^{-E/kT}\,\mathrm{d}\Omega\bigg/\int \mathrm{e}^{-E/kT}\,\mathrm{d}\Omega.\end{aligned} \tag{13.18}$$

We now use the fact that E is separable, and thus split off the integration over $\mathrm{d}q_j$ from the integration over the remaining phase space variables, $\mathrm{d}\Omega'$. We find that

$$\begin{aligned}\overline{\alpha q_j^2}&=\frac{\int_{-\infty}^{\infty}\alpha q_j^2\,\mathrm{e}^{-\alpha q_j^2/kT}\,\mathrm{d}q_j\int \mathrm{e}^{-E'/kT}\,\mathrm{d}\Omega'}{\int_{-\infty}^{\infty}\mathrm{e}^{-\alpha q_j^2/kT}\,\mathrm{d}q_j\int \mathrm{e}^{-E'/kT}\,\mathrm{d}\Omega'}\\ &=kT\int_{-\infty}^{\infty}x^2\,\mathrm{e}^{-x^2}\,\mathrm{d}x\bigg/\int_{-\infty}^{\infty}\mathrm{e}^{-x^2}\,\mathrm{d}x,\end{aligned}$$

since the two integrals over $\mathrm{d}\Omega'$ cancel out. Using an integration by parts we deduce that

▶ $$\overline{\alpha q_j^2}=\tfrac{1}{2}kT, \tag{13.19}$$

the energy associated with q_j is equal to $\frac{1}{2}kT$. An exactly similar derivation would apply if the separable squared term involved a momentum variable. This result is the *equipartition theorem*. It is sometimes loosely stated by saying that 'there is $\frac{1}{2}kT$ of energy for every degree of freedom'. This statement is too vague, however. The theorem must be understood in the following way:

(i) The energy term of interest must be *separable* in terms of a proper set of Hamiltonian variables q_i and p_i.
(ii) The energy term of interest must be *squared* (proportional to q_i^2 or p_i^2).
(iii) The 'degrees of freedom' concerned are the $6N$ phase space co-ordinates and include momentum components as well as spatial degrees of freedom.
(iv) The theorem only applies when the classical limit is valid.

Let us consider a few examples to illustrate the range of applicability of this idea. Take the kinetic energy of the diatomic molecule, for instance. We have already seen that its mean translational kinetic energy is $\frac{3}{2}kT$, corresponding to three squared momentum terms. We have also seen that, in the classical limit its mean rotational kinetic energy is kT, corresponding to two squared angular momentum terms. (Note that in both of these cases we are using generalised co-ordinates – centre of mass co-ordinates and angles of rotation – and the squared momentum terms involve *canonical* momenta – centre of mass and angular momenta.) But it is *also* true that the kinetic energy of each atom in the molecule is separable, and its mean value is $\frac{3}{2}kT$. Indeed, this is *always* true in classical theory, whatever the forces involved. Note, for instance, the paradoxical result that for a gas in equilibrium in a gravitational field, the mean translation kinetic energy for molecules at great height is still $\frac{3}{2}kT$, the same as for those at sea level, in spite of the fact that a molecule rising through the gas must lose kinetic energy as it rises. (The paradox is resolved when we realise that of the molecules at sea level only the most energetic will rise to great heights: thus the *density* at great heights is very low, but it turns out that the mean kinetic energy which the energetic molecules have left when they reach great heights is still $\frac{3}{2}kT$, the loss in energy as they rise just compensating for the fact that only the most energetic molecules can get there.) In the same way, intermolecular attractions may slow down molecules approaching the boundary of a gas. This has the effect of reducing the pressure of the gas. But note carefully that the kinetic energy of those molecules which are close to the wall is still $\frac{3}{2}kT$ on average. The pressure is reduced not by a reduction in the mean kinetic energy of those striking the wall but by a reduction in their *number*. In fact, in classical theory all atoms in thermal equilibrium obey the Maxwell–Boltzmann velocity distribution, wherever they may be.

§§5.2, 12.6

§9.1.2

If we turn to potential energy, we might consider the harmonic oscillator in one dimension. This has a separable squared momentum term in the kinetic energy, and a separable squared position term in the potential energy. The average of each of these terms will be $\frac{1}{2}kT$, corresponding to a mean energy of kT for the oscillator as a whole.

§9.1.1

Note carefully two cases to which the equipartition theorem does *not* apply. We know already that at low temperatures the total energy of an oscillator and the rotational energy of a diatomic molecule are quenched, and are *not* equal to their equipartition values of kT. The reason is that at low temperatures the spacing between the energy levels in these systems is greater than kT – *they are not in the classical limit.* As a second case, consider the gravitational potential energy of a molecule in the atmosphere. This is a separable potential energy, but its mean value is *not* $\frac{1}{2}kT$ but kT, as is easily seen by finding how the number density varies with height. The equipartition theorem does not apply in this case because the *separable energy is not a squared term.* I hope that these examples will show the importance of understanding the limits within which the equipartition theorem applies.

§§9.1.1, 9.1.2

Problems

13.1 In Fig. 13.2 a small box of representative points is initially near A, and bounded by the two trajectories and by two lines parallel to the x axis. What happens to its shape and area as it moves from A to C? If the energies of the two trajectories are in the ratio 0.9:1, what shape will the box have reached after 10 000 cycles of the outer trajectory?

13.2 A classical particle is in equilibrium in a potential $\phi(r)$ at temperature T. Show, using (13.14), (i) that its probability density in real space is proportional to $\exp(-\phi/kT)$, and (ii) that it obeys the Maxwell–Boltzmann distribution (5.8) at all positions. Estimate roughly the atmospheric composition at the height of the noctilucent clouds about 70 km above the tropopause, where mixing by convection ceases. [Air at ground level is 78% N_2, 21% O_2 and 0.010% H_2 by volume. Mean temperature of the stratosphere = 240 K.]

13.3 An asymmetric molecule such as CO normally has a permanent electric dipole moment of magnitude p_0. Explain why at room temperature the probability of finding it at an angle between θ and $\theta + d\theta$ to an external electric field of strength $\mathscr{E}_E$ is proportional to $\exp(p_0 \mathscr{E}_E \cos\theta/kT)\, 2\pi \sin\theta\, d\theta$. Deduce that the mean dipole moment parallel to $\mathscr{E}_E$ is given by

The Langevin theory of electric susceptibility of polar gases

$$p = p_0 \mathscr{L}(p_0 \mathscr{E}_E/kT), \quad \text{where } \mathscr{L}(x) = \coth x - 1/x.$$

Sketch $p(x)$, and compare the result with Fig. 9.4.

§9.2.1

13.4 Use the equipartition theorem to predict at room temperature (i) the rotational heat capacities of CO_2 (a linear molecule) and H_2O (a non-linear molecule), (ii) the rms angular displacement of a mirror suspended vertically from a wire of torsion constant 10^{-16} N m radian^{-1}, and (iii) the rms thermal noise voltage across a 20 pF capacitor in an LRC circuit.

13.5 Show that S calculated as $k \ln \gamma$ for a classical system of N non-interacting particles of mass m in volume V is approximately equal to $Nk \ln[(emE/3N\pi\hbar^2)^{\frac{3}{2}} V]$. Gibbs observed that this expression is not *extensive*, as S should be for a pure gas. After comparing it with (12.26), can you see how and why it should be corrected? [The volume of a hypersphere of radius p in $3N$ dimensions is $\pi^{3N/2} p^{3N}/(3N/2)!$].

Gibbs' paradox
§12.6

14

The problem of the equation of state

For any system of identical particles the pressure exerted by the system is a function of the number density of particles n and the temperature T. This relation between Pn and T is known as the *equation of state*. As we have seen,
§5.4
for an ideal classical gas the equation of state takes the form $P = nkT$. In the classical gas, however, we ignore the forces acting between the particles. In this chapter we shall have to take these forces into account. This will inevitably make the statistical thermodynamics much harder, since we shall no longer be dealing with independent particles, and shall have to consider the statistics of the system as a whole. In addition, we know that at certain number densities and temperatures the system is unstable, and breaks up into mixtures of the *phases* which we know as gases, liquids and solids. Indeed, a *complete* theory of the equation of state will include within it theories of the three separate phases and the phase transitions. An example of how P depends upon n and T for a
p.248
real substance is given in Fig. 18.1. Notice that the function shows *singularities* at the points where phase transitions occur. Obviously the complete theory of the equation of state will not be a straightforward matter. In this chapter I shall simplify the treatment in several respects. First, I shall consider only the
§13.1
classical limit. We saw in the previous chapter that this is adequate for gases and most liquids, and it is also adequate for solids at temperatures above the
§10.4
Debye temperature. Secondly, I shall consider only the simplest types of interaction force. Thirdly, I shall refer only briefly here to the instabilities which lead to phase transitions: they will appear again in Chapter 18. Even with these simplifications, the subject remains a difficult one, and this chapter will be a survey of salient ideas and useful approaches and not in any sense a systematic treatment.

14.1 Intermolecular forces

The important forces between atoms and molecules are electrical in origin and depend in a complex way on the quantum dynamics of the electron shells of the atoms. A good deal is known about them both theoretically and experimentally. Two forces are always present. There is a *hard core repulsion* between all atoms which is negligible at large separations, but rises very rapidly as soon as the electron wave functions of the atoms start to overlap. There is also the *van der Waals attraction.* This is an effect which arises when the spontaneous quantum fluctuations of the electric dipole moment of one atom induce corresponding dipole fluctuations in a second neighbouring atom. The force between the correlated dipole moments is attractive. The interatomic potential energy due to the van der Waals attraction can be calculated in terms of the atomic polarisabilities, and at short ranges it takes the form $-B/r^6$, where r is the separation between the atomic centres. The hard core repulsion is harder to calculate from first principles, but rises so rapidly at short range that the precise form is not very important. It is often, rather arbitrarily, represented by a term A/r^{12}. If these two terms are combined we have the *Lennard–Jones 6–12 potential*

The Lennard–Jones potential

$$u(r) = A/r^{12} - B/r^6$$
$$= 4\varepsilon[(r_0/r)^{12} - (r_0/r)^6], \tag{14.1}$$

shown in Fig. 14.1. This approximate form is a surprisingly good fit to the potentials measured for simple atoms such as inert gas atoms. It also works well for small molecules which, because they rotate rapidly, even in the liquid state, behave as though they were spherical. Note that it takes two constants to

Fig. 14.1. The Lennard–Jones 6–12 potential, showing the so-called 'hard core radius' r_0 (i.e. the separation of centres when in contact at zero potential, better described as the hard core diameter), and the pair binding energy ε. The intermolecular attractive force is the gradient of this curve, and is zero at the minimum, at $r^* = 1.12r_0$. The inset shows the values of ε/k and r_0 for various gases, obtained by fitting the second virial coefficient.

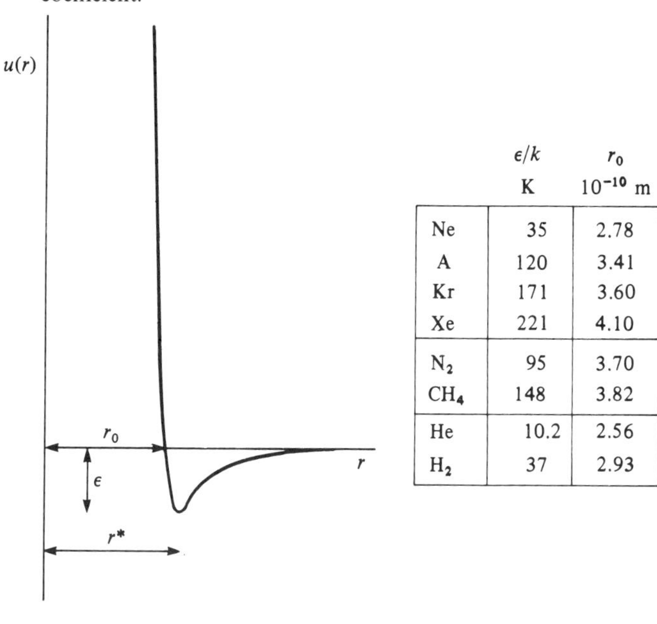

	ϵ/k K	r_0 10^{-10} m
Ne	35	2.78
A	120	3.41
Kr	171	3.60
Xe	221	4.10
N_2	95	3.70
CH_4	148	3.82
He	10.2	2.56
H_2	37	2.93

specify a Lennard–Jones potential. We may think of them as the amplitudes A and B of the hard core and van der Waals' terms, but I have indicated how the formula may be rewritten in terms of the *hard core radius* r_0 and the *pair binding energy* ε shown in the diagram. This form is usually more useful.

Other forces

Not all interatomic forces are so simple. If the atoms are linked by a *covalent bond* there will be in addition strong, directional bond forces. In an ionic substance such as sodium chloride there will be strong, long-range *Coulomb forces* acting between the ions, which depend, of course, on the charge of the ion. Molecules such as H_2O which have a permanent electric dipole moment will have both *dipole forces* and *dipole couples* acting between them. To simplify the discussion I shall ignore all such complications and limit our consideration to substances such as noble gases or gases of small molecules in which the only important forces are the hard core repulsion and the weak van der Waals attraction. Even in this case we have to remember that, strictly speaking, we should take account of the fact that the force between any two atoms may be affected by the presence of other atoms in the neighbourhood. For instance, if a third atom passes between a pair of atoms, we expect it to reduce their mutual van der Waals attraction by dielectric screening. Again, I shall ignore such complications and suppose for simplicity that all atoms or molecules interact by simple two-body potentials which are unaffected by the presence of other bodies, of the type shown in Fig. 14.1.

Three-body forces

14.2 The virial theorem

There are two commonly used approaches to the problem of finding an equation of state in the classical limit. From the point of view of the development which we have used so far, the most obvious procedure would be to calculate the classical partition function (or alternatively the classical grand partition function) and thence the free energy and the pressure, and we shall indeed use this approach in § 14.5. But it is in many ways more illuminating to follow a different route, first suggested by Clausius. In this route we first show that the pressure is related to a quantity known as the *radial distribution function*, $g(r)$. We therefore have to find procedures for computing $g(r)$ rather than Z or Ξ from first principles. In this section we shall establish the connection between P and $g(r)$.

We first introduce a new quantity known as the *virial* $\mathscr{V}$, defined as follows:

$$\mathscr{V} = -\tfrac{1}{2}\sum_i \mathbf{f}_i \cdot \mathbf{r}_i, \tag{14.2}$$

where $\mathbf{f}_i$ is the force on the ith particle in the system, and $\mathbf{r}_i$ is its position vector. Clausius pointed out that the mean virial is connected with the kinetic energy as well as with the interparticle potentials and the pressure. To see this, note first that using Newton's second law we may rewrite $\mathscr{V}$ as

$$\mathscr{V} = -\frac{1}{2}\sum_i m_i \frac{\mathrm{d}}{\mathrm{d}t}(\mathbf{v}_i)\cdot\mathbf{r}_i$$

$$= -\frac{1}{2}\sum_i m_i \frac{\mathrm{d}}{\mathrm{d}t}(\mathbf{v}_i\cdot\mathbf{r}_i) + \frac{1}{2}\sum_i m_i v_i^2.$$

Now, if the energy in the system is fixed, and the system is bounded so that the particles are never more than a certain distance from the origin, the quantity $\mathbf{v}_i\cdot\mathbf{r}_i$ can only fluctuate between certain finite limits. Thus, if we average over a long enough time the first term on the right-hand side must average zero. This leaves the result

$$\blacktriangleright \quad \bar{\mathscr{V}} = \frac{1}{2}\sum_i m_i \overline{v_i^2}, \tag{14.3}$$

the mean virial is equal to the mean kinetic energy. This result is *Clausius' virial theorem* in its general form. (It is worth noting that the result is true provided the averages are taken over times long enough to make the term in $(\mathrm{d}/\mathrm{d}t)(\mathbf{v}_i\cdot\mathbf{r}_i)$ negligible. This may be much shorter than the time required to achieve full thermal equilibrium and the virial theorem has been applied, for instance, to the internal motions of spiral galaxies which are not in thermal equilibrium.)

Problem 14.2

To apply the result to our problem of the equation of state we notice that the mean virial is *also* related to the forces acting. In fact, we may split the virial into two parts

$$\mathscr{V} = \mathscr{V}_{\text{ext}} + \mathscr{V}_{\text{int}}$$

$$= -\frac{1}{2}\sum_i \mathbf{f}_i^{\text{ext}}\cdot\mathbf{r}_i - \frac{1}{2}\sum_i \mathbf{f}_i^{\text{int}}\cdot\mathbf{r}_i, \tag{14.4}$$

where the external part $\mathscr{V}_{\text{ext}}$ is associated with external forces due to the pressure exerted by the container, and the internal part $\mathscr{V}_{\text{int}}$ is associated with the internal forces acting between the particles. If we imagine a rectangular container (Fig. 14.2), then, if we consider the forces exerted on the particles by the x faces of the container, we find that they contribute $\frac{1}{2}PV$ to $\mathscr{V}_{\text{ext}}$ on average. Thus, taking all three pairs of faces, we find that

$$\bar{\mathscr{V}} = \tfrac{3}{2}PV + \bar{\mathscr{V}}_{\text{int}}. \tag{14.5}$$

Fig. 14.2. Two contributions to the virial, $-\frac{1}{2}\sum_i \mathbf{f}_i\cdot\mathbf{r}_i$. For the attractive forces between particles 1 and 2 we have $-\frac{1}{2}(\mathbf{f}\cdot\mathbf{r}_1 - \mathbf{f}\cdot\mathbf{r}_2) = \frac{1}{2}\mathbf{f}\cdot(\mathbf{r}_2 - \mathbf{r}_1) = \frac{1}{2}fr$. For the external pressure of the x faces of the container on the molecules colliding with the wall we have $\frac{1}{2}PA \times L = \frac{1}{2}PV$.

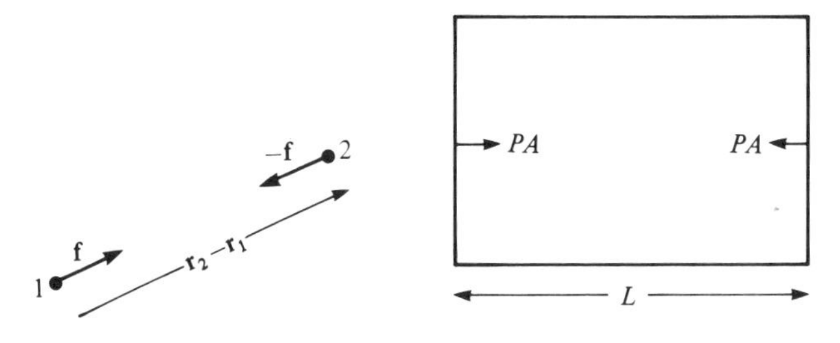

§13.5

In thermal equilibrium we know from the equipartition theorem that the mean kinetic energy for all the particles is $\frac{3}{2}NkT$. Thus on combining (14.3) with (14.5) we find that in thermal equilibrium

$$\tfrac{3}{2}PV = \tfrac{3}{2}NkT - \bar{\mathscr{V}}_{\text{int}}. \tag{14.6}$$

At this point we introduce a new distribution function. If the particles had been non-interacting particles with uncorrelated positions then the probability density per unit volume of finding a particle at distance r from the centre of any particular particle would have been simply $n = N/V$, the number density. In practice, the forces between the molecules lead to correlations in their positions. To describe these correlations we write the probability density per unit volume of finding a second particle at distance r from the particle of interest as $g(r)n$, where $g(r)$ is known as the *radial distribution function*. Some examples of $g(r)$ are shown in Fig. 14.3. At large values of r, $g(r)$ always tends to

The radial distribution function $g(r)$

Fig. 14.3. The radial distribution function $g(r)$ for the 6–12 potential. The upper diagram shows the results of Monte-Carlo calculations for $T = 1.06\varepsilon/k$, which lies below T_c. The curves are labelled with the value of V/V^* where $V^* = Nr_0^3$. The arrows at $r = r_s$, $\sqrt{2}r_s$ and $\sqrt{3}r_s$ correspond to the first, second and third neighbours in the classical close-packed solid at $T = 0$ and zero pressure. The solid curves correspond to crystalline states. The state with $V/V^* = 1.2$ is liquid, and the state with $V/V^* = 20$ is vapour (after Wood, 1968). The lower diagram compares the Monte-Carlo result at the higher temperature $T = 2.74\varepsilon/k$, above T_c, with the Yvon–Born–Green and Percus–Yevick theories for a dense fluid with $V/V^* = 0.9$ (after Broyles, Chung & Sahlin, 1962).

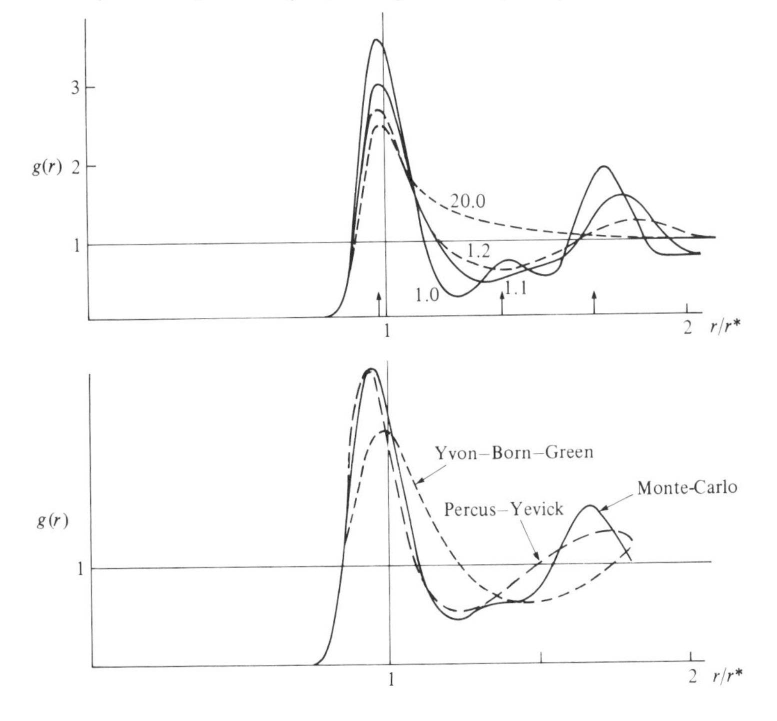

unity, while for small values of r it always tends to zero because the hard core repulsions keep the particles apart. The function $g(r)$ can be *measured* from the X-ray or neutron diffraction pattern for elastic scattering.

$\mathscr{V}_{int}$ is made up of contributions from pairs of atoms at separation r (Fig. 14.2). If the attractive force is $f(r)$, the corresponding contribution to $\mathscr{V}_{int}$ is $\frac{1}{2}f(r)r$. On averaging $\mathscr{V}_{int}$ over the positions of all the atoms we find for its mean value

$$\bar{\mathscr{V}}_{int}=\frac{1}{4}N\int_0^\infty f(r)rg(r)n\,4\pi r^2\,dr, \tag{14.7}$$

where an extra factor of $\frac{1}{2}$ has been included to avoid counting each contribution twice. On substituting this result in (14.6) we find that

$$P=nkT-\frac{n^2}{6}\int_0^\infty 4\pi r^3 f(r)g(r)\,dr. \tag{14.8}$$

The virial equation of state

This is the *virial equation of state.* It expresses the pressure in terms of the radial distribution function $g(r)$ and the interparticle attraction $f(r)$. If there are no interparticle forces, we recover from it the usual equation of state for the perfect gas. If we know $g(r)$ as a function of n and T, we may deduce from (14.8) the equation of state in the general case.

14.3 The second virial coefficient

It is not hard to find a formal expression for $g(r)$. According to the classical Boltzmann distribution, the probability of any particular configuration is given by

§ 13.4

$$dp=\frac{e^{-\phi/kT}\,dV_1\,dV_2\ldots dV_N}{\int e^{-\phi/kT}\,dV_1\,dV_2\ldots dV_N}, \tag{14.9}$$

where $\phi(\mathbf{r}_1,\mathbf{r}_2,\ldots,\mathbf{r}_N)$ is the potential energy of the configuration. The probability of finding the first and second particles in definite positions is therefore

$$dp_{12}=\frac{dV_1\,dV_2\int e^{-\phi/kT}\,dV_3\,dV_4\ldots dV_N}{\int e^{-\phi/kT}\,dV_1\,dV_2\ldots dV_N}. \tag{14.10}$$

Thus the probability density of finding *some* other particle at vector displacement $\mathbf{r}$ from the first particle is given by

$$ng(r)=\frac{N\left[\int e^{-\phi/kT}\,dV_1\,dV_3\,dV_4\ldots dV_N\right]_{\mathbf{r}_2=\mathbf{r}_1+\mathbf{r}}}{\int e^{-\phi/kT}\,dV_1\,dV_2\ldots dV_N}. \tag{14.11}$$

Formal expression for the radial distribution function

In *principle,* this result solves the problem of the equation of state. Unfortunately it is extremely difficult to evaluate the integrals in the expression.

In the limit of low density, however, it is easy to see from (14.11) what the first-order expression for $g(r)$ should be. At low densities, when we integrate over the positions of particles $3, 4, \ldots, N$, we may safely assume that in almost all configurations these particles are distant from particles 1 and 2, and also distant from each other, so, for almost all configurations we may replace ϕ by $u_{12}(r)$, the interaction potential of particles 1 and 2. Indeed, in the denominator particles 1 and 2 are *also* far apart in almost all configurations, so we can ignore u_{12} also. In this approximation, then, we have

$$ng(r) \simeq \frac{N\,\mathrm{e}^{-u_{12}(r)/kT}\,V^{N-1}}{V^N}, \tag{14.12}$$

so that $g(r) \simeq \mathrm{e}^{-u_{12}/kT}$, as it would have been if only two particles had been present. If we insert this approximation into (14.8) we find that

$$P/kT \simeq n + \left[(-1/6kT)\int_0^\infty 4\pi r^3 (\mathrm{d}u/\mathrm{d}r)\,\mathrm{e}^{-u/kT}\,\mathrm{d}r\right] n^2. \tag{14.13}$$

We evidently have here the first two terms of the expansion of P/kT in powers of the number density n, known as the *virial expansion*. The term in square brackets is the *second virial coefficient*, usually written $B_2(T)$. This term may conveniently be rewritten as

The exact expression for the second virial coefficient

$$B_2(T) = \frac{2\pi}{3}\int_0^\infty \frac{\mathrm{d}}{\mathrm{d}r}(\mathrm{e}^{-u/kT}-1)r^3\,\mathrm{d}r, \quad \text{or}$$

▶ $$B_2(T) = -\int_0^\infty 2\pi r^2(\mathrm{e}^{-u/kT}-1)\,\mathrm{d}r, \tag{14.14}$$

after integration by parts. Higher-order terms in the virial expansion are due to the parts of the integrals in (14.11), ignored in the approximation just used, in which more than two particles may be clustered close together, as we shall see later.

§ 14.5

Values of the second virial coefficient are easily obtained from measurements of the pressures of gases at low densities, and $B_2(T)$ may be calculated from (14.14) for any assumed intermolecular potential $u(r)$. Fig. 14.4 shows a comparison, expressed in reduced units, of the experimental values for $B_2(T)$ for some noble gases and small molecules with the theoretical curve for the Lennard–Jones potential. The fit is clearly excellent. (However, the most recent results show that this fit is less precise at low temperatures, and that the interatomic potential for argon, for instance, must be deeper and steeper than the 6–12 potential.)

14.4 The physical behaviour of imperfect gases

Gases at moderate densities whose behaviour is affected to some extent by the intermolecular forces are known as *imperfect gases*. In this section we shall examine and try to account for some of their properties, concentrating particularly on densities low enough to make the first two terms of the virial expansion a good approximation,

$$P \simeq nkT + B_2(T)n^2kT. \tag{14.15}$$

Note first that, as is apparent from Fig. 14.4, there is a temperature known as the *Boyle temperature* T_B at which B_2 is zero. At this temperature the gas is most like an ideal gas at low densities. At higher temperatures B_2 is positive: the gas is *harder* to compress than an ideal gas because the effect of the hard core repulsion is dominant. At lower temperatures, on the other hand, B_2 is negative: the gas is *easier* to compress than an ideal gas, because the effect of the van der Waals attraction is dominant.

The Boyle temperature T_B

Consider next the *Joule expansion*, the expansion of the gas through a valve into an evacuated container. We have already noted that the temperature of an ideal gas does not change in a Joule expansion. For an imperfect gas there is a small temperature change, and we may find an expression for it by using a thermodynamic argument. In a Joule expansion the internal energy of the gas is conserved, so we may write the rate of change of temperature with volume as

§5.6

Fig. 14.4. The second virial coefficient $B_2(T)$ in reduced units. The solid lines are calculated for the 6–12 potential in the classical limit. In the upper diagram this is compared, using a logarithmic temperature scale, with data for five gases. Helium does not fit the prediction at low temperature because it is not in the classical limit; the remaining four gases fit very well (after Hirschfelder, Curtis & Bird, 1967). In the lower diagram, B_2 for the 6–12 potential is compared with B_2 for van der Waals' equation, with values of a and b fitted at the Boyle temperature T_B. The Joule–Kelvin inversion temperature occurs where the radius from the origin is a tangent to the curve, and is at $1.9T_B$ for the 6–12 potential.

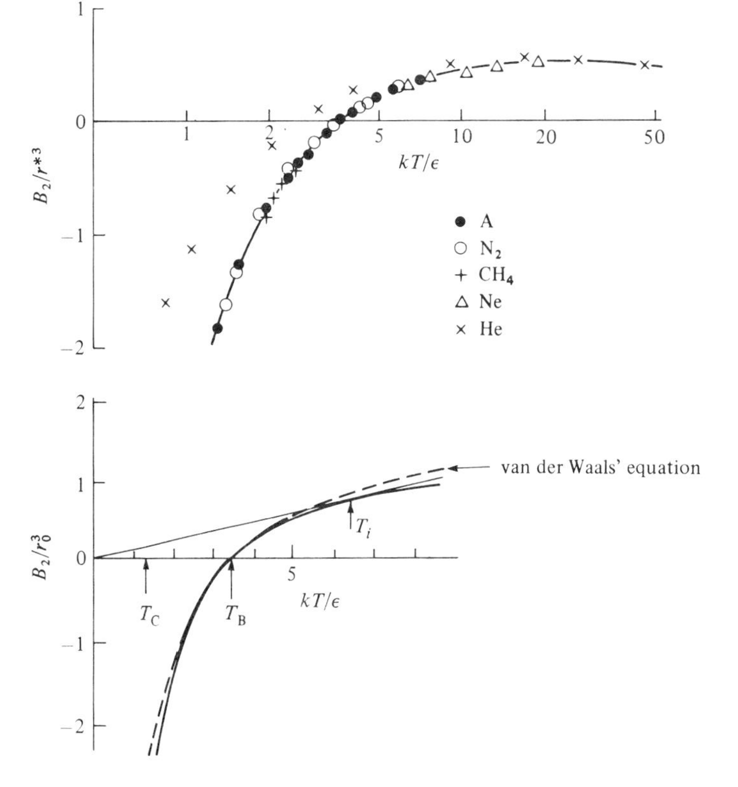

The cooling in a Joule expansion

$$(\partial T/\partial V)_E = -(\partial T/\partial E)_V(\partial E/\partial V)_T$$
$$= -(1/C_V)[T(\partial S/\partial V)_T - P]$$
$$= -(1/C_V)[T(\partial P/\partial T)_V - P], \tag{14.16}$$

§§8.1, 6.8, 8.2

using successively (8.3), the expression for dE (6.18) and a Maxwell relation (8.6). This shows that the temperature change may be written in terms of the heat capacity C_V and the equation of state. Using (14.15) we deduce that at low densities

▶ $$(\partial T/\partial V)_E = -(1/C_V)[n^2kT^2\, dB_2/dT]. \tag{14.17}$$

As will be apparent from Fig. 14.4, this expression is negative, except at very high temperatures: real gases normally *cool* slightly in a Joule expansion. The physical reason is easy to understand. As the gas expands, the molecules move further apart, and do work against the intermolecular attractions. This raises the potential energy. Since the total energy is conserved, the translational and internal energies must fall, corresponding to a fall in temperature.

The different type of expansion which occurs when a gas flows continuously from high to low pressure through an insulated throttle valve is known as a *Joule–Kelvin expansion.* Since this is a flow process and no process energy enters the gas it occurs at constant *enthalpy*. The temperature change which occurs may be found by a thermodynamic argument similar to that which we used to discuss the Joule expansion. We may write the change of temperature with pressure as

§7.3

The temperature change in a Joule–Kelvin expansion

$$(\partial T/\partial P)_H = -(\partial T/\partial H)_P(\partial H/\partial P)_T$$
$$= -(1/C_P)[T(\partial S/\partial P)_T + V]$$
$$= -(1/C_P)[-T(\partial V/\partial T)_P + V], \tag{14.18}$$

where we have used (7.7), (8.3), the expression for dH (7.2), and a Maxwell relation. In this case we find from (14.15) that

$$-T(\partial V/\partial T)_P + V = -V\left[\frac{1+nB_2+nT\,dB_2/dT}{1+2nB_2}\right]+V$$
$$= N[B_2 - T\,dB_2/dT],$$

to first order in n. Thus

▶ $$(\partial T/\partial P)_H = -(N/C_P)[B_2 - T\,dB_2/dT], \tag{14.19}$$

at low densities. This expression does *not* always predict cooling on expansion, for the term in square brackets changes sign as the temperature falls. At high temperatures it is positive, and the gas *warms* as the pressure falls, essentially because the hard core effects are dominant, and this means that $P_1V_1 > P_2V_2$ as the gas passes through the valve. The surrounding flow therefore does net work on the process box, which warms the gas. At low temperatures both terms in the square bracket are negative, and the gas *cools* as the pressure falls, essentially because work is being done against the intermolecular attractions. The temperature at which the effect reverses sign is known as the *Joule–Kelvin inversion temperature*, T_i. Its value may be obtained from $B(T)$ by the

§7.3

The inversion temperature T_i

construction shown in Fig. 14.4. For the 6–12 potential T_i is $1.9T_B$ and this is approximately true for many real gases also. The cooling on expansion below T_i was put to use in early helium liquefiers, to avoid the need for mechanical moving parts, which are difficult to lubricate at temperatures of a few K. Compressed helium gas passed through a counter-flow heat exchanger, being thus cooled before it reached an expansion valve. At the valve it was cooled by Joule–Kelvin expansion, and it was returned to the compressor via the heat exchanger, so providing the coolant for the incoming gas. Once the process was started, the temperature at the valve fell steadily, until eventually drops of liquid were formed there. However, since the inversion temperature of helium is only 51 K, the compressed gas had to be cooled in liquid hydrogen before the process could start. Though inefficient in use of energy, such liquefiers were simple and reliable.

14.5* The virial expansion

The next step in the theory of the imperfect gas is to find the higher-order coefficients B_n in the virial expansion,

$$P/kT = n + B_2(T)n^2 + B_3(T)n^3 + \cdots. \quad (14.20)$$

Mayer first found a general procedure for determining these coefficients. They may be found by expanding in powers of n the general expression (14.11) for $g(r)$, and using the virial equation of state (14.8). For our present purposes it is more convenient to proceed by expanding instead the classical partition function Z_{cl} from which the pressure may be obtained using the relation $P/kT = (\partial \ln Z_{cl}/\partial V)_T$.

§13.4

§9.2

The classical partition function for N identical particles in the classical limit is

$$Z_{cl} = \frac{1}{N!}\int e^{-E/kT}\, d\Omega,$$

where the integral is taken over the phase space of the N particles. The factor $1/N!$ is included to allow for the effects of particle identity. (In the classical limit we may safely assume that all particles are in different quantum states in all important configurations. In that limit we overcount the number of distinct states by a factor of $N!$ when we integrate over phase space.) Now $d\Omega$ is $\prod_i (dp_i\, dx_i/h)$, and

Problem 13.5

$$E = \sum_i p_i^2/2m + \phi(\mathbf{r}_1, \mathbf{r}_2, \ldots, \mathbf{r}_N),$$

so the phase space integral may be broken up into the product

$$Z_{cl} = \frac{1}{N!}\prod_i\left[\int_{-\infty}^{\infty}\frac{dp_i}{h}\, e^{-p_i^2/2kTm}\right]\int e^{-\phi/kT}\, dV_1\, dV_2 \ldots dV_N$$

$$= \frac{1}{N!}(2\pi mkT/h^2)^{3N/2} I, \quad (14.21)$$

where I is the *configuration integral*

$$I=\int e^{-\phi/kT}\,dV_1\,dV_2\ldots dV_N, \tag{14.22}$$

taken over *configuration space*, the $3N$-dimensional space of the position co-ordinates of the N particles. If we can find this integral, we can find Z_{cl} and hence P, and the problem is solved. (Inspection of (14.22) shows that I is closely related to the expression for $g(r)$, (14.11).) For instance, for non-interacting particles $\phi=0$ and thus $I=V^N$; we deduce then that $P=kT(\partial \ln Z_{cl}/\partial V)_T=kT(\partial \ln I/\partial V)_T=NkT/V$, the usual result for an ideal gas.

§ 14.3

Mayer chose to rewrite I as follows:

$$\begin{aligned}I&=\int e^{-\phi/kT}\,dV_1\,dV_2\ldots dV_N\\&=\int\prod_{\text{pairs}}(e^{-u_{ij}/kT})\,dV_1\,dV_2\ldots dV_N\\&=\int\prod_{\text{pairs}}(f_{ij}+1)\,dV_1\,dV_2\ldots dV_N,\end{aligned} \tag{14.23}$$

where $f_{ij}=\exp(-u_{ij}/kT)-1$, the product is taken over all possible pairs of atoms i, j, and $u_{ij}(\mathbf{r}_i-\mathbf{r}_j)$ is the intermolecular potential for atoms i and j. The reason for writing I in this way is that the *pair function f_{ij} is only non-zero when the atoms are close together.* If N is large there will be a very large number of pairs, and the product when written out will actually contain *all possible products of pair functions.* For instance, there will be a contribution to I of the form

$$C=\int f_{12}f_{23}f_{13}f_{45}f_{78}f_{89}\,dV_1\,dV_2\ldots dV_N,$$

amongst many others. As an aid to the memory, Mayer attached to each such contribution a corresponding *diagram.* The diagram for contribution C is shown in Fig. 14.5. Each atom is represented by a labelled dot, and each factor f_{ij} by a line joining atoms i and j. In such a diagram, a group of atoms

Fig. 14.5. The diagram for a typical contribution in Mayer's expansion of the configuration integral I. In this diagram, cluster B is of first order, and clusters A and C of second order. A and B are *irreducible*, but C is not: it can be split into two by a cut at atom 8.

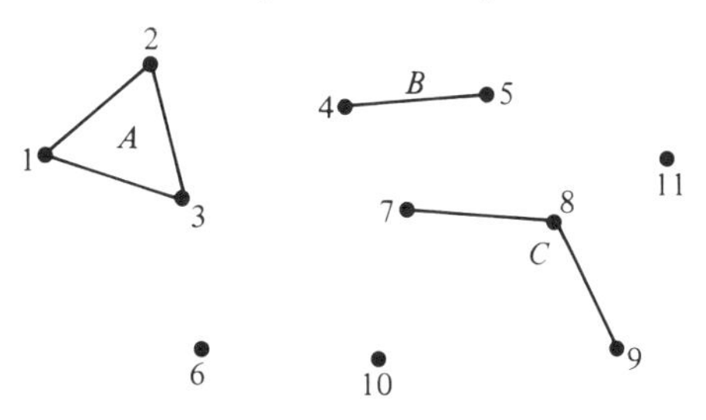

linked directly or indirectly by lines is called a *cluster*. If the cluster contains k atoms we say that the *order* of the cluster is $j=k-1$. The diagram in the figure contains two second-order clusters, one first-order cluster, and the rest of the atoms are in zero-order clusters of one atom only. We notice that our contribution C may be written as a product of integrals, one for each cluster:

$$C=\int f_{12}f_{23}f_{13}\,\mathrm{d}V_1\,\mathrm{d}V_2\,\mathrm{d}V_3\times\int f_{45}\,\mathrm{d}V_4\,\mathrm{d}V_5$$

$$\times\int f_{78}f_{89}\,\mathrm{d}V_7\,\mathrm{d}V_8\,\mathrm{d}V_9\times\int \mathrm{d}V_6\times\int \mathrm{d}V_{10}\ldots.$$

We are now in a position to recognise the terms in the expansion of I in powers of the density. The point is that *each of the integrals in the above product is proportional to V.* For instance, we may write the first integral as

$$\int f_{12}f_{23}f_{13}\,\mathrm{d}V_1\,\mathrm{d}V_2\,\mathrm{d}V_3=\int \mathrm{d}V_1\left[\int f_{12}f_{23}f_{13}\,\mathrm{d}V_2\,\mathrm{d}V_3\right]_{\mathbf{r}_1=\text{const}}$$

$$=VI_\mathrm{c},$$

where I_c represents the term in square brackets, known as a *cluster integral*. In the cluster integral we hold one atom fixed, and integrate over the positions of the others. Since $f_{12}f_{23}f_{13}$ drops to zero unless atoms 2 and 3 are within a few atomic radii of molecule 1, this cluster integral will be of order v_a^2, where v_a is a typical atomic volume, and will be *independent of the volume of the container.* Thus if the diagram contains clusters of order j, the corresponding contribution to I will be of order $V^{N-r}v_\mathrm{a}^r$, where $r=\sum j$. It follows that we may expand I in powers of $1/V$ as

$$I=V^N[1+C_1(1/V)+C_2(1/V)^2\ldots],\tag{14.24}$$

where *C_r is just the sum of all possible products of cluster integrals such that $\sum j=r$.* This gives the required expansion of I in powers of the density.

From this point onwards the argument becomes very technical and I shall omit the details. First, we must convert the expansion of I in powers of $1/V$ into a corresponding expansion of $\ln Z_\mathrm{cl}$ (for instance, by using the expansion for $\ln(1+x)$ in powers of x, though more elegant methods are better). Then we get the expansion for P/kT by writing it as $(\partial \ln Z_\mathrm{cl}/\partial V)_T$. In the resulting virial expansion

$$P/kT=n+B_2n^2+B_3n^3\ldots$$

it is clear that the virial coefficient B_s must have dimensions of $(\text{volume})^{s-1}$ and can only involve products of cluster integrals such that $\sum j=s-1$, each of which will be of order v_A^{s-1}. Mayer proved the surprising result that, when all the various transformations are performed much cancellation of terms occurs, and *the only surviving contributions are those corresponding to a special type of diagram, the single irreducible clusters of order $s-1$.* He showed in fact that

▶ $$B_s=-\frac{s-1}{s}\frac{1}{(s-1)!}\sum I_s,\tag{14.25}$$

where the sum is taken over the cluster integrals for all the *irreducible* clusters containing s distinct atoms. A cluster is said to be irreducible when it cannot be split into two by cutting the diagram at a single atom (see Fig. 14.5).

We therefore have a definite prescription for writing down the sequence of virial coefficients, and this provides an excellent and accurate theory for the equation of state of imperfect gases at moderate densities. You may like to check that Mayer's expression

$$B_2 = -\frac{1}{2}\left[\int f_{12}\, dV_2\right]_{\mathbf{r}_1 = \text{const}} \tag{14.26}$$

§ 14.3

is equivalent to our earlier result (14.14). For the third virial coefficient we have only one irreducible cluster of three molecules, which gives

$$B_3 = -\frac{1}{3}\left[\int f_{12} f_{23} f_{13}\, dV_2\, dV_3\right]_{\mathbf{r}_1 = \text{const}}. \tag{14.27}$$

But you will find that there are 10 distinct irreducible clusters of four molecules, and for the higher coefficients the number of integrals involved increases very rapidly.

14.6* Attempts to analyse the high density limit

At first sight, Mayer's virial expansion provides a complete theory of the equation of state, but this is not so. As the density rises, higher-order virial coefficients are required, and it becomes increasingly difficult to compute them as the order rises. At present it seems unlikely that coefficients beyond B_7 will be calculated, because of the problems of accounting for the possible diagrams. (For seven particles there are 21 pairs and 2^{21} – more than two million – possible diagrams, not all irreducible, however.) This is enough to give about 1% accuracy for temperatures well above the critical temperature and densities as high as half the solid density. For higher densities the method breaks down completely, because, since $B_s \approx v_a^{s-1}$, *as soon as n approaches $1/v_a$ the terms of the virial expansion cease to decrease with s, and the series cannot converge.* (This is also apparent from the observation that $(\partial P/\partial n)_T$ is discontinuous at a phase transition, as may be seen in Fig. 18.1, for a convergent power series cannot have a discontinuous derivative.) To put it another way, the virial expansion is a method which starts with the ideal gas, and allows for the effects of small clusters. It cannot cope with the case of a liquid or solid, where *all* the molecules are aggregated into a single, giant cluster.

Failure of the virial expansion

p. 248

A more hopeful approach is to try to find *self-consistent equations*, usually for the radial distribution function $g(r)$, which are satisfied at *any* density. As an illustration of this idea, consider the *Kirkwood integral equation*, put forward in 1935. We introduce a hierarchy of distribution functions, g. The nth such distribution function is proportional to the joint probability of finding n particles each at a particular position, and all the functions are scaled so that they tend to unity when the particles are far apart. Using the classical Boltzmann distribution we find that the two-body distribution function g_{12} is given by

$$g_{12}=g(\mathbf{r}_1,\mathbf{r}_2)=V^2\frac{[\int e^{-\Sigma u_{ij}/kT}\,dV_3\,dV_4\ldots dV_N]_{\mathbf{r}_1,\mathbf{r}_2\text{ fixed}}}{\int e^{-\Sigma u_{ij}/kT}\,dV_1\,dV_2\ldots dV_N}$$

while (14.28)

$$g_{123}=V^3\frac{[\int e^{-\Sigma u_{ij}/kT}\,dV_4\,dV_5\ldots dV_N]_{\mathbf{r}_1,\mathbf{r}_2,\mathbf{r}_3\text{ fixed}}}{\int e^{-\Sigma u_{ij}/kT}\,dV_1\,dV_2\ldots dV_N},$$

and so on. Kirkwood pointed out that if we differentiate g_{12} with respect to $\mathbf{r}_1$, since

$$\nabla_1(\sum u_{ij}/kT)=\nabla_1(u_{12})/kT+\sum_{j>2}\nabla_1(u_{1j})/kT,$$

we find that

$$-kT\,\nabla_1 g_{12}=g_{12}\,\nabla_1 u_{12}+(N-2)V^2$$
$$\times\int\nabla_1 u_{13}\,dV_3\frac{\int e^{-\Sigma u_{ij}/kT}\,dV_4\,dV_5\ldots dV_N}{\int e^{-\Sigma u_{ij}/kT}\,dV_1\,dV_2\ldots dV_N},$$

or

▶ $$-kT\,\nabla_1 g_{12}=g_{12}\,\nabla_1 u_{12}+n\int g_{123}\,\nabla_1 u_{13}\,dV_3, \tag{14.29}$$

where we have treated $N-2$ as equal to N. A similar relation connects g_{123} with an integral over g_{1234}, and so on. The advantage of this hierarchy of equations is that the set of distribution functions all refer to the actual high density situation of interest – we are no longer starting from the low density limit. If we can find a self-consistent solution of this set of equations, the high density problem is solved.

The problem of finding an *exact* solution is as intractable as ever, though, and some sort of approximation is necessary. Kirkwood suggested writing $g_{123}\simeq g_{12}g_{23}g_{31}$, known as the *superposition approximation.* It works well at short and long ranges, but how good it is in the crucial medium range of separation is hard to judge. With this approximation (14.29) becomes a self-consistent equation for $g(r)$ which may be solved numerically, giving $g(r)$ as a function of n and T. The equation of state can then be found, using the virial theorem, from (14.8).

Self-consistent equation for $g(r)$

§ 14.2

A number of other integral equations for $g(r)$ have been developed. The *Yvon–Born–Green* equation is similar to Kirkwood's, and also uses the superposition approximation. The *Percus–Yevick* equation was derived in a quite different way, by first translating the problem into collective variables, corresponding to the amplitudes of density waves in the fluid. This is analogous to the treatment of a solid in terms of phonons, and the hope was that the different density wave amplitudes would be less strongly coupled than the position and momentum variables for the molecules, and that this would make suitable approximations easier to recognise: this hope was partly realised. A closely related approach leads to the *hypernetted chain* equation.

All of these approaches involve self-consistent equations for $g(r)$. Since they are self-consistent, they are designed to cope with high densities, and this is reflected in the fact that when the corresponding pressure is expanded as a virial expansion, they are found to contain *some* of the required diagrams to all orders. But none of them contains *all* of the required diagrams and, indeed, they all cease to be exact beyond the third virial coefficient. From this point of view these approaches are various attempts to find an approximately self-consistent integral equation which includes *enough* of the important diagrams to give hope of being a good approximation at high density.

All of these self-consistent integral equations are better than the virial expansion as the high density region is approached. For the 6–12 potential, for instance, they give a reasonable account of the equation of state up to about 0.7 of the solid density. The best of them show errors of order 10% at the solidification pressure. *None* of them convincingly predicts the solid–liquid phase transition, however. Fig. 14.3 gives a good idea of their lack of precision in the high density regime. The diagram shows predictions of $g(r)$ at about twice the critical temperature, and at a density very close to the liquid–solid transition line – in other words, for a very dense system at a relatively high temperature – compared with the known form obtained from a Monte-Carlo calculation. Clearly, the Percus–Yevick theory is the most successful, but it still shows serious errors in the important region between the first and second peaks.

§ 14.8

It may be that approximate integral equations of this type will eventually give a good account of the whole of the equation of state, including the solid state. But at present analytic theories are very far from this goal. We must therefore examine some other approaches to the problem.

14.7 The law of corresponding states

Since at present we have to abandon exact analytic methods in the high density region it is useful to ask whether there are any *general* principles which govern the equation of state. In this section we examine an inexact generalisation which can be very useful.

Suppose we have two different heavy inert gases, for both of which the Lennard–Jones potential is a good approximation. In other words, for the two gases the interatomic potential is the same *shape*, and differs only in *scale*. Let us then adopt a special system of units, based on the atomic mass m, the hard core radius r_0 and the pair binding energy ε. In this system we have the following units:

§ 14.1

Quantity:	*Unit:*	*Quantity:*	*Unit:*
Mass	m	Energy	ε
Length	r_0	Pressure	εr_0^{-3}
Time	$(mr_0^2/\varepsilon)^{\frac{1}{2}}$	Temperature	ε/k

Now, in this special system of units the two gases have the *same* interatomic forces and *same* equations of motion. They would thus, in these units and in the classical limit, be indistinguishable, and must therefore have *the same equation of state*. If P, n and T are written in *ordinary* units, this equation of state would take the form

$$P/P_0 = f(n/n_0, T/T_0), \tag{14.30}$$

where $P_0 = \varepsilon r_0^{-3}$, $n_0 = r_0^{-3}$ and $T_0 = \varepsilon/k$. In other words, *if the equation of state is written in reduced form*, in terms of a characteristic pressure, number density and temperature which are related to r_0 and ε for the gas, *the two equations of state will be identical*. In practice, this conclusion is a useful generalisation for a wide range of substances, and not just for the noble gases: *when expressed in a suitable reduced form the equations of state of all substances are much the same.* This generalisation is known as the *law of corresponding states*. It works because the intermolecular potential curves for almost all substances are at least roughly similar in form, though they differ in scale. Note that the atomic mass does not appear in (14.30). The pressure is, in fact, *always* independent of the atomic mass in classical calculations of the equation of state: for *any* interatomic potential, if, for instance, the mass is multiplied by 4 while n and T are held fixed, the velocities are all halved and the time scale of the collisions is doubled, but the rates of change of momentum and all forces, including the pressure force, are unchanged.

The extent to which the law of corresponding states is obeyed depends partly on how similar the potentials involved are in form, and partly on how sensitive the property of interest is to the details of the potential. Generally speaking, the law works well for gases, and less well as we approach the solid state. For instance, in Fig. 14.4, which was plotted in reduced units, we saw that the values of the second virial coefficient for the range of imperfect gases plotted were impressively consistent. But the law holds also for more complex phenomena which are much harder to analyse.

§ 14.4

In Chapter 18, for instance, we shall discuss the phase separation between liquids and gases. This only occurs at temperatures below a critical temperature T_c and pressures below a critical pressure P_c. At higher temperatures, increasing the pressure moves the system from a low density, gas-like state to a high density, liquid-like state, but the system remains homogeneous, and no separation into phases occurs. At temperatures below T_c, on the other hand, as the volume is reduced a stage is reached where liquid separates from the gas as a phase of greater density. Over a certain range of volumes both phases exist together, gas condensing into liquid as the volume is reduced. A plot against temperature of the densities of the two phases when in mutual equilibrium is known as the *co-existence curve*. The co-existence curve for a number of noble gases and small molecules is shown in reduced units in Fig. 18.11. Once again, the agreement between the different gases is excellent.

p. 278

Table 14.1 gives some idea of the effectiveness of the law of corresponding states for a wider range of molecules. According to the law the four characteristic ratios shown on the left of the table, whose meaning is explained below, should be the same for all substances. We see that for the heavier noble gases the law works very well in all cases. Indeed, it can be applied to them as a group in considering many other properties such as compressibility and expansion coefficient in the solid state, surface tension, and transport properties such as thermal conductivity and viscosity. The law only fails seriously for them in the solid state below the Debye temperature, where quantum effects become important. If we turn to other substances, in the first row we have values of the *ratio of* kT_B, *the energy corresponding to the Boyle temperature, to the binding energy* ε *of the measured 6–12 potential.* The Boyle temperature is the point at which $B_2(T)=0$ (see Fig. 14.4), and is a property of gases at low density and relatively high temperature. It is not particularly sensitive to the details of the potential, and we see that kT_B/ε for the small molecules N_2 and CH_4 agrees well with the theoretical value for the 6–12 potential, which is 3.42. In CO_2 and H_2O the potential is too far from 6–12 form for ε to have much meaning, and I therefore quote no value for the ratio. In He and H_2, on the other hand, the 6–12 potential is quite good, but since these are light particles at low temperatures they are not in the classical limit. It is clear from the table that this has a substantial effect on the value of kT_B/ε.

The second entry in the table is *the ratio of the critical temperature for liquid–gas separation* T_c *to the Boyle temperature* T_B. Here we are comparing a property of the high density critical state (which is very difficult to predict theoretically) with a low density, high temperature property. The nature of the critical state depends more strongly on the shape of the interatomic potential, and in this ratio differences between the effectively spherical molecules and the noble gases are beginning to appear. The non-spherical shape of the CO_2 molecule and the dipole forces in the water molecule have increased T_c relative to T_B, because at lower temperatures these asymmetric molecules will bind

Table 14.1. *Illustrations of the precision of the law of corresponding states (data from a wide range of sources)*

Charac-teristic ratios	Heavy noble gases				Effectively spherical molecules		Longer molecule	Dipole forces	Quantum effects	
	Ne	A	Kr	Xe	N_2	CH_4	CO_2	H_2O	He	H_2
kT_B/ε	3.43	3.42	3.44	3.43	3.43	3.42	—	—	2.3	2.8
T_c/T_B	0.364	0.368	0.361	0.367	0.386	0.37	0.53	0.47	0.23	0.32
$P_c v_c/kT_c$	0.287	0.292	0.291	0.290	0.292	0.290	0.29	0.23	0.30	0.30
T_{trip}/T_c	0.553	0.555	0.554	0.557	0.50	0.48	0.71	0.42	No triple point	0.42

together more strongly than equivalent spherical atoms, as correlations between their orientations become significant. The quantum effects in He and H_2, on the other hand, tend to counteract condensation, and keep T_c relatively low. The third entry is the *critical ratio* $P_c v_c / k T_c$, where v_c is the volume per particle at the critical point. This is still surprisingly uniform for all molecules, though the effects of the dipole forces in H_2O are apparent.

Problem 14.6

When the solid state is involved, there are many obvious respects in which the law fails. We show in the table the ratio of the *triple-point temperature* T_{trip} (the temperature at which solid, liquid and vapour can be in equilibrium together) to the critical temperature T_c. The 'effectively spherical' molecules are no longer effectively spherical at this lower temperature and the asymmetries of CO_2 and H_2O now have a large effect. The quantum effects in helium are so strong that there is *no* triple point, the system remaining liquid down to $T=0$ at normal pressures. The law also obviously fails in so far as solids do *not* all have the cubic close-packed structure of the solid inert gases, and do *not* all expand by 15% on melting. (When ice melts to form water the density *increases*.)

§ 18.1

Clearly, though the law of corresponding states is very useful, and often provides an excellent guide to the behaviour to be expected from a substance, it must be used with some awareness of where it works best and where it is likely to fail.

14.8 Direct computation for small numbers of molecules

Because the analytic theory of dense systems has not so far proved very successful, and itself involves lengthy computation, it is natural to ask whether a powerful computer could be used to calculate the properties of a fluid *directly*. It turns out that this approach works well, though it is expensive in computer time.

Two methods are used. In both we explore the distribution of configurations for a finite number of molecules, subject to some assumed law of force. At present the number of molecules is limited to a few hundred, by restrictions in available computer power. In the *method of molecular dynamics* what is done is in effect to perform an experiment in the computer. We start the molecules in some definite configuration, each with some definite speed. The computer then simply calculates, using an assumed force law, the subsequent trajectory of each molecule. A record of the molecular motions is kept and, by time averaging once equilibrium has been established, quantities such as $g(r)$ and hence the equation of state may be computed. Some trajectories obtained in such a calculation are shown in Fig. 14.6. The method may also be used to compute non-equilibrium properties such as thermal conductivity or viscosity, so long as the time required does not exceed a few hundred collision times.

The method of molecular dynamics

The second approach uses a *Monte-Carlo method* to find the distribution of configurations in equilibrium. We know from the Boltzmann

The Monte-Carlo method

distribution that the probability of any particular configuration at temperature T is proportional to $\exp(-\phi/kT)$, where ϕ is the potential energy of the configuration. At first sight, we should be able to find quantities such as $g(r)$ by simply making the computer take the average over all possible configurations. This cannot be done, however. To get an accurate form for $g(r)$ we need to define the position of each molecule to within at least $\frac{1}{10}$th of the core radius. For 100 molecules at typical density each atom would have at least 10^6 different positions, so 100 atoms would have at least 10^{600} configurations. This is too large a number for even a very large computer to explore. But we are saved by two facts. First, when the density is high almost all the configurations have one or more pairs of cores overlapping, and therefore have negligible probability. Secondly, we do not need to average over all configurations, but only a suitable random sample of configurations – about 10^5 is enough to give effective averaging for most quantities. The Monte-Carlo method is designed to take advantage of these facts. We start with the molecules in some definite configuration, and then proceed to move them, one at a time, through a suitable *randomly chosen* vector displacement, according to the following rule.

Fig. 14.6. Trajectories calculated by the molecular dynamics method (reprinted by permission from Alder & Wainwright, 1959). The photographs show the complete cell: particles leaving on one side are treated as entering on the opposite side. The trajectories are projected onto the x–y plane. The trajectories correspond to the density $V/V_s = 1.525$ in Fig. 14.8, near the lower end of the solid branch of the hard sphere equation of state. The computed system makes occasional changes between a 'solid' state, shown on the left, and a 'fluid' state, shown on the right. Once formed, these states are persistent – both photographs cover 3000 collisions. The system contains 32 particles. (In the solid state the projection used means that they are superimposed in pairs.)

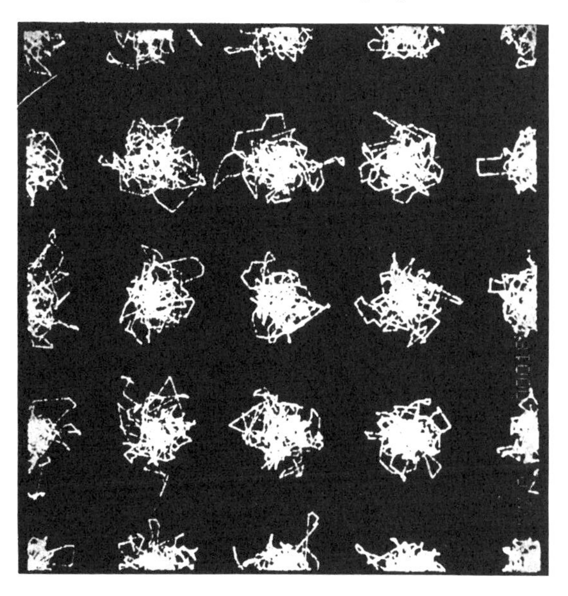

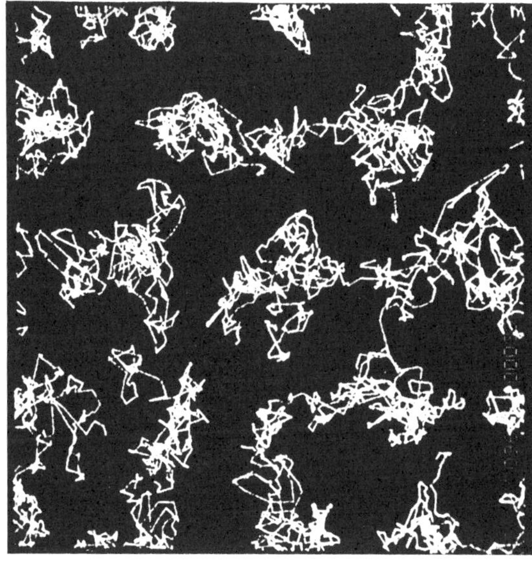

The computer calculates the change in potential energy $\Delta\phi$ which the displacement produces. If $\Delta\phi$ is negative, the jump is allowed to occur. If $\Delta\phi$ is positive, the computer is programmed to make the jump, on a random basis, with probability $\exp(-\Delta\phi/kT)$. (If the computer chooses not to make the upward jump, the calculation is ignored and the computer tries another randomly chosen displacement.) It is not hard to see that *with this rule in operation, the system will reach statistical equilibrium when the probability of each configuration is the required Boltzmann probability*, for in this statistical state we have *detailed balance* – on average every change of configuration will occur as often as the reverse change. Thus the Monte-Carlo method, like the method of molecular dynamics, generates a sequence of configurations over which equilibrium properties such as $g(r)$ can be averaged. If the size of the random displacement vector is suitably chosen, the Monte-Carlo method converges considerably faster than the method of molecular dynamics. But it cannot be used to calculate non-equilibrium properties, or, without considerable modification, equilibrium time-dependent properties such as fluctuation frequencies. The two methods are therefore complementary.

In both methods the limitation to a few hundred particles means that, in a cubic cell, the cell will be less than ten particles wide. At first sight this suggests that *surface effects* might lead to serious errors. If the molecules were treated as being confined within a rigid box, this would certainly be true. To get round this difficulty we imagine that the cell is surrounded not by rigid walls but by copies or 'ghosts' of itself. Thus, when a molecule leaves the cell, an equivalent molecule enters through the opposite boundary. The calculation takes into account forces due to 'ghost' molecules in neighbouring cells. Calculations with cells of different sizes have confirmed that in most situations this device makes the effects of the finite number of particles negligible. (Exceptions to this statement include investigations of melting, where care has to be taken that the lattice of cells chosen neither inhibits nor encourages the formation of the crystalline lattice, and critical point phenomena, where density fluctuations occur on a very large scale.) § 18.9

Generally speaking, both methods have an accuracy of about 1%, at present. They reproduce, using the 6–12 potential, many of the experimental results on the heavy noble gases quite impressively, and, unlike the analytic methods used so far, they predict the solid–liquid transition. However, the question of how far they agree with experiment is rather beside the point: in a sense they *are* experiments, done in the computer on small groups of molecules having rather special interatomic potentials. Enough of such work has been done to establish that such experiments yield consistent results. If they disagree with real argon, it is almost certainly because real argon has different interatomic forces, involving three-body potentials. From this point of view it seems best to treat the computed results as experimental data for simple and precisely known laws of force, against which analytical theories can be tested. Considered as experimental data the computed results can give many quantities (such as high-order correlation functions and velocity correlations) much more accurately than we can hope to measure them on a real system. Considered as a piece of theory, the computed results are less satisfying. They may allow us to confirm that the application of Newton's laws to particles with known forces acting between them *does* lead to an appropriate equation of state – but few physicists doubt that, anyway. They do *not* give any direct insight into *why* a particular equation of state or phase transition occurs. On the other hand, they can be extremely useful in suggesting new approaches, or in confirming or refuting conjectures. For instance, computer models do confirm that in some régimes Kirkwood's superposition approximation is quite sound. They also provide a basis for the rather simple generalisations with which we shall close this chapter.

14.9 The hard sphere fluid

In the *solid* phase a substance such as argon has an obvious and simple *structure* – it forms a regular close-packed crystalline lattice. The reason for this is obvious: the hard core repulsions keep the atoms apart, and at high density the only possible arrangement which can accommodate the atoms within the required volume is the regular crystalline structure. At $T=0$ we may think of the atoms (in the classical limit) as being at rest, and perfectly ordered. But solids may also exist, if enough pressure is applied, at very *high* temperatures. In such states the crystalline order persists, but the hard cores are no longer in contact: each atom rattles around within a cage formed by its neighbours (Fig. 14.6). In both cases we notice that *the structure is due to the hard cores.* Modifying the attractive part of the potential has little effect upon it.

The random close-packed structure

It is perhaps less obvious that a dense fluid also has a structure, and that *this structure is also determined mainly by the hard core interactions.* The liquid structure is in fact very similar to *random close-packing*, as Bernal has emphasised. The idea of random close-packing may be understood in two dimensions by placing a large number of identical coins on a flat table, and pushing them together. They do *not* readily form a hexagonal crystalline close-packed structure, each surrounded by six neighbours. They usually form a more random pattern in which each coin is typically in contact with four or five others, and no coin is free to move (Fig. 14.7). Pursuing this idea in three dimensions, Bernal and others have analysed the structure of aggregates of balls pressed together in rubber balloons. An aggregate in which all balls are held rigidly by their neighbours and there are no voids large enough to contain a ball is said to have *dense random close-packing.* Such an aggregate provides a remarkably good model of the structure of a liquid at low temperatures. For instance, its volume is 1.16 times the close-packed volume: when argon melts at the triple point, its volume increases by a factor of 1.15. The radial distribution function for random close-packing fits well the measured function for dense liquids at low temperatures, showing a mean co-ordination number of about 8 (as opposed to 12 for crystalline close-packing) in good agreement with the measured values. There are other similarities between solids and liquids associated with this picture. Measured over short times, for instance, the atomic motion in a dense liquid at finite temperature is very much like that in the solid, the individual atom rattling around in a cage formed by its neighbours. The rearrangements which give the fluid its fluidity in fact only occur very occasionally (Fig. 14.6). Indeed, it is well established that, when measured over a short enough time interval a dense fluid has a well-defined and large *shear* modulus, as well as being hard to compress. Elastic waves can travel quite easily in dense fluids. At higher temperatures both the solid and the liquid structure develop vacancies or voids, and the increasing motion of the atoms means that *there is no striking difference between the radial distribution functions for the two phases*, as Fig. 14.3 shows: the main difference is a fall in the *number* of atoms at the peaks corresponding to nearest and third-nearest neighbours, but the peaks have only changed their positions slightly. Evidently the break-up of the long-range crystalline order has rather little effect on the short-range structure.

p. 157

These considerations suggest that many important properties of liquids are due almost entirely to the hard cores. Machine calculations support this idea. Monte-Carlo studies show that at high densities $g(r)$ for the 6–12

Diffusion and viscosity are controlled by the hard cores

Fig. 14.7. An example of dense random close-packing in two dimensions. All the atoms are held fixed by their neighbours, but if the volume were slightly increased, atoms A, B and C would be able to make a substantial movement. It is such atoms which allow flow to occur when the packing is not dense.

potential is very similar to $g(r)$ for a fluid of simple hard spheres at an equivalent density. It has also been realised that transport processes such as diffusion and viscous flow in fluids can be explained by a hard sphere model. The rates of such processes increase rapidly with temperature, and this was formerly interpreted as due to thermal activation over an energy barrier. It is now known from experiment and computer studies that this interpretation is not correct: the determining factor is not the temperature but the number of sites in the fluid at which structural rearrangement can occur. It has been found, in fact, that if the ratio of the volume to the effective hard core volume of the fluid is kept constant (by applying enough pressure) the rates of diffusion and viscous flow do *not* increase appreciably with temperature. The rapid increase previously observed was due to *thermal expansion*, which had the effect of dramatically increasing the small number of structural rearrangement sites in the fluid.

§16.6

These discoveries suggested that it might be fruitful to explore theoretically the properties of an idealised fluid of hard spheres, and a good deal of effort has been spent in the required molecular dynamics and Monte-Carlo calculations. In considering the results note that the law of corresponding states takes a different form for hard spheres. Since we have no binding energy ε, there is no natural energy scale in the potential. We are therefore free to use kT as the unit of energy, and this has the effect that *all temperatures become equivalent*: if we multiply T by 4, we double the velocities in ordinary units but, since the unit of time is simultaneously halved, in *reduced* units the velocities, the time between collisions, etc., are all unchanged, and the system appears identical to the original system. Thus formally, on modifying (14.30) in the required manner we find that

§14.7

The equation of state for hard spheres

$$P/n_0kT = f_h(n/n_0, T/T) = f_h(n/n_0), \tag{14.31}$$

and all isotherms are equivalent. A scaling of T simply produces a corresponding scaling of P, and there can be no critical point. In Fig. 14.8 we therefore show a single reduced isotherm. This isotherm is revealing. At large volumes the pressure is proportional to $1/V$, as for an ideal gas. On compression there is no vapour–fluid transition. (This strongly suggests that for systems in which liquefaction does occur it must be a density instability brought about by intermolecular attractions: for hard spheres there are no attractions, and therefore no instability, and no distinction between liquid and vapour.) But there *is* a fluid–solid transition, with two different branches of the equation of state at high density. The solid phase has a regular, close-packed lattice. As the volume approaches the closed-packed volume V_s, the pressure diverges, being proportional to $(V-V_s)^{-1}$ as one would expect.

14.10 The modified hard sphere fluid

We have seen that the hard sphere model gives quite a good account of the liquid–solid transition, the radial distribution function $g(r)$, and the rates of diffusion and viscous flow in dense fluids. However, it does not account at all for the vapour–liquid transition, which is due to the attractions between the molecules. The fact that the hard sphere model *does* give a good account of $g(r)$ suggests a simple modification of the hard sphere approach which might give a good semi-quantitative description of most of the phase diagrams. Let us start with the hard core equation of state, as shown in Fig. 14.8, and ask what happens if we add a relatively weak attractive interaction for $r > r_0$. We shall refer to this model potential as the *van der Waals model*. According to the virial equation of state (14.8) the effect of the attractive potential is simply to add an extra pressure given by

van der Waals' model

§ 14.2

$$P_{\text{att}} = \frac{-n^2}{6} \int_{r_0}^{\infty} 4\pi r^3 f(r) g(r)\, \mathrm{d}r, \tag{14.32}$$

where $f(r)$ is the new attractive force. Let us now assume that the liquid structure and hence $g(r)$ is determined by the hard cores. If this is correct we may insert into (14.32) the radial distribution $g_n(r)$ calculated for the hard core case, which is a function of density only, and can be obtained from the Monte-Carlo calculation. In this approximation $P_{\text{att}} = -n^2 a(n)$ where

$$a(n) = \frac{1}{6} \int_{r_0}^{\infty} 4\pi r^3 f(r) g_n(r)\, \mathrm{d}r. \tag{14.33}$$

Fig. 14.8. The fluid and solid branches of the hard-sphere equation of state at high densities compared with van der Waals' approximation $P = NkT/(V - Nb)$ with $b = r_0^3$, a value nearer to the high density limit than the low density limit. Here V_s is the close-packed volume, $Nr_0^3/\sqrt{2}$. The solid branch diverges at $V = V_s$. The fluid branch appears to diverge at the density of dense random close-packing, $V = 1.16V_s$ (after Wood, 1968).

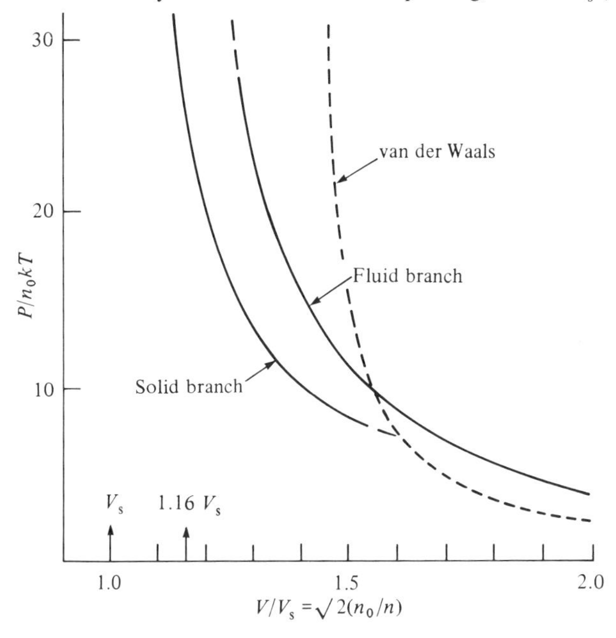

If we add this extra pressure to the hard core pressure we have

▶ $$P = n_0 kT f_h(n/n_0) - n^2 a(n). \qquad (14.34)$$

This is the *modified hard core equation of state*. It gives quite a good account of both the solid–fluid and vapour–liquid transitions and is perhaps the best general equation of state which we have available, though its accuracy is not very high.

See, however, § 18.9

The modified hard core equation of state is closely related to *van der Waals' equation*,

van der Waals' equation

▶ $$(P + N^2 a/V^2)(V - Nb) = NkT, \qquad (14.35)$$

which was proposed in 1873 as a model of the vapour–liquid transition. The simple ideas which underlie it may be seen most easily by rewriting it as

$$P = \frac{NkT}{V - Nb} - n^2 a, \qquad (14.36)$$

a form analogous to the modified hard core equation (14.34). The *first* term, like the first term in (14.34), is meant to represent the hard core equation of state. Van der Waals assumes simply that the system behaves like an ideal gas, except that the volume available is reduced by an *excluded volume* Nb, which is related in some way to the fact that any one molecule is excluded from certain parts of the volume by the hard cores of the others. As Fig. 14.9 shows, this simple picture with a constant value of b is exactly correct in one dimension – and in one dimension there is only a single branch to the hard sphere equation of state, and no discontinuous transition from fluid to solid as the density increases. But, as we may see in Fig. 14.8, the van der Waals hard core approximation is in *three* dimensions a rather poor fit to the fluid branch of the exact equation of state. In fact, the effective value of b *decreases* from $2\pi r_0^3/3$

Fig. 14.9. van der Waals' hard sphere assumption in one dimension. In one dimension the motion of N hard spheres of diameter D along a line of length L is identical with the corresponding motion of N hard point particles along a line of length $L-4D$, with the same spaces between them. The system exerts a force on the boundaries equal to $nkT/(L-ND)$. But this simple idea cannot be extended to three dimensions (see Fig. 14.8).

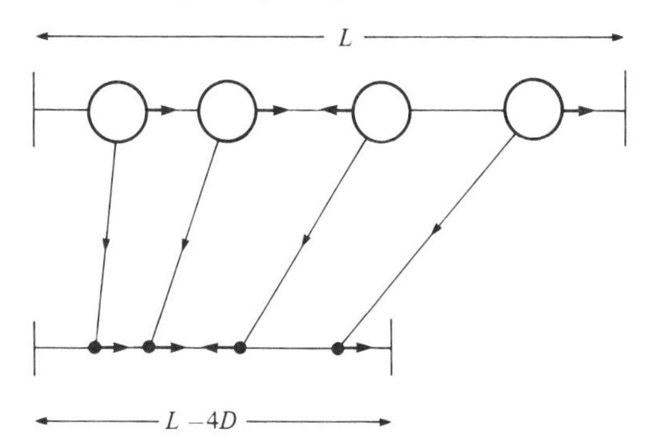

(four times the hard sphere volume) at low densities to $1.16r_0^3/\sqrt{2}$ at the highest densities (the volume per atom in random close-packing). The reason is that at low densities we can safely assume that volume excluded by one atom will not be excluded by another, but at high density the same volume may be excluded by more than one atom at once, so the total volume excluded is less. The *second* term in van der Waals' equation represents a reduction in pressure due to intermolecular attractions. The idea here is that we might expect a lowering in potential energy of the gas of the form $-aN^2/V$, on the basis that each near neighbour lowers the energy by a fixed amount on average, and we have N atoms each with a number of near neighbours proportional to n. When the volume changes the work done in changing this energy will be $(aN^2/V^2)\,\mathrm{d}V$, and if we equate this with the work $-P_{\mathrm{att}}\,\mathrm{d}V$ done against a 'pressure due to attractions' we find that $P_{\mathrm{att}}=-n^2a$. If we compare this term with the second term of (14.34) we see that van der Waals' equation in effect ignores the variation of $g(r)$ with density. Fig. 14.3 shows that this is only a crude approximation. In fact $a(n)$ increases substantially with density, although not as much as one might have expected, for the peaks and troughs in $g(r)$ at high densities tend to cancel one another out in the integral (14.33).

p. 157

The van der Waals isotherms are shown in Fig. 18.4, and may be compared with the real isotherms of argon, shown in Fig. 18.1 (and we shall discuss the van der Waals isotherms more fully when we come to investigate liquefaction). The equation is clearly qualitatively successful. But, because it treats a and b as constants, it does not make accurate quantitative predictions. By making a and b functions of the density we can make van der Waals' equation identical with the more accurate modified hard core equation. It is of some interest to derive the appropriate values of a and b in the low density limit, for instance. For van der Waals' equation the second virial coefficient is given by

p. 259
p. 248
§§ 18.4, 18.5

Values of a and b at low densities

$$B_2(T)=b-a/kT. \tag{14.37}$$

If we apply the general formula for B_2 (14.14) to the van der Waals' model potential, we find that

$$\begin{aligned} B_2(T)&=\int_0^{r_0} 2\pi r^2\,\mathrm{d}r-\int_{r_0}^{\infty} 2\pi r^2(\mathrm{e}^{-u/kT}-1)\,\mathrm{d}r \\ &\simeq 2\pi r_0^3/3-\left[\int_{r_0}^{\infty}-2\pi r^2 u(r)\,\mathrm{d}r\right]\Big/kT, \end{aligned} \tag{14.38}$$

in the approximation $u \ll kT$. Thus we may identify b as $2\pi r_0^3/3$, as we noted earlier, and a as the integral in square brackets, which is positive since u is negative. (This last result agrees with (14.8) if we set $g(r)=1$.) The van der Waals second virial coefficient fits the data on the noble gases quite well, as Fig. 14.4 shows; and for van der Waals' equation $T_i/T_{\mathrm{B}}=2$, whereas the ratio for the 6–12 potential and the noble gases is 1.9. It must be remembered, however, that the values of a and b just quoted are only good approximations

p. 160

in the low density limit, and for other applications different values may be more appropriate.

14.11 Summary

Let us summarise our present understanding of the equation of state. For gases at moderate densities the virial expansion is a precise analytical theory, but it fails to converge at high densities. For fluids at high density the various self-consistent integral equations give a fairly good account of the pressure, but do not give $g(r)$ accurately. Nor do they give a good account of the phase transitions, or even predict the existence of the solid state. Once we know that the crystalline solid state exists, we can give a good account of its properties in terms of phonons, vacancies, etc. The very crude van der Waals equation gives at least a qualitative account of the liquid–vapour transition, and with the help of Monte-Carlo calculations on hard spheres, the modified hard sphere equation gives a reasonable account of both phase transitions. For reasons which will appear later none of these theories works well near the liquid–vapour critical point. The Monte-Carlo and molecular dynamics methods are generally successful. They have pointed to the importance of the hard sphere picture of solids and dense liquids in understanding structure and transport properties as well as the equation of state, but they do not give much insight into why the solid–liquid transition occurs. Thus, in general, though we have a good working knowledge of the equation of state, we are at present far from possessing effective, precise analytical methods which account for the whole of it, even for the simplest systems of noble gases.

§§ 10.4, 9.3.1

§§ 18.9, 18.10

Problems

14.1 Solid argon has a face centred cubic lattice. If the hard core repulsion acts only between nearest neighbours but the van der Waals attraction acts between all atoms, show that the potential energy of the solid with the atoms at rest may be written as $24N\varepsilon[(r_0/r)^{12} - \alpha(r_0/r)^6]$, where α may be written down as a series. (Ignore surface energies.) Deduce that, as Fig. 14.3 shows, the classical equilibrium spacing between nearest neighbours r_s is slightly less than r^*. Discuss the value of the latent heat of sublimation of solid argon at low temperatures, which is 1.28×10^{-20} J atom^{-1}. [Use the data of Fig. 14.1. θ_D for solid argon is 92 K.]

Latent heat of sublimation

14.2 Use the virial theorem (14.3) to show that the mean gravitational energy of an isolated galaxy is equal to minus twice its mean kinetic energy.

14.3 An imperfect gas at low density undergoes a small isothermal expansion. Use a Maxwell relation with (14.15) to find how much heat enters. Deduce that the potential energy of the gas may be written as $-(N^2kT^2/V)\,dB_2/dT$, and confirm this result using (14.14). Show that for a gas obeying van der Waals' equation (14.36) the heat entering in any isothermal expansion is independent of a, and check that its potential energy is equal to $-N^2a/V$.

Potential energies of imperfect gases

14.4 $B_2(T)$ for helium varies as follows:

T/K	10	20	30	40	50	60	70
$B_2/10^{-29}\,\mathrm{m}^3$	−3.9	−0.7	0.4	0.9	1.27	1.48	1.63

Find T_B and T_i for helium. A vessel containing helium at 26 K and 1 atmosphere situated within a much larger evacuated container is ruptured. If no heat is exchanged with the containers, what is the drop in temperature of the helium? (Use question 14.3.)

14.5 Neon has critical pressure 27 atmospheres and critical temperature 44 K. Its vapour pressure P in units of torr is given in terms of its temperature in K by the relation $\log_{10}(P/\mathrm{torr}) = 8.75 - 0.0437(T/K) - 127(T/K)^{-1}$. Rewrite this relation in terms of P/P_c and T/T_c, and then use the law of corresponding states to deduce the vapour pressure of argon at 135 K. For argon $P_c = 48$ atmospheres and $T_c = 151$ K.

14.6 Dimensional analysis shows that when quantum effects are important the law of corresponding states (14.30) must be rewritten in the more general form $P/P_0 = f(n/n_0, T/T_0, \alpha)$ where α is the dimensionless ratio $h/(m\varepsilon r_0^2)^{\frac{1}{2}}$. The ratio kT_c/ε, instead of being the same for all substances which have similar interatomic force laws, then becomes a function of α. With this in mind, use the data of Fig. 14.1 and Table 14.1 to predict T_c for the light isotope of helium, ^{3}He. [The cryogenic properties of ^{3}He were successfully predicted, using arguments of this type, by de Boer in 1949, when there was no reliable theory of quantum fluids, and too little ^{3}He available for measurement.]

de Boer's dimensional predictions for ^{3}He

14.7 Show that the classical partition function Z_h for a gas of hard spheres is equal to $(2\pi mkT/h^2)^{3N/2} I(N,V)/N!$, where $I = \int_a dV_1 dV_2 \ldots dV_N$ and the integral is taken over the available parts of configuration space in which no two spheres overlap. Writing the corresponding free energy, pressure and heat capacity at constant volume as F_h, P_h and C_h, deduce that $C_h = 3k/2$ and that P_h/kT is a function of density only, as in (14.31).

Hard sphere and modified hard sphere partition functions

If we change to a modified hard sphere gas by adding a weak attraction for $r > r_0$, show that the corresponding partition function Z_{mh} may be written to first order in ϕ/kT as $Z_h(1 - \bar{\phi}/kT)$, where $\bar{\phi}(N, V) = \int_a \phi \, dV_1 \, dV_2 \ldots dV_N/I$ and ϕ is the potential energy of the configuration. Deduce that at sufficiently high temperatures $P_{mh} = P_h - d\bar{\phi}/dV$, as in (14.34).

Show that at low densities $I \simeq (V - Nb)^N$, where b is four times the volume of a hard sphere, and that $\bar{\phi} \simeq -N^2 a/V = -\frac{1}{2}Nn \int_{r_0}^{\infty} 4\pi r^2 u(r)\, dr$, and hence obtain van der Waals' equation in the low density, high T limit of the modified hard sphere model.

15

Electric and magnetic systems

I have avoided discussing electric and magnetic systems in any depth up to this point because they present some difficulties of convention, concerning the treatment of the *field energies* of the electric and magnetic fields. These difficulties present no intractable problems of principle – it is simply a question of keeping a clear head. But, since there are more than a dozen possible combinations of convention, many of them used by one author or another, and since even within a single convention it is easy to get confused, I have set out our point of view at some length in the first two sections and the last section of this chapter. The other sections describe two magnetic systems, of some interest in their own right, by way of illustration.

This chapter assumes a working knowledge of the energies of configurations of electric charge and electric currents, and of atomic quantum numbers.

15.1 Electrical energy and electrical work

Any configuration of electrical charges has an *electrostatic field energy*, equal to the work done in assembling the charges when they are initially separated by great distances. The field energy is independent of the order in which the charges are assembled. It may be thought of as stored in space with a density of $\frac{1}{2}\varepsilon_0\mathscr{E}^2$, where $\mathscr{E}$ is the local electric field, as is shown in texts of electromagnetism. We shall use the symbol $\mathscr{F}$ for the field energy.

When we are dealing with the thermodynamics of an electrical material subject to an *externally applied field* $\mathscr{E}_E$, it is necessary to distinguish between the *internal charges*, the charges of the material itself, and whatever *external charges* we suppose to be producing the external field $\mathscr{E}_E$. In general, the medium will be polarised, and the internal charges will generate an *internal field* or *polarisation field* $\mathscr{E}_P$. The total field $\mathscr{E}$ at any point is equal to $\mathscr{E}_E + \mathscr{E}_P$.

In such a situation, we may think of the total field energy as the sum of three parts. We may imagine that we first assemble the internal charges alone, in the required polarised configuration. The work required to do this is the *internal self-energy*, $\mathscr{F}_I$, equal to $\int \frac{1}{2}\varepsilon_0\mathscr{E}_P^2\,\mathrm{d}V$. We then imagine that, quite

The three parts of the field of energy

separately, we assemble the external charges. The work required for this is the *external self-energy*, $\mathscr{F}_E$, equal to $\int \frac{1}{2}\varepsilon_0 \mathscr{E}_E^2 \, dV$. Finally we bring the two distributions into the required relative positions. The work required to do this is the *mutual energy*, $\mathscr{F}_M$. Since the total field energy at the end must be $\int \frac{1}{2}\varepsilon_0(\mathscr{E}_E + \mathscr{E}_P)^2 \, dV$, we see that the mutual energy is equal to $\int \varepsilon_0 \mathscr{E}_E \cdot \mathscr{E}_P \, dV$. We also note that, since the work done in placing a dipole moment $\mathbf{p}$ in a fixed external field $\mathscr{E}_E$ is $-\mathbf{p}\cdot\mathscr{E}_E$, the mutual energy may alternatively be written as $-\sum \mathbf{p}\cdot\mathscr{E}_E$, where the sum is taken over the dipole moments within the polarised dielectric.

We now face a problem of definition. Do we regard these three contributions to the field energy as part of the internal energy of the dielectric, or not? Different authors have adopted different conventions at this point. In this book, however, I have made it clear that I identify the *internal energy* with the *quantum energy*: I regard the possible values of the internal energy E as being the quantum energies E_i of the dielectric system. Now, in quantum mechanics there is a well-defined procedure for handling charged particles which move in an external field. When we write down Schrödinger's equation we *include* in the Hamiltonian both their electrostatic interaction energy (the internal self-energy) and their potential energy in the external field (the mutual energy), but we do *not* include the self-energy of the external field. Note that the mutual energy has to be included because the particles exchange energy with the external field as they move, and if we omitted the mutual energy the state energy E_i of the particles would not be well defined. Consequently we shall in this book regard E as *including* the parts $\mathscr{F}_I$ and $\mathscr{F}_M$ but *not* the part $\mathscr{F}_E$ of the electrostatic field energy.

With this definition of E, what do we mean by the *external work* done on the system dw? It is perhaps natural to think of 'the work' as the work done on the polarisation charges by the field. But, as Fig. 15.1 makes clear, if we regard $\mathscr{F}_I$ and $\mathscr{F}_M$ as contributing to E, the work done on the polarisation charges by the field does not represent *external* work – it represents a transfer from one form of internal energy to another. We noted, however, that the mutual energy $\mathscr{F}_M$ may be written as $-\sum \mathbf{p}_i \cdot \mathscr{E}_E$. Thus in general any *change* in $\mathscr{F}_M$ may be written as $-\sum d\mathbf{p}_i \cdot \mathscr{E}_E - \sum \mathbf{p}_i \cdot d\mathscr{E}_E$. These two terms represent the two different ways in which energy can be stored in the mutual field. The first term is the work done by the internal forces of the dielectric medium when the polarisation charges move in the external field. It follows by subtraction that the second term represents the work done by the *observer* when, in changing $\mathscr{E}_E$, he has to move the free charges in the polarisation field. Since the mutual energy is part of the system but the observer is not, this term *does* represent external work done on the system. It is marked with an asterisk in Fig. 15.1, which makes clear that this is the *only* point at which external electrical work can be done on the system as here defined. So we must identify dw by writing

▶ $$dw = -\sum \mathbf{p}_i \cdot d\mathscr{E}_E \tag{15.1}$$

for the dielectric system. Note that, with the definitions which we have used, $\mathscr{E}_E$ is a *constraint* in the sense of Chapters 1–4: it can be held fixed by the observer, and when it is held fixed the quantum energies of the system are well defined. The above expression is the electrical analogue of the mechanical work term $dw = -P\,dV$.

15.2 Magnetic energy and magnetic work

I shall take the view that *all* magnetic moments and magnetic fields are due to circulating currents. (This raises a question concerning the treatment of *spin* moments to which we shall return later.) Any configuration of circulating currents has a *magnetic field energy*, which is the work which we have to do on the currents if we build them up *in situ* from zero. (The work is

§ 15.3

Fig. 15.1. Energy flux in electric and magnetic systems, showing the conventions used in defining which parts of the field energy are to be regarded as inside the system. In each case, work can only be done on or by the 'system' at the point marked with an asterisk.

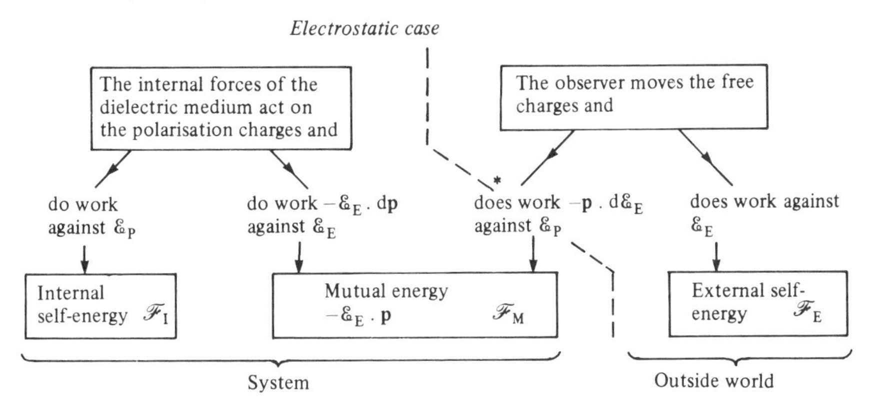

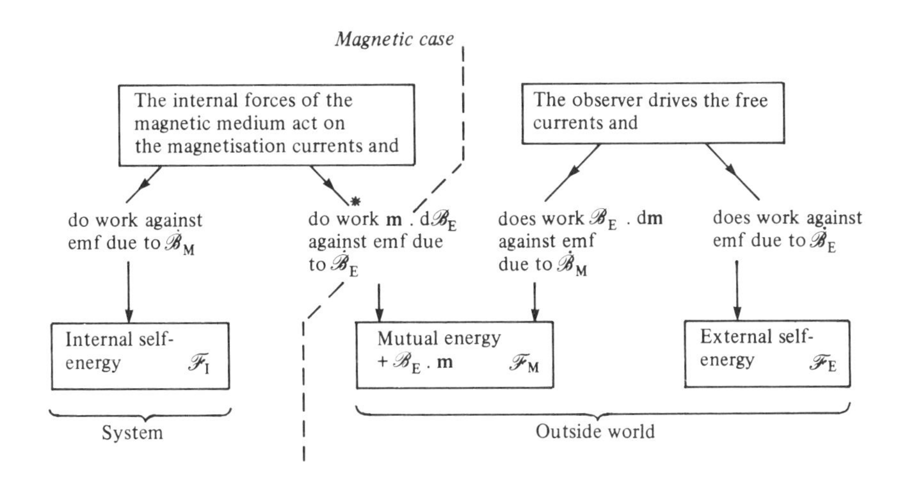

done against the Faraday back emf which acts in the current loops while the magnetic fields are changing.) As in the electrical case we distinguish between the *magnetisation field* $\mathscr{B}_M$ due to the internal *magnetisation currents*, and the *external applied field* $\mathscr{B}_E$, which we suppose to be due to *external currents* flowing in some convenient magnetisation coil. The total field energy may be thought of as stored in space with a density $\frac{1}{2}\mathscr{B}^2/\mu_0$, where $\mathscr{B}$ is the total field $\mathscr{B}_M + \mathscr{B}_E$.

Once again, we have an *internal self-energy* $\mathscr{F}_I$ equal to $\int (\mathscr{B}_M^2/2\mu_0)\,dV$. The external currents have an equivalent *external self energy* $\mathscr{F}_E$, and there is a *mutual energy* $\mathscr{F}_M$ equal to $\int (\mathscr{B}_E \cdot \mathscr{B}_M/\mu_0)\,dV$. In this magnetic case we may rewrite the mutual energy in a convenient form by supposing that we establish the magnetisation current *first*. If we subsequently build up the external current there will be a back emf within each loop of magnetisation current equal to $\mathbf{S}\cdot(d\mathscr{B}_E/dt)$, where $\mathbf{S}$ is the area of the loop. The mutual energy is then the total work which we have to do in maintaining the magnetisation currents in the face of this emf while $\mathscr{B}_E$ is changing, which is $\sum I\mathbf{S}\cdot\mathscr{B}_E$, or $+\sum \mathbf{m}\cdot\mathscr{B}_E$, where I is the current and $\mathbf{m}$ is the magnetic moment of the loop, and we sum over all loops. (Note that this formula has the opposite sign to the equivalent electrical one.)

The three parts of the magnetic field energy

The procedure for writing down Schrödinger's equation in the presence of magnetic fields may be unfamiliar, but the conclusion is not surprising. The magnetic Hamiltonian contains a term corresponding to the internal magnetic self-energy *but no term corresponding to the mutual energy*. The reason is fairly obvious. So long as the applied field is constant, it generates no back emf in the magnetisation loops and therefore there can be no energy exchange between the magnetisation currents and the mutual energy. (The mutual energy does change, but the work is done on the external currents, and the energy exchange is between the *observer* and the mutual energy, not the *medium* and the mutual energy.) To put it another way, there can be no energy exchange between moving charges and a fixed *magnetic* field, because the magnetic force acting on any charge is always perpendicular to its velocity. Thus in a magnetic field, the quantum energies of the magnetic material do *not* include the mutual energy, and we shall accordingly include only $\mathscr{F}_I$, and not $\mathscr{F}_M$ or $\mathscr{F}_E$ in our definition of E.

The definition of E

If we adopt this definition of E, then external work can only be done when energy is exchanged between the medium and the mutual field (Fig. 15.1). This occurs when the external field changes, and there is a back emf in the magnetisation loops. Since dw is the work done *on* the loops *by* the field, we have $dw = -d\mathscr{F}_M$, or, since $\mathscr{F}_M = \sum \mathbf{m}\cdot\mathscr{B}_E$,

$$dw = -\sum \mathbf{m}\cdot d\mathscr{B}_E. \tag{15.2}$$

Note that, although derived quite differently, this formula is analogous to the electrical one (15.1), and $\mathscr{B}_E$, like $\mathscr{E}_E$, is a true constraint.

(It may be helpful to mention here that a different formula for the work done appears in texts of electromagnetism. Written in the usual electromagnetic notation for fields it states that $dw = \int (\mathbf{E} \cdot d\mathbf{D} + \mathbf{H} \cdot dB)\, dV$. This is a formula for the total work done by the observer on the free charges and currents, which is not what we mean by dw in this book. It is sometimes, loosely, said to represent a change in 'field energy'. More accurately, since it represents the external work done on a system consisting of all the fields *plus all the media*, it should properly be regarded as a change in the *free* energy of that entire system, provided the change is made reversibly at constant temperature. For other changes it does not represent a function of state at all.)

15.3 Spin systems

The paramagnetic moments of atoms, ions and nuclei

Many atoms and ions have a permanent paramagnetic moment, whose magnitude is usually written as $g\mu_B[J(J+1)]^{\frac{1}{2}}$, where μ_B is the *Bohr magneton*, an elementary magnetic moment equal to $e\hbar/2m_e$, g is a constant lying between 1 and 2 known as the *Landé splitting factor*, and J is the quantum number for total angular momentum. (The same is true for many nuclei, except that μ_B is replaced by the smaller *nuclear magneton* $\mu_N = e\hbar/2m_p$, and J is replaced by the nuclear spin quantum number I.) In the presence of a magnetic field $\mathscr{B}$ in the z direction the paramagnetic moment can only lie at certain angles to the field, so that the component of magnetic moment in the z direction takes one of the quantised values

$$m_z = M_J g \mu_B, \quad \text{where } -J \leqslant M_J \leqslant J. \tag{15.3}$$

The internal energy of a magnetic atom in an external field

We need to know the internal energy of the atom when its permanent moment lies at some definite angle to the applied field. The internal energy of the atom changes when the flux linked with its circulating current changes: in such a situation the back emf does work on the circulating current, and this work is stored inside the atom. The linked flux can change in two ways. If the moment is at rest and the external field changes, the mutual field does work on the circulating current which, as we have seen, is equal to $-\mathbf{m} \cdot d\mathscr{B}_E$. If, on the other hand, the external field is fixed, and some external couple *rotates* the dipole in the magnetic field, this external couple does work equal to $-d\mathbf{m} \cdot \mathscr{B}_E$. A moment's thought shows that this work is stored inside the atom and not as field energy: the external field is constant, so the mutual field has done no net work on the charges, and $-d\mathbf{m} \cdot \mathscr{B}_E$ is easily seen to be the work done on the atom by the back emfs as the linked flux changes when the moment rotates. Thus in this process the mechanical work done by the external couple is converted *directly* into internal energy of the atom. Combining these two contributions we may write for the total change in the internal energy of the atom $dE = -d(\mathbf{m} \cdot \mathscr{B}_E)$, and it follows that in an external field an atom of fixed moment has an extra energy

$$\Delta E = -m_z \mathscr{B}, \tag{15.4}$$

if $\mathscr{B}$ is the applied field in the z direction. (It is worth reemphasising that, unlike the equivalent electrical term, this is *not* a mutual field energy – and has the opposite sign to the mutual energy. It is energy stored inside the atom. In a simple Bohr model, for instance, it takes the form of a kinetic energy term, associated with the Larmor precession of the atom in a magnetic field. For a spinning electron it can be identified as a change in rest mass.)

Problem 15.3

We discussed the magnetisation of a simple paramagnetic salt having $J=S=\frac{1}{2}$ and $g=2$ in §9.2.1, using the partition function. To vary the approach, let us now calculate the mean magnetisation of a single paramagnetic ion in the general case by using the Boltzmann distribution. If we calculate $\bar{m}$ as $\sum m_{M_J} p_{M_J}$ we find that

The mean magnetisation of a paramagnetic ion in a given field

$$\bar{m}=\sum_{M_J} M_J g\mu_B \, e^{M_J g\mu_B \mathscr{B}/kT} \Big/ \sum_{M_J} e^{M_J g\mu_B \mathscr{B}/kT}.$$

With some algebraic effort this may be rewritten as

$$\bar{m}=gJ\mu_B\left[\frac{2J+1}{2J}\coth\left(\frac{2J+1}{2J}x\right)-\frac{1}{2J}\coth\left(\frac{x}{2J}\right)\right], \tag{15.5}$$

Problem 15.4

where $x=Jg\mu_B\mathscr{B}/kT$. The expression in square brackets is the *Brillouin function* of order J, and represents a generalisation of the tanh x of (9.14). It is similar to the tanh function, tending to unity for large x, and to $(J+1)x/3J$ for small x. Thus for small values of applied field *Curie's law*, Equation (9.15), still stands, if we identify m_0^2 as $g^2\mu_B^2 J(J+1)$. Some comparisons of the Brillouin theory with measurements on dilute paramagnetic salts at low temperatures are shown in Fig. 15.2. The theory clearly works very well.

§9.2.1

15.4 Cooling by adiabatic demagnetisation

We have already noted that a paramagnetic salt may be substantially cooled by reversibly removing a strong magnetic field. This is an example of the *magnetocaloric effect*. In any thermally isolated system the change of temperature with applied field $\mathscr{B}_E$ is given by

§9.2.1

$$\begin{aligned}(\partial T/\partial \mathscr{B})_S &= -(\partial T/\partial S)_{\mathscr{B}}(\partial S/\partial \mathscr{B})_T \\ &= -(T/C_{\mathscr{B}})(\partial M/\partial T)_{\mathscr{B}}.\end{aligned} \tag{15.6}$$

The magnetocaloric effect

Here $\mathscr{B}$ is the component of field in the z direction and we are, for convenience, assuming a geometry in which the magnetisation field is zero, so that we can write $\mathscr{B}_E=\mathscr{B}$. The total magnetic moment of the system is M, which we assume to be in the z direction, and we have used the Maxwell relation $(\partial S/\partial \mathscr{B})_T = (\partial M/\partial T)_{\mathscr{B}}$, the analogue of (8.6). This result shows that *in any system whose magnetisability varies with temperature we can expect to see a change of temperature when the field is altered.*

§8.2

To see how this works in a paramagnetic salt, we show in Fig. 15.3 a plot of the magnetic entropy as a function of temperature for several different fields. (The diagram is based on Equation (9.16) but the behaviour for values of J other than $\frac{1}{2}$ is very similar.) A term proportional to T^3 has been added to represent the entropy due to lattice vibrations at low temperatures, on the Debye model. When a large field is removed reversibly and the salt is thermally

§10.4

isolated, S remains constant, as shown in the diagram. Notice that, if we start at a point such as A, the cooling when the field is completely removed is not very large. But if we start at a point such as B where the initial value of S is less than $k \ln 2$, the system should cool, on this simple model, to absolute zero. For reasons which we shall examine below, this cannot actually occur, but the

Fig. 15.2. Magnetisation of three dilute paramagnetic salts in large applied fields at low temperatures compared with the corresponding Brillouin functions (after Henry 1952). The samples were spheres of $CrK(SO_4)_2 \cdot 12H_2O$, $FeNH_4(SO_4)_2 \cdot 12H_2O$ and $Gd_2(SO_4)_3 \cdot 8H_2O$, three salts commonly used for cooling by adiabatic demagnetisation. Note that data for four different temperatures fit the same curve for each salt. $L=0$ for Gd^{3+} and Fe^{3+}, while for Cr^{3+} the orbital angular momentum is quenched by crystal field effects, so for each of these ions in effect $J=S$ and $g=2$.

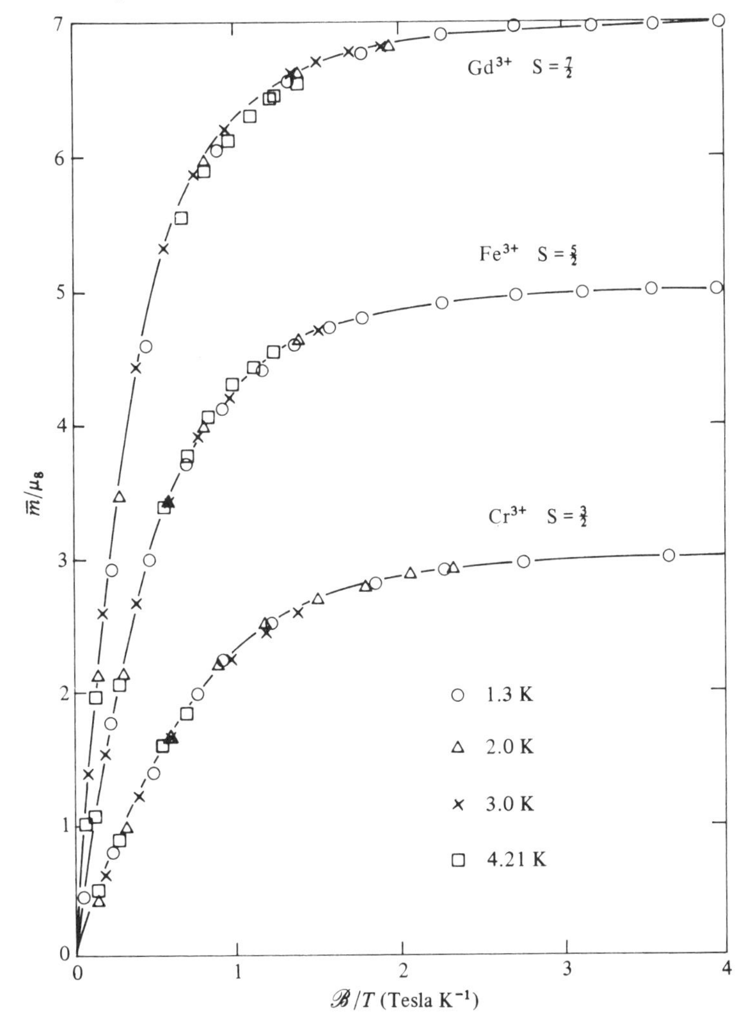

cooling is nevertheless very effective. To reach a starting position such as B requires a large magnetic field and a low starting temperature: in a practical apparatus the starting temperature may be about 1 K (produced by cooling in liquid helium which is evaporating at reduced pressure) and the starting field about 1 T.

In fact, *we cannot in any system reach $T=0$ in any finite set of processes*. The argument leading to this conclusion is based on the third law. If some finite process for reaching $T=0$ existed, the final stage would have to be some operation on a thermally isolated system, which would be most efficient if reversible. In other words, the final step would have to be some operation in which S was constant. But at any finite temperature it is obvious from its definition that the canonical entropy must be non-zero. The third law, on the other hand, requires that $S \to 0$ as $T \to 0$ for all states. Thus there cannot be an adiabatic which takes any system from a finite temperature directly to $T=0$, and therefore no finite set of processes leading to $T=0$ can exist.

The inaccessibility of absolute zero

§4.3

It follows that there must be an error in our treatment of the paramagnetic salt. In fact, the error is obvious: Fig. 15.3 is inconsistent with the third law – the curve for $\mathcal{B}=0$ as drawn has a range of non-zero entropies at $T=0$. This error has arisen because we have assumed that the spins interact with nothing but the applied field. As we noted in discussing the third law, this cannot be correct at sufficiently low temperatures. In fact, each spin must eventually become governed by weak internal fields due, for instance, to the

§4.5

Fig. 15.3. Schematic plot of $S(T)$ for a paramagnetic salt at low temperatures in various applied fields $\mathcal{B}$, including the magnetic entropy and the lattice entropy. The horizontal lines starting at A and B represent adiabatic demagnetisation from the maximum available field $\mathcal{B}_{max}$ at two different starting temperatures. It appears that demagnetisation from B leads to $T=0$, but there are always residual internal fields which prevent this. For a residual internal field equivalent to $0.02\mathcal{B}_{max}$, for instance, the final temperature is T_f. The dashed line represents the lattice entropy alone.

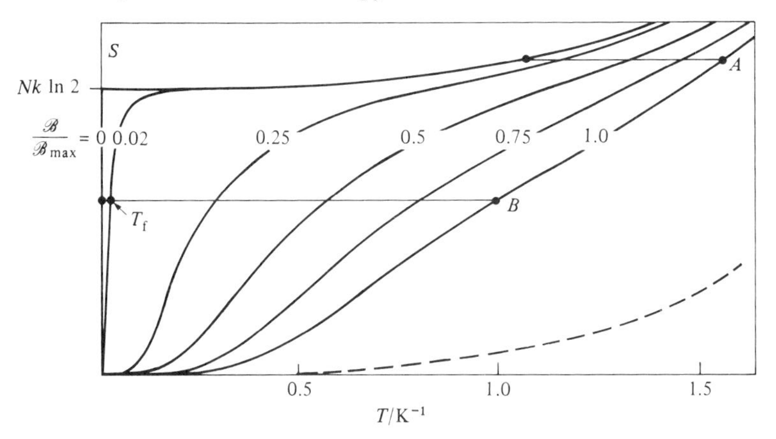

magnetic effects of its neighbours, anisotropic electric fields of the lattice or interaction with the spins of the nuclei. At the lowest temperatures the system is always subject to some such effective residual internal field which splits the levels with different M_J, and ensures that the entropy tends to zero as T tends to zero. The magnitude of this effective residual field, which cannot be removed, may be of order 5×10^{-3} T in a good salt. (A salt such as $Ce_2Mg_3(NO_3)_{12}\cdot24H_2O$ is particularly useful, for the Ce^{3+} ion has $J=\frac{5}{2}$ and therefore has a usefully large saturated spin entropy of $k\ln 6$, while the ions are widely spaced in the lattice, which keeps the residual field low.) Thus the plot of $S(T)$ is not a step function at $T=0$, but more like the curve for $\mathscr{B}=0.02\mathscr{B}_{max}$ in the diagram, drawn for a field equal to the residual field. By demagnetising cerium magnesium nitrate from 1.2 K from a field of 1.5 T a temperature of about 4 mK can be reached.

The limiting temperature

Having reached such low temperatures, we have the problem of measuring them. A simple method of doing so is to measure the magnetisability of a dilute salt such as cerium magnesium nitrate in a very small field. This is the basis of the CMN thermometer, commonly used at temperatures below 0.3 K. If Curie's law holds, the magnetisability should be proportional to $1/T$: temperatures measured in this way are designated T^*. But we have no guarantee that Curie's law holds, and indeed we expect it to fail at the lowest temperatures, so T^* will not be perfectly accurate, and we need some *absolute* measure of T. A way of getting one is to measure the heat capacity of the salt in T^* units, as

§9.2.1

Absolute measurement of very low temperatures

$$C^*=T(\partial S/\partial T^*)_{\mathscr{B}=0}, \tag{15.7}$$

by adding a small amount of heat and measuring the rise in T^*. We then aim to determine $\partial S/\partial T^*$ separately, and so find T from the ratio. To find $(\partial S/\partial T^*)_{\mathscr{B}=0}$ we proceed as follows. We first measure the magnetisability of the salt over a range of relatively high temperatures near 1 K which can be established by gas thermometry. This gives us M as a function of $\mathscr{B}$ and T, from which we can determine $(\partial M/\partial T)_{\mathscr{B}}$ as a function of $\mathscr{B}$ at some convenient known temperature T_0. Then, using the Maxwell relation

$$(\partial S/\partial\mathscr{B})_T=(\partial M/\partial T)_{\mathscr{B}},$$

we note that we have obtained $(\partial S/\partial\mathscr{B})_T$ as a function of $\mathscr{B}$ at the known temperature T_0. By integrating with respect to $\mathscr{B}$ we may thus determine S as a function of $\mathscr{B}$ at the same temperature, apart from an arbitrary constant. We now perform a series of adiabatic demagnetisations from known fields at temperature T_0, recording the apparent temperature T^* reached in zero field in each case. Since S is constant in an adiabatic demagnetisation and we know the starting value of S (apart from a constant) we may deduce S as a function of T^* in zero field (apart from the same constant). We then deduce $(\partial S/\partial T^*)_{\mathscr{B}=0}$, and hence T from (15.7). This method is not easy – it requires precise measurements of four types: low temperature susceptibility measurements to determine T^*, heat capacity measurement at low temperature to find C^*, susceptibility measurements at a range of known temperatures near T_0 to determine $M(\mathscr{B}, T)$ and hence $S(\mathscr{B})_{T_0}$, and finally adiabatic demagnetisation to determine $S(T^*)_{\mathscr{B}=0}$ at low temperatures. It illustrates the considerable problems in establishing the absolute temperature scale in regions where gas thermometry is not available.

It is possible to reach still lower temperatures by demagnetising *nuclear* spins. The nucleus of copper, for instance, has a spin quantum number $I=\frac{3}{2}$. Because the nuclear magneton is 1840 times smaller than the Bohr magneton, it is much harder to reach the required starting condition; in practice, we require a field of about $5T$ and a starting temperature of about 0.01 K, produced by a separate demagnetisation of a paramagnetic salt in a different part of the apparatus, which cools the copper by thermal conduction. But once the required starting condition has been reached, the final temperature is much improved, for, although the residual field is not much less than in the case of the salt, the corresponding energy, and hence the final temperature, are much smaller. Nuclear demagnetisation can bring the nuclear spin system itself to temperatures of 1 μK or less. It is, however, extremely difficult to get effective thermal contact between the nuclear spins and the lattice vibrations at these temperatures, and there are very few experiments in which the sample *as a whole* has been brought within range of 1 μK.

Nuclear demagnetisation

15.5 The superconducting critical field

A number of metals become *superconducting* at temperatures below 20 K, and this phenomenon provides an interesting illustration of the thermodynamic behaviour of a magnetic system.

For simplicity, let us imagine that we have a long, uniform rod of superconductor, parallel to a uniform applied field (Fig. 15.4). In this geometry the fields are all parallel and uniform and we may drop the vector notation. The rod will, in general, be uniformly magnetised with magnetisation $\mathscr{M}$ measured in the direction of the applied field $\mathscr{B}_E$, associated with a magnetisation current which consists of a surface sheet of current flowing around the rod. The magnetisation field $\mathscr{B}_M$ will be equal to $\mu_0\mathscr{M}$ inside the rod, and zero outside.

A superconductor is a *perfect conductor*: it carries current without resistance. Using Faraday's law one can show that in a perfect conductor magnetic fields are *perfectly screened*: if we attempt to change the magnetic field inside the system persistent eddy currents are immediately set up which prevent the field from changing. Thus, if we start with the rod in zero field, and increase the applied field, current will circulate on the surface of the rod which ensures that $\mathscr{B}$ remains zero inside. This implies that $\mathscr{B}_M = -\mathscr{B}_E$, and hence that

The diamagnetic behaviour

$$\mathscr{M} = -\mathscr{B}_E/\mu_0. \tag{15.8}$$

The magnetisation opposes the applied field: the superconductor behaves as a *perfect diamagnet* (Fig. 15.4).

In many superconductors the perfect diamagnetism persists up to the *thermodynamic critical field* $\mathcal{B}_c$, and at this field the superconducting state collapses, the screening currents disappear and the magnetic field enters the metal. At higher fields the material behaves like normal metal, showing only the usual very weak net paramagnetism (which we shall ignore, for simplicity). This is known as *type I behaviour*. The details of the behaviour at the critical field show, however, that a superconductor is more than an extremely good conductor. In an ordinary conductor the magnetic self-energy lost when screening currents decay appears as heat. But when the superconductor enters the normal state and loses its self-energy heat is *absorbed*, showing that the internal electronic structure of the material has taken up the self-energy and extra heat energy as well. Moreover, the transition is *reversible*. If we reduce the applied field through the critical field $\mathcal{B}_c$, the field is not *trapped* inside the material, as it would have been if the material had simply become highly

Type I transition in a magnetic field

Fig. 15.4. Magnetisation of superconductors. The upper drawing shows the direction in which the screening currents flow on the surface of a type I superconducting rod in a field $\mathcal{B}_E$ applied parallel to the rod. The corresponding magnetisation plot is shown on the left. $\mathcal{B}_c$ is the thermodynamic critical field, and the shaded area is equal to $F_N - F_S$. The lower curve shows the magnetisation plot for an equivalent type II superconductor having the same value of $F_N - F_S$. Both plots refer to ideal materials in which there is no flux pinning, and the magnetisation is reversible.

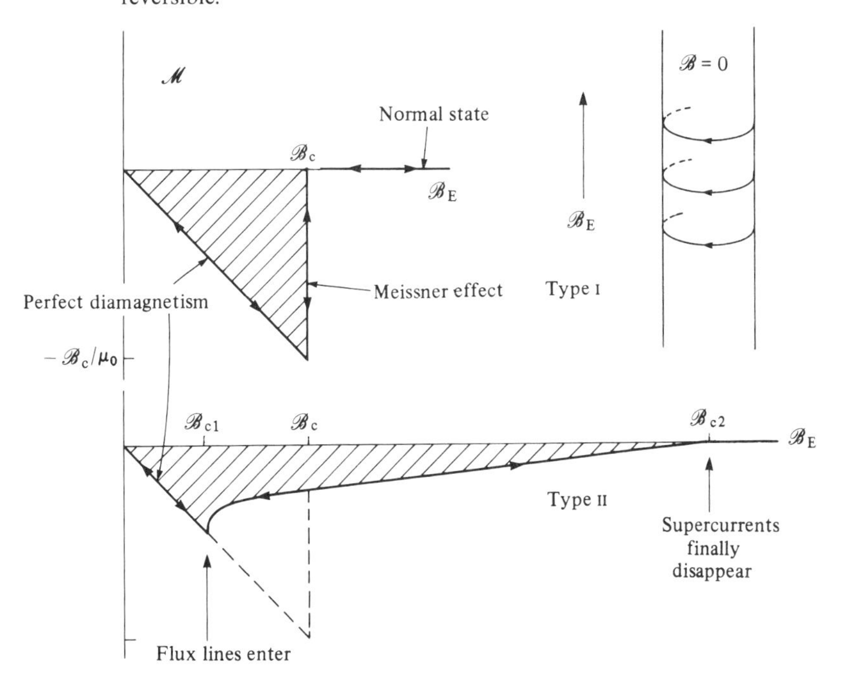

conducting, but *expelled*, so that $\mathscr{B}=0$ inside once again, and the magnetic self-energy reappears. This remarkable phenomenon is called the *Meissner effect*. Evidently there must be strong internal forces which act on the screening currents. At the moment when the flux is expelled the magnetic self-energy rises from zero to $\mathscr{B}_M^2/2\mu_0$ per unit volume. The work done against the back emfs in creating this energy is done by internal forces acting on the screening currents, and is associated with a fall in the non-magnetic part of the internal energy as the superconducting condensation occurs.

The Meissner effect

In fact, the type I superconducting–normal transition at the thermodynamic critical field $\mathscr{B}_c$ is an ordinary, fully reversible *first-order phase transition*, with a definite latent heat. Thus the whole of the upper magnetisation plot shown in Fig. 15.4 is reversible, and we can use this to find the difference $F_N - F_S$ between the free energy in the normal state (which is virtually independent of the field) and the free energy in the superconducting state in zero field $F_S(0)$, which I write as F_S for short. At constant temperature we have

§18.1

$$F_N - F_S = \int_S^N \mathrm{d}w = \int_0^{\mathscr{B}_c} -\mathscr{M}\,\mathrm{d}\mathscr{B}_E$$
$$= \mathscr{B}_c^2/2\mu_0, \qquad (15.9)$$

Problem 15.6

where F_N and F_S refer to unit volume. This quantity is shown as a function of temperature for tin, a typical type I superconductor, in Fig. 15.5. We see that at the *critical temperature*, T_c, the highest temperature at which the system is superconducting in zero field, the two free energies are equal. At lower temperatures $F_S < F_N$.

We can use this result to find the entropy difference between the two states. Since $(\partial F/\partial T)_{\mathscr{B}_E} = -S$ we have

$$S_N - S_S = -(\mathrm{d}/\mathrm{d}T)(\mathscr{B}_c^2/2\mu_0). \qquad (15.10)$$

The measured entropies S_N and S_S are also plotted in Fig. 15.5, and they agree with this prediction. Note that $S_N - S_S$ vanishes, like $F_N - F_S$, at T_c. It also vanishes at $T=0$, as required by the third law. We know that a Maxwell relation requires that $(\partial S/\partial \mathscr{B}_E)_T = (\partial \mathscr{M}/\partial T)_{\mathscr{B}_E}$. In a type I superconductor the right-hand side of this relation is zero, so we may deduce that S_S is independent of $\mathscr{B}_E$. Thus (15.10) holds right up to the transition field, and tells us the entropy change and hence the latent heat of the superconducting–normal transition: the latent heat is positive, as we noted earlier. (Equation (15.10) is the magnetic analogue of the Clausius–Clapeyron equation.)

§4.5

Latent heat

§8.3.1

By differentiating the expression again we can find the difference in heat capacities. Since $C = T(\partial S/\partial T)_{\mathscr{B}_E}$ we have

$$C_N - C_S = -T(\mathrm{d}^2/\mathrm{d}T^2)(\mathscr{B}_c^2/2\mu_0). \qquad (15.11)$$

The measured heat capacities for tin are also shown in Fig. 15.5. Note that at the lowest temperatures $C_N \propto T$, as usual for a condensed Fermi system, but C_S is very small (in fact, because an energy gap appears in the electronic density of

§12.4

states at the Fermi level in a superconductor, and when this gap is larger than kT thermal excitation is quenched). At higher temperatures, however, $C_S > C_N$, and there is a *discontinuity* in the heat capacity in zero field at T_c.

The heat capacity anomaly at T_c: see also § 18.9

Not all superconductors show type I behaviour. The type II superconductors have a different type of magnetisation curve, shown in Fig. 15.4. In low fields they exhibit the same perfect diamagnetism as the type I materials, but at a definite *lower critical field* $\mathscr{B}_{c1}$ flux enters the material and $\mathscr{M}$ rises sharply at first, and then more gradually as the applied field continues to rise. At the *upper critical field* $\mathscr{B}_{c2}$ the magnetisation tends smoothly to zero. At this point the property of perfect conductivity vanishes. It is known that this behaviour is due to the entry of *quantised flux lines* into the material at $\mathscr{B}_{c1}$.

Type II behaviour in an applied field

Fig. 15.5. Thermodynamic properties of superconducting tin. The quantities $\mathscr{B}_c$, $F_N - F_S$, $S_N - S_S$ and $C_N - C_S$ are closely related, as is explained in the text. (For F, S and C the figures refer to 1 mole of tin, which occupies a volume of 1.63×10^{-5} m^3.)

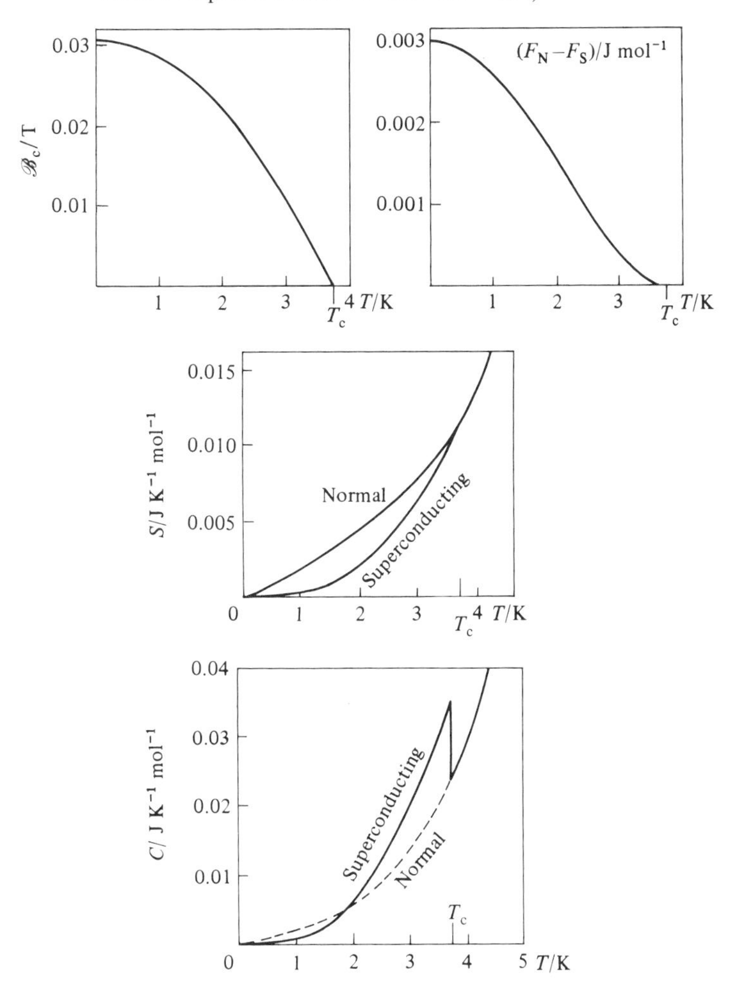

A quantised flux line is a vortex of supercurrent which encloses a normal core and contains a thread of magnetic flux of magnitude $\Phi_0 = h/2e$, running parallel to the axis of the vortex. The flux line has a definite positive free energy per unit length f, which is the work needed to drive the core normal and to create the self-energy of the core field. But when we consider the interaction between the vortex currents and the screening currents flowing on the surface of the superconductor we find that their mutual energy contributes $-\Phi_0\mathscr{B}_E/\mu_0$ per unit length to the internal energy of the system. (Since both of these currents are magnetisation currents their mutual energy is part of the internal energy of the superconductor.) Thus if $\mathscr{B}_E > \mu_0 f/\Phi_0$ it is energetically favourable for flux lines to enter the superconductor, and this is what occurs at $\mathscr{B}_{c1} = \mu_0 f/\Phi_0$.

In a carefully prepared type II material the magnetisation curve is reversible, or almost so, so we may still determine ΔF as

$$F_N - F_S = -\int_0^{\mathscr{B}_{c2}} \dot{\mathscr{M}}\,d\mathscr{B}_E, \tag{15.12}$$

and in a formal sense we may still equate this to $\mathscr{B}_c^2/2\mu_0$, if we interpret $\mathscr{B}_c$ as the critical field the system *would* have had if it had been of type I. This helps us to understand the distinction between type I and type II materials. The material will be of type II if $\mathscr{B}_{c1}$, defined as $\mu_0 f/\Phi_0$, is less than $\mathscr{B}_c$, defined as $[2\mu_0(F_N - F_S)]^{\frac{1}{2}}$, so that flux lines enter before the thermodynamic critical field is reached. We also notice that if the material is *extreme* type II, so that $\mathscr{B}_{c1} \ll \mathscr{B}_c$, we expect to find that $\mathscr{B}_{c2} \gg \mathscr{B}_c$, for otherwise it will not be possible to have the right area above the magnetisation curve. This has important technical consequences. For large electromagnets superconducting windings have the great advantage that maintaining the field costs us no power. But $F_N - F_S$ is never very large, and the greatest known value of $\mathscr{B}_c$ is only about 0.2 T. However, by using an extreme type II material we can arrange that $\mathscr{B}_{c2}$ is of order 20 T. It is thus possible to wind superconducting coils which will operate very well in fields of 15 T – much larger than can be obtained in conventional coils without burning them out, and much larger than the thermodynamic critical field.

Low $\mathscr{B}_{c1}$ implies high $\mathscr{B}_{c2}$

High field superconducting magnets

15.6 The electrochemical potential

Ordinary dielectric insulating materials whose charges cannot leave the medium may be treated straightforwardly using the ideas of § 15.1, and we shall not discuss them further. (Some related problems appear at the end of the chapter.) But the case of freely moving charges needs more attention. We know that when the particles are free to move the chemical potential μ must be equalised in all parts of the system in equilibrium. What does this mean when we are dealing with free *charged* particles such as the electrons in a metal or semi-conductor?

Problems 15.1, 15.2

We notice at once that the chemical potential μ, which in this context is more often referred to as the *electrochemical potential*, includes a contribution due to the electrostatic energy of the electron. Consider the metal shown schematically in Fig. 15.6(*a*). The heavy line shows the electrostatic potential energy as a function of position. Inside the metal this is not independent of position but oscillates as we pass the individual positive ion cores. The horizontal line marks the energy of an electron at the Fermi level. (We assume T to be low so that we may identify ε_F with μ.) The *total* energy of

Fig. 15.6. Work function and contact potential. (*a*) Energy diagram for electrons near the edge of a metal. (*b*) Metals *A* and *B* uncharged and at a distance. (*c*) Metals *A* and *B* in contact: electrons transferred from *A* to *B* set up a large field in the few Å which separate them, bringing the Fermi levels into coincidence. (*d*) The same, showing the surface charge density and electric field outside the metals.

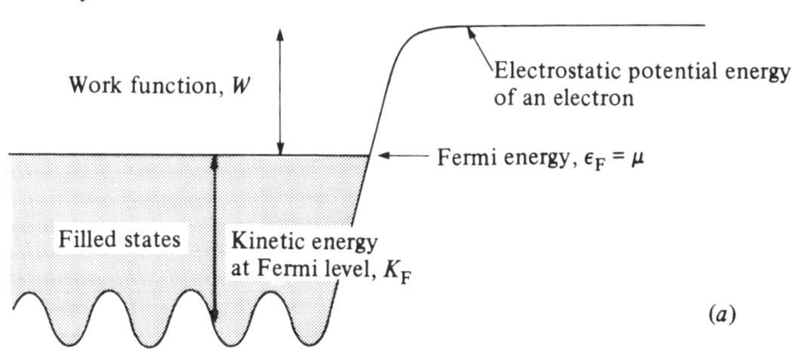

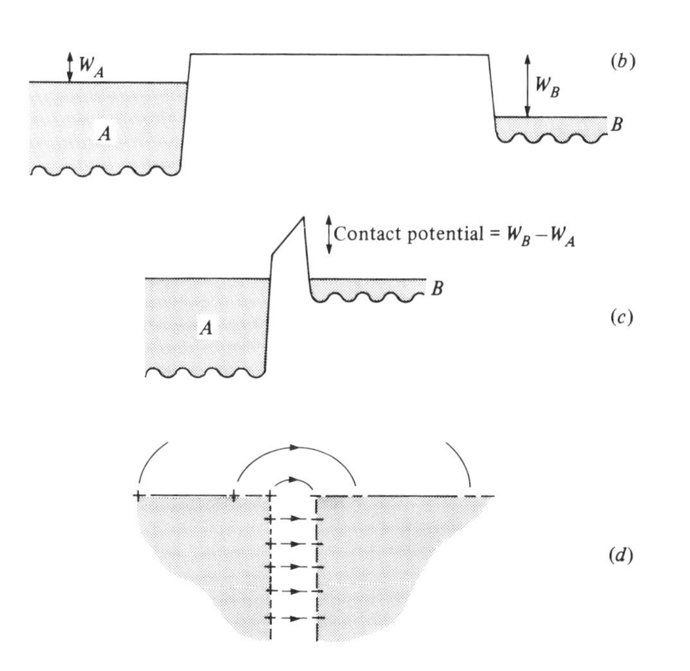

this electron is constant, but the potential and kinetic parts of its energy vary with position. At the metal boundary the potential energy of the electron rises rapidly by several eV. (This is due to two effects. First, because the electron wave functions extend a short distance outside the metal, there is a permanent electrical double layer at the surface with the negative charge on the outside: this generates a local field which drives electrons inwards. Second, each electron in the metal drives other electrons away from it, and this leaves each electron surrounded by a net positive charge due to unscreened ion cores, which lowers its energy inside the metal. When the electron is removed from the metal, extra work has to be done to separate it from its net positive screening charge.) The work done in taking an electron from the Fermi level to a state at rest just *outside* the metal is known as the *work function W*.

Contributions to the work function W

This picture has some surprising implications. For instance, in electricity texts it is often assumed that, if ϕ is the electrostatic potential, the electric field $-\nabla\phi$ is zero in a metal in equilibrium, and ϕ is continuous at the metal boundary. Both assumptions are incorrect. Since

$$\mu = K_F - e\phi \tag{15.13}$$

where K_F is the kinetic energy of an electron at the Fermi level, and $\nabla\mu$ is zero in equilibrium, then if K_F varies with position (as it certainly does between ion cores, or between one metal and another in electrical contact with it) the electric field is *not* zero in equilibrium. *It is μ and not ϕ which determines the flow of electrons.* Ohm's law should properly be written not as $\mathbf{J} = -\sigma\nabla\phi$ (where σ is the conductivity) but as

$\nabla\mu$, not $\nabla\phi$, controls electron flow

$$\mathbf{J} = \sigma\nabla(\mu/e). \tag{15.14}$$

Ordinary voltmeters, which measure the tendency of electrons to flow from one metal to another, in fact measure differences in $-\mu/e$ and *not* differences in ϕ.

We shall discuss two consequences of this idea in particular. Observe first that when we measure the *emf of a cell* using a conventional voltmeter what we really measure is the difference in $-\mu/e$ between the plates on open circuit. We may use this idea to relate the emf of the cell to the chemical potentials of the reagents. For instance, in a Daniell cell zinc and copper electrodes dip into acidified copper sulphate solution, and the following reaction may occur:

Electrode potentials

$$\mathrm{Zn} + \mathrm{Cu}^{++} + 2\mathrm{e}^-_{\mathrm{Cu}} \rightarrow \mathrm{Zn}^{++} + \mathrm{Cu} + 2\mathrm{e}^-_{\mathrm{Zn}}$$

where $\mathrm{e}^-_{\mathrm{Cu}}$ and $\mathrm{e}^-_{\mathrm{Zn}}$ represent electrons on the copper and zinc electrodes. In chemical equilibrium the chemical potentials of the reagents and the products must be equal, so

§11.2

$$\mu_{\mathrm{Zn}} + \mu_{\mathrm{Cu}^{++}} + 2\mu_1 = \mu_{\mathrm{Zn}^{++}} + \mu_{\mathrm{Cu}} + 2\mu_2, \tag{15.15}$$

where μ_1 and μ_2 are the electrochemical potentials of $\mathrm{e}^-_{\mathrm{Cu}}$ and $\mathrm{e}^-_{\mathrm{Zn}}$. The emf η of the cell is just $(\mu_2 - \mu_1)/e$, so

$$\blacktriangleright \quad \eta = E_{Zn} - E_{Cu}, \tag{15.16}$$

where $E_{Zn} = (\mu_{Zn} - \mu_{Zn^{++}})/2e$ and $E_{Cu} = (\mu_{Cu} - \mu_{Cu^{++}})/2e$. These quantities are known as the *absolute electrode potentials* for zinc and copper with respect to the particular electrolyte employed. This expression for the emf in terms of electrode potentials is often useful.

Contact potentials

Secondly, consider the *contact potential* which arises when two different metals or semiconductors are placed in electrical contact. In this situation μ must be the same in both materials, but ϕ in general is not. By contact potential we mean the difference in ϕ which exists between a point just *outside* one metal and a point just outside the other. (It cannot be measured by an ordinary voltmeter, but can be found by electrostatic techniques: one method is to measure the charge which flows onto a small earthed plate which is moved from a position very near to but not touching the surface of one metal to an equivalent position near the other.) To see how the contact potential arises, let us imagine two metals A and B with different work functions which are initially neutral and far apart, as in Fig. 15.6(b). Since the metals are neutral there is no field outside them, and the potentials immediately outside them will be equal. Suppose that $W_B > W_A$. Then $\mu_A > \mu_B$, and electrons must flow from A to B as soon as contact is established. The transferred charge will set up an electrical double layer which lifts the electrostatic energy and hence the Fermi energy of B with respect to A, until the two Fermi levels are equal, when the flow will cease. In practice the charge transfer needed to produce the change in potential is minute, so the electron densities in the two metals will be effectively unchanged. Thus *the effect is simply to lift the complete pattern of occupied states of B with respect to those of A by an amount equal to* $W_B - W_A$, and this is the value of the contact potential. It is often of the order of several volts, and though quite undetectable by an ordinary voltmeter, has important effects, for instance, on the rectifying properties of a p–n semiconductor junction and on measurements of the photoelectric effect.

Problem 15.7

Problems

Langevin's theory of polar gases

15.1 In zero electric field the permanent electric dipole moment $\mathbf{p}_0$ of a molecule in a polar gas is equally likely to point in any direction. If an electric field of magnitude $\mathscr{E}_E$ is now applied parallel to the z axis, show using classical statistics that $f(\theta)\,d\theta \propto 2\pi \sin\theta \exp(p_0 \mathscr{E}_E \cos\theta/kT)\,d\theta$, where $f(\theta)$ is the distribution function for the angle θ between $\mathbf{p}_0$ and the z axis. Hence show that the mean component $\bar{p}$ of moment parallel to the field at temperature T is given by

$$\bar{p} = p_0 \mathscr{L}(p_0 \mathscr{E}_E/kT),$$

where $\mathscr{L}$ is the *Langevin function*, $\mathscr{L}(x) = \coth x - 1/x$. Sketch this function and compare it with (9.14). Why is it valid to use classical statistics in this calculation but not in the equivalent calculation for paramagnetic materials in § 15.3?

15.2 By expanding $\coth x$ in powers of x show that in small electric fields the quantity $\bar{p}$ of the previous question is approximately equal to $p_0^2\mathscr{E}_E/3kT$. Dry steam at 200 °C is subjected to an electric field of 10^4 V m^{-1}. If no heat enters the steam, by how much does its temperature change? [Dipole moment of $H_2O = 6.1 \times 10^{-30}$ Cm. Molecular heat capacity of steam at constant volume at 200 °C = 3.2 k.]

The electrocaloric effect

15.3 When an atom without angular momentum ($J = L = S = 0$) is placed in an external magnetic field $\mathscr{B}_E$ its electrons have superimposed on their original motions a *Larmor precession* at angular velocity $e\mathscr{B}_E/2m_e$. Deduce that the atom has a small *diamagnetic magnetisability* $\alpha_m = -(Ze^2/4m_e)\overline{a^2}$, where $\overline{a^2}$ is the mean square distance of the electrons from the precession axis. With the help of (15.2) show that in an external field $\mathscr{B}_E$ such a diamagnetic atom has an extra internal energy equal to $-\frac{1}{2}\alpha_m\mathscr{B}_E^2$, and show also that this energy takes the form of kinetic energy.

Atomic diamagnetism and magnetic energies

An atom with $S=0$, $L \neq 0$ will also undergo Larmor precession in an applied field. Show that in this case the extra kinetic energy is equal to $-m_z\mathscr{B}_E - \frac{1}{2}\alpha_m\mathscr{B}_E^2$, where m_z is the component of orbital magnetic moment parallel to the applied field.

15.4 Derive equation (15.5), noting that $\bar{m}$ may be written as $gJ\beta(\mathrm{d}S/\mathrm{d}x)/S$, where $S = \sum \exp(M_J x/J)$ is the sum of a finite geometric series. Show that for large J the Brillouin function reduces to the Langevin function of problem 15.1. By expanding the coth functions in powers of x confirm that for any J the Brillouin function approximates to $(J+1)x/3J$ for small x.

15.5 In zero magnetic field the alum $CrK(SO_4)_2 \cdot 12H_2O$ has a Schottky anomaly with the heat capacity maximum at 0.08 K. Estimate roughly the effective residual field in the alum, and show that it is rather too large to be accounted for by magnetic interactions between the Cr^{3+} ions. (It is chiefly a crystal field effect.) Use Fig. 15.3 to explain why there is little point in trying to cool the alum below 0.04 K by demagnetisation. What field would be needed to reach 0.04 K from 1.2 K? Show that if 0.1 mole of the alum is in thermal contact with 1 mole of copper, the cooling is still almost as good. [The Cr^{3+} ions in the alum are about 8×10^{-10} m apart. They have $S = \frac{3}{2}$, and the effective value of L is zero. Heat capacity of copper $= BT^3 + \gamma T$ where $B = 6.5 \times 10^{-5}$ J mol^{-1} K^{-4} and $\gamma = 7.0 \times 10^{-4}$ J mol^{-1} K^{-2}.]

§9.2.1

15.6 The heat capacities per unit volume of normal and superconducting tin are given very approximately by the formulae $C_n = bT^3 + \gamma T$ and $C_s = aT^3$. Using these formulae show that (i) $T_c = [3\gamma/(a-b)]^{\frac{1}{2}}$, (ii) $\mathscr{B}_c = \mathscr{B}_{c0}(1 - t^2)$, where $t = T/T_c$ and $\mathscr{B}_{c0} = (\frac{1}{2}\mu_0\gamma)^{\frac{1}{2}}T_c$, (iii) at $T=0$, $E_n - E_s = \frac{1}{4}\gamma T_c^2$, and (iv) $E_n - E_s$ has a maximum at $T = T_c/\sqrt{3}$. [Since $\frac{1}{2}\gamma T^2$ is the electronic thermal energy in the normal state, the third result supports the suggestion that superconductivity involves a process of condensation among the conduction electrons.]

15.7 Two free electron metals A and B with Fermi energies $\varepsilon_A = 4$ eV and $\varepsilon_B = 3$ eV have work functions $W_A = 1$ eV and $W_B = 1.2$ eV. One face of a cube of A, of side 1 cm, is placed in direct contact with a face of a corresponding cube of B. If the electrical double layer at the interface is 10^{-10} m thick, what fraction of the electrons in A flow spontaneously into B? If a very small current of electrons is driven from B into A at $T=0$, discuss the energy changes which occur. [What happens at finite temperature is discussed in §17.5.]

16

Fluctuations and the approach to equilibrium

This and the following chapter bring together brief treatments of a number of important topics which are in fact closely related, although often regarded as belonging to different areas of physics. The first five sections are concerned with the distribution and power spectrum of the equilibrium fluctuations of thermodynamic variables. The remaining two sections and Chapter 17 examine different aspects of the question of how a thermodynamic system approaches equilibrium. In most of the sections we shall be concerned not with the detailed kinetics but with general properties. Some of the links between these different topics will become apparent as we proceed, but the two chapters should be regarded as a brief preliminary exposure to a large field, and not as a systematic treatment which displays all the relationships involved.

16.1 The equilibrium distribution of fluctuations: general principles

Any variable x of a thermodynamic system which is not constrained will fluctuate around its mean value. How is the distribution of such fluctuations to be calculated? In answering this question we shall assume that, when the system is in a given quantum state i, the parameter x takes a definite value x_i. If we know the probability of finding the system in each state i it is then a simple matter to find the distribution of x. We must be careful here, however. The probabilities $\tilde{p}_i$ used in this book represent the *ensemble* distribution, which is not necessarily the same as the *fluctuation* distribution – they represent the probability of finding the system in i at a fixed moment in time and not the probability of finding a single system in a given state when averaged over its fluctuations. However, we noted earlier that *when in thermal equilibrium* the two distributions are identical. Thus if $\tilde{p}_i$ represents the *equilibrium* ensemble probability for state i we may write the probability that x lies, in the course of its equilibrium fluctuations, in some narrow range between x and $x + \mathrm{d}x$ as

§16.2

§2.5

$$f(x)\,\mathrm{d}x = \sum_{\mathrm{d}x} \tilde{p}_i, \tag{16.1}$$

where $f(x)$ is the *equilibrium fluctuation distribution* of x, and the sum is taken over all states i for which x lies in the range of width dx. In classical statistical mechanics the sum is replaced by an integral over the equivalent phase space volume. Equation (16.1) is our basic principle. (In the next section I refer briefly to some cases in which the application of this principle raises special difficulties.)

The three general expressions for $f(x)$

The values of $\tilde{p}_i$ to be used in (16.1) depend on the type of equilibrium involved. For a system in contact with a heat bath they are the canonical Boltzmann probabilities. If the system is also large, the sum may be replaced by an integral, and $f(x)$ may be written in the form

$$f(x) = \int g(x, E) p(E) \, dE, \qquad (16.2)$$

where $p(E)$ is the Boltzmann probability, and $g(x, E)$ is a *two-dimensional density of states*, defined so that the number of states in which x lies in a narrow range dx and E simultaneously lies in the narrow range dE is equal to $g(x, E)\, dx\, dE$. For a large thermally isolated system, on the other hand, E is conserved and we must take $\tilde{p}_i$ to be the microcanonical probability, which is constant over a narrow accessibility range δE and zero outside it. For such a large thermally isolated system, then, we find simply that

§4.3

$$f(x) \propto g(x, E). \qquad (16.3)$$

The formulae (16.2) and (16.3) apply equally to classical statistics, if we interpret $g(x, E)$ as a density of phase space volume in energy.

16.2* Special difficulties in certain cases

Before we proceed to make calculations it may be helpful to mention certain conceptual difficulties concerning fluctuations which are commonly found confusing. It is, for instance, tempting to assume that the fluctuations in a quantity are no more dependent upon the details of the constraints than its mean value is. This is not always so, however. For instance, the pressure of a gas acting on the walls of its container is the sum of a large number of pressure *pulses* due to the impacts of the molecules on the walls. It is not hard to show that in this situation the fractional fluctuation in the pressure acting on a given area of wall is of order $N_P^{-\frac{1}{2}}$, where N_P is the number of molecules which strike the area on average *in the duration of one pulse*. This fluctuation depends both on the area involved and on the length of the pulse, which in turn depends on such details as the rigidity of the wall, and the range of the forces acting between the molecule and the wall. Thus, unlike the mean pressure, *the distribution of pressure fluctuations depends on the area and nature of the walls.*

Fluctuations may depend on details of the constraints

A second and important difficulty becomes apparent if we consider the same pressure fluctuations from the quantum point of view. Imagine a box of fixed volume containing a single atom, which is thermally isolated (Fig. 5.1). In the absence of collisions the atom will remain in the same quantum state, i.

p. 48

§4.4

The value of the pressure in this state is, according to (4.10), $-\mathrm{d}E_i/\mathrm{d}V$. Thus the pressure is, apparently, *constant* and shows no fluctuations. In the equivalent classical state, however, as the diagram shows, the atom is bouncing off the walls and the pressure due to its impacts is clearly *not* constant. What is the reason for this discrepancy? The answer to this question concerns the phenomenon of *quantum fluctuations*. In quantum theory, when the operator for some quantity does not commute with the Hamiltonian the quantity concerned does not take a definite value in a given energy state, i. If we measure the quantity when the system is in state i we may find one of several values. We cannot tell which of these values will occur, but we can calculate the statistical distribution of values. The mean value of this distribution is called the *expectation value*, and it is this mean value which is usually quoted as the 'value' of x in state i. For our single atom the instantaneous pressure on the walls is a function of the position variable which does not commute with the Hamiltonian. Thus the so-called 'value' of the pressure $-\mathrm{d}E_i/\mathrm{d}V$ is only an expectation value, and the pressure actually observed will show 'quantum fluctuations' around it. In fact, it may be shown that, if we perform the rather tedious calculation of the quantum fluctuations in the pressure for our single atom, we recover the classical expression for the pressure fluctuations to a good approximation. Now the general method discussed in the preceding section clearly assumes that x takes a single, definite value in state i. Thus, *if x does not commute with the Hamiltonian, equations (16.1), (16.2) and (16.3) must be modified to allow for the effect of quantum fluctuations.* This difficulty does not arise in classical statistics. We shall not have space to examine it further in this book and simply warn the reader that cases do arise in which quantum fluctuations can become important – for instance, in discussing some types of electrical noise at low temperatures. (The more general question of how non-commuting variables should be treated in quantum statistics will be discussed briefly later, however.)

Quantum fluctuations

§19.4

A difficulty of a different type concerns the variables T, μ and S, which do not have well-defined values attached to particular quantum states. *These variables are properties of the equilibrium probability distribution over the states, and not of the states themselves.* Without some knowledge of the distribution in energy we cannot describe the temperature, and without knowledge of the range of states over which the probability is spread we know nothing of the entropy. Since T, μ and S do not have identifiable values attached to each state, equation (16.1) cannot be applied to them. Whether we regard them as fluctuating for a system in equilibrium is partly a matter of attitude. In one sense the distribution is *constant*: we may say that the fluctuations *sample* but do not *alter* the equilibrium distribution. In this sense T, μ and S do not change in the course of equilibrium fluctuations. In a different sense we may regard a fluctuation as a measurable departure from the equilibrium distribution. For instance, if the energy of a large system in contact with a heat bath is known to have increased in the course of a fluctuation, we

Special difficulties with fluctuations of T, μ and S

may describe this by saying that the peak of its energy distribution has moved to a larger value. In this sense the system is hotter. To discuss the fluctuations in T, μ and S it would clearly be essential to begin by deciding how T, μ and S were to be defined in the course of a fluctuation. We shall not tackle this delicate question in this chapter. The question of whether S *decreases* during fluctuations is of some fundamental significance, however, and we shall return to it later.

§ 19.7

Finally, it is as well to be reminded that *the usual equilibrium relations between functions of state do not hold in the course of thermal fluctuations.* It is, for instance, tempting, having calculated the energy fluctuations in a gas held at fixed volume and temperature, to try to deduce the pressure fluctuations by using the usual equilibrium relation between P and E at fixed V, or, having calculated the volume fluctuations of a system held at fixed pressure to attempt to deduce the pressure fluctuations at fixed volume by imagining an equilibrium compression to fixed volume superimposed on the volume fluctuation. A little thought shows that both such attempts must be fallacious, since the pressure fluctuations depend on the nature of the wall, but the energy and volume fluctuations do not. Indeed, the first method is quite obviously fallacious, for if we applied it to a thermally isolated gas held at fixed volume, E would be *constant* and we should be forced to conclude that the pressure did not fluctuate, which is clearly nonsensical. It is equally invalid to attempt to derive formulae for the fluctuations in S and T from known fluctuations in E, V, etc., by using the equilibrium relations between them.

Equilibrium relations cannot be used to calculate one fluctuation from another

With these warnings behind us, let us in more positive mood examine some detailed calculations of fluctuations.

16.3 Example: fluctuations of magnetisation in a simple paramagnetic salt

The simplest way to calculate a fluctuation distribution is to use (16.1), or the related results (16.2) or (16.3), directly. This may always be done when the microstates of the system and their energies are known, so long as quantum fluctuations are not important. The details differ slightly for classical and quantum calculations, and for different combinations of constraint. Here we shall consider, as an example, the fluctuations in magnetisation of a quantum system, a simple paramagnetic salt held at temperature T in an applied field $\mathscr{B}$ in the z direction. Other cases are left as exercises.

Problems 16.1, 16.2

Suppose that the salt contains N non-interacting ions each with $S=\frac{1}{2}$. If the number of spin up ions is $N_\uparrow$ and the number of spin down ions is $N_\downarrow$ we have $N=N_\uparrow+N_\downarrow$, while the total magnetic moment m for the whole sample is equal to $m_0(N_\uparrow-N_\downarrow)$. Let us calculate the distribution of fluctuations in m. When we apply (16.1) to this problem we find that this is a particularly simple case because E happens to be a function of m: in fact, $E=-m\mathscr{B}$. Thus for any particular value of m only one value of E contributes to the sum, and we find that the fluctuation distribution of m is given by

§ 15.3

Fluctuations of m in a paramagnetic salt

$$f(m) = W(m)p(E), \tag{16.4}$$

where $W(m)$ means the number of configurations of spin for which the total magnetic moment is m, and $p(E)$ is the Boltzmann probability calculated for $E = -m\mathscr{B}$. For fixed m we observe that $N_\uparrow$ and $N_\downarrow$ are fixed, and W is just the number of ways of choosing the $N_\uparrow$ up spin ions from the total of N, which is $\binom{N}{N_\uparrow}$. Thus we find that

$$f(m) \propto \frac{N!}{N_\uparrow!\,N_\downarrow!}\,\mathrm{e}^{m_0\mathscr{B}(N_\uparrow - N_\downarrow)/kT}. \tag{16.5}$$

Obviously $f(m)$ will be a maximum at some value of $N_\uparrow$ which is the most probable value. To find how m is distributed it is convenient to expand $\ln f$ in powers of the departure $\Delta N_\uparrow$ of $N_\uparrow$ from this most probable value. If $N_\uparrow$ is increased by unity $f(m)$ is *multiplied* by $[N_\downarrow/(N_\uparrow + 1)] \exp(2m_0\mathscr{B}/kT)$, so, writing $N_\uparrow + 1$ as $N_\uparrow$ on the assumption that $N_\uparrow$ is large, we find that

$$\mathrm{d}\ln f/\mathrm{d}N_\uparrow = \ln(N_\downarrow/N_\uparrow) + 2m_0\mathscr{B}/kT.$$

This must be zero at the maximum, so $N_\downarrow/N_\uparrow = \exp(-2m_0\mathscr{B}/kT)$. We deduce that $m = m_0(N_\uparrow - N_\downarrow)$ is equal to $Nm_0 \tanh(m_0\mathscr{B}/kT)$, in agreement with (9.14). On differentiating again we find that

§9.2.1

$$\mathrm{d}^2\ln f/\mathrm{d}N_\uparrow^2 = -1/N_\uparrow - 1/N_\downarrow.$$

To second order in $\Delta N_\uparrow$, then, we find that $\ln f = \ln f_0 - (1/N_\uparrow + 1/N_\downarrow)\frac{1}{2}\Delta N_\uparrow^2$, by Taylor series expansion about the most likely value. To this order of approximation, writing $\Delta N_\uparrow$ as $\frac{1}{2}\Delta m/m_0$, we find that f takes the form

$$\blacktriangleright \quad f(m) = f_0\,\mathrm{e}^{-\frac{1}{8}\Delta m^2(1/N_\uparrow + 1/N_\downarrow)/m_0^2}, \tag{16.6}$$

§3.4

a Gaussian distribution whose mean square deviation is given by

$$\overline{\Delta m^2} = 4m_0^2/(1/N_\uparrow + 1/N_\downarrow). \tag{16.7}$$

In zero field we have $N_\uparrow = N_\downarrow = \frac{1}{2}N$, so the rms fluctuation in m is equal to $m_0 N^{\frac{1}{2}}$. In the general case we find that the rms fluctuation is equal to $m_0 N^{\frac{1}{2}}/\cosh(m_0\mathscr{B}/kT)$, showing that the fluctuations decrease substantially as the field increases and the system approaches saturation.

16.4 Relations between fluctuations and thermodynamic response functions

§3.4

We saw in Chapter 3 that the fluctuations in energy of a system in contact with a heat reservoir at temperature T are given by the relation

$$\blacktriangleright \quad \overline{\Delta E^2} = kT^2 C, \tag{16.8}$$

where C is the heat capacity of the system. This is an example of a type of relationship which often, but not always, exists between the fluctuations of a quantity (in this case the energy E) subject to a controlling restriction (the reservoir temperature T), and the corresponding *thermodynamic response function* (the heat capacity $C = \mathrm{d}E/\mathrm{d}T$) which measures the rate of change of the equilibrium value of the quantity as the controlling restriction is varied.

We shall see shortly that a similar relation connects the volume fluctuations of a system subject to a fixed external pressure with the response function $\mathrm{d}V/\mathrm{d}P$, which may be found by measuring the compressibility of the system. Such relations between a fluctuation distribution and the corresponding measurable response function are often more useful and more illuminating than detailed microscopic fluctuation calculations of the sort considered in the preceding section. In what follows we shall explore them in some detail.

The simplest situation to consider is the fluctuation of a variable x inside a system which is completely isolated from the external world, in cases where we are free to fix x, if we wish, by applying an extra constraint to the system, without disturbing the microstates significantly. In such a situation (16.3) tells us that when x is *not* constrained its fluctuation distribution $f(x)$ is proportional to $g(x, E)$. On the other hand, if x *is* constrained to a particular value, or to a small range near this value, we may use the large system approximation to write the corresponding equilibrium entropy of the system $S(x, E)$ as $k \ln g(x, E)$. On combining these two relations we find that

The entropy expression for the fluctuation distribution in an isolated system

§4.3

▶ $$f(x) \propto \mathrm{e}^{S(x, E)/k}. \tag{16.9}$$

In this formula for the fluctuation distribution for x when *un*constrained, $S(x, E)$ stands for the equilibrium entropy which the system has when x *is* constrained, at the value of interest. Obviously, the most likely value $x = x_0$ occurs where $S(x)$ is a maximum. If we expand S around this point in a Taylor series to second order in $\Delta x = x - x_0$ we find that $S(x) \simeq S_0 + \frac{1}{2}(\partial^2 S/\partial x^2)_E\, \Delta x^2$. (I leave it as an exercise to show that the higher-order terms are normally negligible in large systems.) Using this expansion we deduce that $f(x)$ is Gaussian to a good approximation,

Problem 16.3

$$f(x) \propto \mathrm{e}^{\frac{1}{2}(\partial^2 S/\partial x^2)_E \Delta x^2/k}.$$

By comparison with the standard Gaussian we see that the mean square fluctuation is given by

§3.4

$$\overline{\Delta x^2} = -k/(\partial^2 S/\partial x^2)_E. \tag{16.10}$$

Using classical thermodynamics it is usually a simple matter to relate $(\partial^2 S/\partial x^2)_E$ to a measurable response function.

For instance, suppose that the isolated system consists of two parts, A and B, in thermal contact, and consider the energy fluctuations of E_A, the energy in part A. We may clearly constrain E_A by inserting some thermal insulation between A and B, so this is a situation to which the above analysis applies, and S means $S(E_A)$, the equilibrium entropy which the joint system has when E_A is constrained. Remembering that $\mathrm{d}E_B = -\mathrm{d}E_A$ for an isolated system we find that

Fluctuations of energy inside an isolated system

$$\begin{aligned}\mathrm{d}S &= \mathrm{d}S_A + \mathrm{d}S_B \\ &= (\partial S_A/\partial E_A)\,\mathrm{d}E_A + (\partial S_B/\partial E_B)\,\mathrm{d}E_B \\ &= (1/T_A - 1/T_B)\,\mathrm{d}E_A.\end{aligned}$$

S is a maximum at the equilibrium point where $T_A = T_B$ and $(\partial S/\partial E_A)_E = 0$. Differentiating again we find that

$$(\partial^2 S/\partial E_A^2)_E = -(1/C_A + 1/C_B)/T^2,$$

where $C = \mathrm{d}E/\mathrm{d}T$ is the corresponding heat capacity. Using (16.10) we deduce that

$$\overline{\Delta E_A^2} = kT^2/(1/C_A + 1/C_B). \tag{16.11}$$

Notice that in this expression it is the *smaller* of the two heat capacities which is dominant. In particular, if part B is much larger than part A, we may ignore C_B. In this limit (16.11) reduces to our earlier result (16.8), as one would expect, for B is now in effect an external reservoir with which A is interacting.

The availability expression for the fluctuation distribution in a system which may be interacting with its surroundings

The limitation to isolated systems is inconvenient, however, for there are many possible combinations of external controlling restriction which may interest us. We may be interested, for instance, in systems which can exchange heat or particles with a reservoir at temperature T_R and chemical potential μ_R. We may be interested in a system whose volume is free to fluctuate, enclosed by surroundings at a fixed pressure P_R. A general formalism which allows for all these possibilities may be set up by applying (16.9) not to the system alone but to the complete assembly consisting of the system, the reservoir and the surroundings. Then $S(x, E)$ must be replaced by $S_{\text{tot}}(x, E_{\text{tot}})$, applied to the assembly as a whole. If S_R is the reservoir entropy we have in any change

$$\begin{aligned} \mathrm{d}S_{\text{tot}} &= \mathrm{d}S + \mathrm{d}S_R \\ &= \mathrm{d}S + \mathrm{d}E_R/T_R - (\mu_R/T_R)\,\mathrm{d}N_R \\ &= (1/T_R)(T_R\,\mathrm{d}S - \mathrm{d}E - P_R\,\mathrm{d}V + \mu_R\,\mathrm{d}N) \\ &= -\mathrm{d}A/T_R, \end{aligned} \tag{16.12}$$

§7.5

where $A(x) = E - T_R S + P_R V - \mu_R N$ is the *availability* of the system, calculated in equilibrium with x constrained, and we have used energy conservation in the third line. Thus (16.9) may be replaced by

▶ $$f(x) \propto \mathrm{e}^{-A(x)/kT_R}, \tag{16.13}$$

and to discuss fluctuations of x we must now expand A instead of S in powers of $\Delta x = x - x_0$ around the equilibrium point. Note carefully that this analysis makes no assumptions about whether E, V and N are *actually* free to fluctuate: the result holds whether they are constrained or not. For instance, we note that for an isolated system with V fixed, E and N are automatically fixed also, and (16.13) reduces to (16.9). But we may equally well use (16.13) to discuss, for instance, volume fluctuations at fixed P and T, or the fluctuations of some internal variable x at fixed V, T and N, or at fixed P, T and μ, etc.

Volume fluctuations at fixed P, T and N

As an example, consider the volume fluctuations of a system subject to the pressure P_R of the surroundings, at fixed T and N. To discuss them we need $A(V)$, which is the equilibrium value of the availability calculated as a function of the constraint on V at fixed N, but with E free to fluctuate, so that $T = T_R$. Then, since $\mathrm{d}A = (T - T_R)\,\mathrm{d}S + (P_R - P)\,\mathrm{d}V$, we find that

$$(\partial A/\partial V)_{T,N} = P_R - P$$

and

$$(\partial^2 A/\partial V^2)_{T,N} = -(\partial P/\partial V)_{T,N}.$$

Thus on expanding A in powers of $\Delta V = V - V_0$ around the equilibrium point to second order, we find using (16.13) that $f(V)$ is Gaussian, and that the mean square fluctuation is given by

▶ $$\overline{\Delta V^2} = -kT(\partial V/\partial P)_{T,N}. \tag{16.14}$$

This is our anticipated formula linking the volume fluctuations at fixed external pressure with the corresponding thermodynamic response function $(\partial V/\partial P)_{T,N}$, which may be found by measuring the isothermal compressibility of the system.

The fractional fluctuations of extensive variables are small in large systems

This result provides a convenient context in which to note the general point that the *fractional* fluctuations in any extensive variable are normally very small in large, homogeneous systems. Because V is extensive but P is intensive, $(\partial V/\partial P)_{T,N}$ is proportional to the size of the system. It follows that the *fractional* fluctuation $(\overline{\Delta V^2})^{\frac{1}{2}}/V$ is proportional to $(\text{size})^{-\frac{1}{2}}$, and may therefore be made very small by taking a sufficiently large system. In the case of an ideal gas, for instance, we note that $(\partial V/\partial P)_{T,N} = -V^2/NkT$. In this case, then, we see using (16.14) that the fractional volume fluctuation is equal to $N^{-\frac{1}{2}}$. It is very small indeed in a sample of everyday size containing, say, 10^{20} molecules. Similar considerations apply to the fractional fluctuations of other extensive variables.

Simultaneous fluctuations of two or more variables

The availability method described above may be extended to cover the simultaneous fluctuations of two or more variables. For instance, consider the fluctuations of energy and volume for the system just considered, which is in mechanical and thermal contact with the reservoir. For this purpose we need to know how the availability changes when we vary simultaneous constraints on the energy and the volume. Writing A as $E - T_R S + P_R V$, since N is fixed in this case, and performing a *double* Taylor expansion with respect to E and V, we find that

$$(\partial A/\partial E)_V = 1 - T_R/T,$$
$$(\partial A/\partial V)_E = -T_R P/T + P_R,$$
$$(\partial^2 A/\partial E^2)_V = (T_R/T^2)(\partial T/\partial E)_V = T_R/C_V T^2 = \alpha_{EE},$$
$$(\partial^2 A/\partial E\,\partial V) = (T_R/T^2)(\partial T/\partial V)_E = -T_R[\partial(P/T)/\partial E]_V = \alpha_{EV},$$
$$(\partial^2 A/\partial V^2)_E = -T_R[\partial(P/T)/\partial V]_E = \alpha_{VV}.$$

At the equilibrium point $T = T_R$ and $P = P_R$, so to second order

$$A = A_0 + \tfrac{1}{2}\alpha_{EE}\,\Delta E^2 + \alpha_{EV}\Delta E\,\Delta V + \tfrac{1}{2}\alpha_{VV}\,\Delta V^2.$$

This leads to a *double* Gaussian distribution,

▶ $$f(E, V) \propto e^{-(\frac{1}{2}\alpha_{EE}\Delta E^2 + \alpha_{EV}\Delta E\Delta V + \frac{1}{2}\alpha_{VV}\Delta V^2)/kT}. \tag{16.15}$$

Note that, in general, $\alpha_{EV} \neq 0$, and this means that *the fluctuations in E and V are correlated*. To find the distributions of the fluctuations in E or V *alone* we must therefore integrate over the other variable. I leave it as an exercise to show that integrating the above distribution over E leads back to expression (16.14) for $\overline{\Delta V^2}$, for instance. For the special case of an ideal monatomic gas we find that $\alpha_{EE}/2kT = 3N/4E^2$, $\alpha_{VV}/2kT = N/2V^2$ and $\alpha_{EV} = 0$. In this special case, then, the fluctuations of E and V turn out to be uncorrelated. The fractional fluctuation in energy $(\Delta E)^{\frac{1}{2}}/E$ is $(2/3N)^{\frac{1}{2}}$, and the fractional fluctuation in volume is $N^{-\frac{1}{2}}$ as we noted earlier.

Problem 16.5

Quasi-Boltzmann distributions for systems at fixed T

For certain combinations of controlling restriction our expression (16.13) in terms of the availability reduces to expressions in terms of the free energy, or other familiar thermodynamic potentials. For instance, for fluctuations of a variable x in a system held at fixed T and N, whose constraints are fixed so that no external work can be done on the system, we have $A = E - TS + \text{const} = F + \text{const}$. Thus

$$f(x) \propto e^{-F(x)/kT}. \tag{16.16}$$

This may be regarded as a *quasi-Boltzmann distribution* for x. It represents yet another respect in which the free energy behaves like a potential energy for a system held at constant temperature. In a similar way, for systems held at fixed T, N and P we find that $A(x) = G(x) + \text{const}$, so that $f(x) \propto \exp[-G(x)/kT]$. Other similar results are summarised in Table 7.1.

§7.2

p. 70

Fluctuating variables which cannot be constrained

Up to this point we have limited our discussion to those fluctuating variables which we are actually able to constrain, if we wish, without materially changing the nature of the available states. This is not possible for all fluctuating variables. For instance, we could only fix the instantaneous magnetic moment of a paramagnetic salt by clamping the individual ions, or by coupling them together in some manner which would completely alter the forces acting on them. We could likewise only fix the instantaneous pressure of a gas on the walls of its container by controlling the number of molecules in contact with the walls at any instant in some way. Since we cannot do either of these, even in principle, without drastically altering the states available to the systems, it is not clear whether results analogous to (16.14) hold in such cases, for we can no longer use the same method of relating $g(x, E)$ to an observable thermodynamic response function. However, our expression (16.3) for the fluctuation distribution still holds. Thus, in a formal sense, if we *define* $S(x, E)$ as $k \ln g(x, E)$, and *define* $A(x)$ as $E - T_R S(x, E) + P_R V - \mu_R N$, we may continue to write $f(x) \propto \exp[-A(x)/kT_R]$. It remains true that $A(x)$ must be minimised at the equilibrium point $x = x_0$ and, as before, we find on expanding $A(x)$ in powers of $\Delta x = x - x_0$ that $f(x)$ is Gaussian, and that $\overline{\Delta x^2} = kT_R/(d^2A/dx^2)$. Our difficulty is that we no longer know how to relate d^2A/dx^2 to an observable thermodynamic response function. Now, although in the cases under consideration here we cannot fix the *instantaneous* value of x, we can usually control its *mean* value $\bar{x}$ by varying an appropriate external 'force', y. Consider, for instance, the fluctuations of magnetic moment m in a paramagnetic salt. We cannot fix m, but we can fix $\bar{m}$ by varying the external magnetic field $\mathscr{B}_E$. It sometimes happens that a small change in the 'force' δy changes the energies of all the microstates by $-x\delta y$, leaving the x values unchanged: in the case of the paramagnetic salt, for instance, we know that the states of given magnetic moment have an energy $-m\mathscr{B}_E$, so that in a change of external field their energies change by $-m\delta\mathscr{B}_E$. For a system held at fixed temperature this simply subtracts $x\delta y$ from the availability: this may be seen most easily by writing A as $F+P_R V-\mu_R N$, and remembering that $F=-kT\ln Z$. Thus the new availability is given by

§15.3

$$A'(x)=A(x)-x\delta y.$$

Now $A'(x)$ must still be minimised at the equilibrium point, so

$$(\mathrm{d}A/\mathrm{d}x)_{\text{new equilibrium}}-\delta y=0,$$

or, since $(\mathrm{d}A/\mathrm{d}x)$ at the old equilibrium point was zero,

$$(\mathrm{d}^2A/\mathrm{d}x^2)\delta\bar{x}-\delta y=0,$$

where $\delta\bar{x}$ is the shift in the equilibrium point produced by δy. This identifies $\mathrm{d}^2A/\mathrm{d}x^2$ as $(\partial y/\partial\bar{x})_T$, and leads to an expression for $\overline{\Delta x^2}$ analogous to (16.14),

$$\blacktriangleright \qquad \overline{\Delta x^2}=kT(\partial\bar{x}/\partial y)_T. \qquad (16.17)$$

In the case of the paramagnetic salt, for instance, we find that

$$\overline{\Delta m^2}=kT(\partial\bar{m}/\partial\mathscr{B}_E)_T,$$

and you may like to check that this result agrees with (16.7), which we obtained by considering the detailed microstates of the system.

The expression (16.17) must be used cautiously, however, observing the requirement that δy must change the energies of all microstates by exactly $-x\delta y$, while leaving the x values unchanged. This applies, as we have just seen, to the fluctuations of m in a paramagnetic salt in fixed external magnetic field. It applies also to the fluctuations of electric dipole moment p, if we identify y as the external electric field: the energy $-p\delta\mathscr{E}_E$ is the mutual field energy in this case, and we note that a small change in field does not affect the dipole moment of a given charge configuration. But the formula does *not* apply to a system of non-interacting diamagnetic atoms, for instance, for in that case the extra magnetic field does not leave the magnetic moments of the microstates unchanged: it changes the magnetic moment of the atom in its ground state. (In fact, if excitation into higher states is ignored, the diamagnetic moment in fixed applied field is a function of the field only, and does not fluctuate, but $(\partial\bar{m}/\partial\mathscr{B}_E)_T$ is non-zero, so (16.17) is clearly incorrect.) It is also worth noting that our result (16.14) for the *volume* fluctuations at fixed pressure may be regarded as a valid example of (16.17), for a change in P_R increases the availability $A(V)$ by $V\delta P_R$, but leaves the V values of the microstates unchanged. Thus $-P_R$ may be identified as the 'force' for the volume fluctuations, and (16.17) is equivalent to (16.14). But the *pressure* fluctuations at fixed volume cannot be handled in this way. Although it is true that, for a

§15.1

Cases in which the standard result is not valid

given configuration of gas molecules, for instance, the potential energy of interaction between the molecules and the walls changes by $-P\delta V$ in a very small movement of the walls, where P is the instantaneous value of the pressure, so that V might be thought to be the appropriate 'force' in this case, changing V also changes P for a given configuration, so that (16.17) is invalid. Indeed, it is obviously invalid in this case because $(\partial\bar{P}/\partial V)_T$ is negative, while $\overline{\Delta P^2}$ must be positive. Clearly, in cases where the fluctuating variable cannot be constrained formula (16.17) must only be used with great caution.

16.5 The power spectrum of thermal fluctuations

If x is a fluctuating variable whose mean value is zero we may Fourier analyse its time dependence $x(t)$ and so find its components at various frequencies. In doing this it is important to understand two facts. First, since $x(t)$ extends over an infinite time, the usual formula for the Fourier transform *diverges*, so we must modify our definition of what we mean by the component of $x(t)$ at a given frequency. Secondly, because of the random nature of $x(t)$, the different frequency components usually have completely random *phase*. For these reasons we usually speak of the *power spectrum* of $x(t)$ rather than its ordinary frequency components. To understand what is meant by the power spectrum let us start by analysing $x(t)$ as a Fourier series over the *finite* period $0 \leqslant t \leqslant \tau$,

The meaning of the power spectrum

$$x(t) = \sum_{-\infty}^{\infty} x_n \, e^{i\omega_n t}, \tag{16.18}$$

where $\omega_n = 2\pi n/\tau$ and the coefficients x_n are, in general, complex. For a variable fluctuating in a random manner it is not hard to show that $|x_n|$ will be proportional to $\tau^{-\frac{1}{2}}$ if τ is large. Consider now the mean square value or 'power' of x. This may be written as

$$\begin{aligned} \overline{x^2} &= (1/\tau) \int_0^\tau x(t)x^*(t)\,dt \\ &= (1/\tau) \sum_{-\infty}^{\infty} x_n x_n^* \tau \\ &= \sum_{-\infty}^{\infty} |x_n|^2. \end{aligned} \tag{16.19}$$

Since $|x_n|^2$ is independent of the phase of x_n, we may take it to be a smooth function of frequency. We also have $|x_{-n}|^2 = |x_n|^2$. Thus, if we count positive frequencies only, the contribution to x^2 associated with a narrow range of frequencies $d\nu$ may be expressed as

▶ $$d(\overline{x^2}) = 2\tau |x_\nu|^2 \, d\nu, \tag{16.20}$$

since there are $\tau\,d\nu$ frequencies in the range $d\nu$. The quantity $2\tau|x_\nu|^2$ is known as the *power spectral density* which describes the *power spectrum* of x. Notice that, since $|x_\nu| \propto \tau^{-\frac{1}{2}}$, the power spectrum is independent of τ.

Now a knowledge of the *distribution* of a fluctuating variable tells us nothing directly about its power spectrum. Consider a simple pendulum immersed in a gas at temperature T. The bombardment of the pendulum bob by the gas molecules will prevent it from coming completely to rest. In fact, since at a small angle of displacement θ the pendulum has potential energy $\frac{1}{2}mgl\theta^2$, we know from the classical Boltzmann distribution that

§13.4

$$f(\theta) \propto e^{-mgl\theta^2/2kT}, \qquad (16.21)$$

where m is the mass of the bob and l the length of the shaft. The mean square displacement $\overline{\theta^2}$ will be equal to kT/mgl, and the mean potential energy will be $\frac{1}{2}kT$, as required by the equipartition theorem. Now any one collision has little effect on the pendulum. If it is lightly damped we expect to find that θ executes harmonic oscillations at the pendulum frequency whose amplitude increases and decreases slowly with time in a random manner. In this case the power spectrum of $\theta(t)$ is clearly very strongly peaked at the pendulum frequency. Suppose, however, that the pendulum is heavily damped by immersion in a light but very viscous liquid. Since this does not affect the formula for the potential energy we are forced to the perhaps surprising conclusion that the *distribution* of θ is unchanged. The *power spectrum* of $\theta(t)$ is, however, completely altered. Because the motion is heavily damped the regular oscillations are replaced by a much more irregular motion having a much broader power spectrum.

A given distribution may have various possible power spectra

This illustration also brings out a second important point. Since in the viscous liquid the damping is much increased but the mean square amplitude is unchanged, it is clear that the random forces which generate the motion must also be greatly increased: *if the damping is larger the random forces which generate the fluctuations must be larger too.* That there should be a microscopic connection between the damping and the random driving forces is not surprising, for these effects are due to the same mechanism, collisions between the bob and the molecules of the fluid. In fact, whenever we have a mechanism which dissipates mechanical energy as heat there is always an inverse mechanism present by which heat can generate a mechanical fluctuation. *There is always a connection between dissipation and fluctuation.* In particular, if the dissipation is negligible, the random driving forces must be negligible, too, for otherwise the system would gradually accumulate more mechanical energy from the driving forces than the equipartition theorem allows.

Fluctuation is connected with dissipation

To determine the power spectrum of fluctuations from first principles involves details of the kinetics which vary from case to case. Rather than considering such details it is more illuminating to establish a *general* formula which gives us the power spectrum of the fluctuations in terms of the dissipation in the system of interest. We shall do the calculation for the special case of an electrical circuit, which provides one of the many important applications of this theory. If part of the circuit has a resistance R this means that there is a mechanism available capable of converting electrical energy into heat. The same mechanism will generate random electrical disturbances from

The thermal noise emf of a resistor

the heat energy: in a metal wire, for instance, the electrons are subject to random bombardment by the elastic lattice waves known as phonons. Thus the wire behaves as though it contained a voltage source, a fluctuating *noise emf* which tries to drive a random current through the wire (Fig. 16.1). If, on the other hand, any part of the circuit has negligible resistance, this means that energy cannot leave it to enter the heat bath. By the principle of detailed balance it follows that energy cannot *enter* it from the heat bath either. The electrical part of its energy is, so to speak, thermally isolated, which means that *there is no noise emf acting on it.*

This last point provides us with a method for determining the power spectrum of the noise emf associated with resistance R. We imagine that we join R in series with an ideal resistanceless coil of inductance L and an ideal resistanceless capacitor of capacitance C. Since these components are resistanceless they do not affect the noise emf $V_N(t)$ acting in the circuit. But the circuit has now become a damped oscillator. By suitably adjusting L and C we may adjust the resonant frequency of oscillation $\omega_0 = (LC)^{-\frac{1}{2}}$ to any value that we like, and we may also keep the bandwidth of resonance R/L as small as we like. In the presence of the noise emf the circuit has the equation of motion

$$L\,\mathrm{d}I/\mathrm{d}t + RI + Q/C = V_N(t), \tag{16.22}$$

the *Langevin equation* for the circuit, where I is the current flowing, Q is the charge on the capacitor and $I = \mathrm{d}Q/\mathrm{d}t$. The solution is the superposition of the solutions for the various Fourier components of $V_N(t)$. Let us take some period $0 \leqslant t \leqslant \tau$ much longer than the time constant of the circuit. We may then safely treat the response as the sum of the long-term responses to the Fourier

Fig. 16.1. Thermal (Johnson) noise in a circuit. (*a*) shows the equivalent circuit of a resistor, which behaves as though it has an internal random noise emf of magnitude $V_N(t)$. (*b*) shows the resistor in an oscillator circuit in which L and C are ideal, loss-free components which have no internal noise source. By adjusting L and C we may vary the natural frequency to any required value while keeping the resonant bandwidth small, and so deduce the power spectrum of $V_N(t)$ by applying the equipartition theorem to the oscillator. (*c*) shows an alternative equivalent circuit for the resistor, which has a parallel noise current source instead of a series noise voltage source.

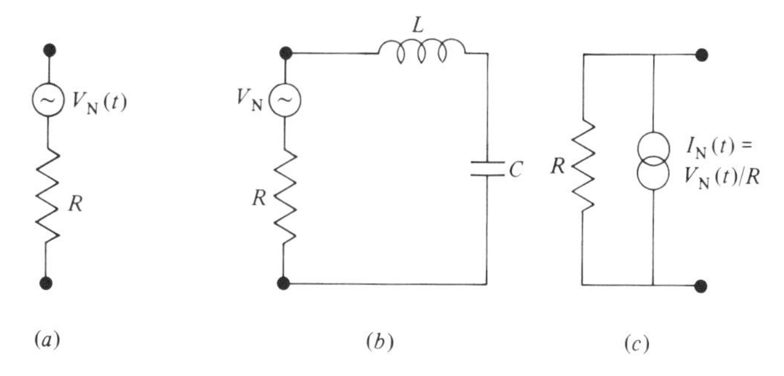

components of $V_N(t)$ measured over this period, ignoring all transient effects at the beginning and end of the period. Thus, if we expand $V_N(t)$ and $I(t)$ over this period as Fourier series we find to a good approximation that

$$I_n = \frac{V_n}{i\omega_n L + R + 1/i\omega_n C}. \tag{16.23}$$

Using (16.20) we deduce that

$$\begin{aligned} \overline{I^2(t)} &= (\tau/\pi) \int_0^\infty |I_n|^2 \, d\omega \\ &= (\tau/\pi) \int_0^\infty \frac{|V_n|^2 \, d\omega}{(\omega L - 1/\omega C)^2 + R^2}. \end{aligned} \tag{16.24}$$

We now make use of the fact that we have chosen L so that R/L is small, and thus the integrand is only large near the resonant frequency ω_0. This means that we may treat $|V_n|^2$ as constant, equal to $|V_{\omega_0}|^2$, and take it outside the integral. It also means that we can approximate $\omega L - 1/\omega C$ as $2\omega' L$ in the important range, where $\omega' = \omega - \omega_0$. Changing the variable to $x = 2\omega' L/R$ we find that

$$\begin{aligned} \overline{I^2(t)} &= (\tau |V_{\omega_0}|^2 / 2\pi L R) \int_{-\infty}^{\infty} dx/(x^2+1) \\ &= \tau |V_{\omega_0}|^2 / 2LR. \end{aligned} \tag{16.25}$$

Now $2\tau|V_{\omega_0}|^2$ is the power spectral density of V, while we know from the equipartition theorem that $\frac{1}{2}L\overline{I^2}$, the mean energy stored in the coil, must be $\frac{1}{2}kT$. Thus on combining (16.20) and (16.25) we find that

$$\blacktriangleright \quad d(\overline{V^2}) = 4RkT \, d\nu. \tag{16.26}$$

This is *Nyquist's formula* for the contribution to the power spectrum of the noise emf associated with a resistance R, for a frequency range $d\nu$. It is one of the forms of a general result known as the *fluctuation–dissipation theorem*, which shows how the noise driving forces in a system are related to its dissipation.

We must note several points about the Nyquist formula. The electrical noise referred to here is the *thermal* or *Johnson noise*, which is inevitably associated with a resistance R. (The circuit may well contain other types of noise, such as shot noise, due to the finite size of the electronic charge, and $1/f$ noise, which is usually due to thermal fluctuations in the resistance of different parts of the circuit.) The Johnson noise is *white noise*: that is to say its spectral density turns out to be independent of frequency. Note carefully that it is the noise emf whose power spectrum we have calculated. Unless the system is on open circuit, this is *not* the same as the voltage which actually appears across the resistor, which must be calculated in the usual way using circuit theory and allowing for the noise sources in all components. We have assumed that R is independent of frequency: if it is not, it is easy to show that the formula still holds if we use at each frequency the appropriate value of $R(\omega)$.

We have also assumed the classical limit, in using the equipartition theorem. This is only valid if $h\nu \ll kT$. At very high frequencies or very low temperatures kT must be replaced by Planck's formula $h\nu/[\exp(h\nu/kT)-1]$, and we see that at very high frequencies there is a cut-off to the Johnson noise, which ceases to be white.

§9.1.1

16.6 Thermally activated processes

The meaning of thermal activation

The rest of this chapter will be concerned with *rates of approach to equilibrium*. In this section we shall examine one such problem which, though not simple, is of considerable importance. It arises when a system is in or near a configuration X at which its potential energy is locally a minimum, but is able to move through an *activated configuration* A of greater energy to a final configuration Y of considerably lower energy (Fig. 16.2). The amount by which the potential energy at A exceeds the minimum energy at X is known as the *activation energy*, ϕ_A, and is typically much greater than kT. If such a system is kept at fixed temperature, its thermal fluctuations will explore the configurations around X, and it will eventually reach the activated configuration, from which it may well fall to Y. The system thus has a definite probability per unit time, R, of escaping from its *metastable state* near X. If the temperature is raised the probability of a fluctuation large enough to reach the activated configuration A becomes rapidly larger, and thus the rate of transfer from X to Y is greatly increased. For this reason such processes are said to be *thermally activated*. There are many important examples of such situations. In most chemical reactions, for instance, the reacting molecules have to move from one configuration to another of lower potential energy. But the original molecules are locally stable, so the system has to pass through an energy barrier corresponding to an intermediate state in which the chemical bonds are stretched or broken. Diffusion in alloys takes place by the movement of atoms into neighbouring vacancies: during such movements the lattice is disturbed and the increased potential energy presents a barrier to the diffusion. In much the same way electrical conduction in insulators is due to the hopping of electrons from one 'trap' to another, or the diffusion of ions from one lattice site to another: both processes involve passing through states of higher energy. We shall see later that the nucleation of droplets in a supersaturated vapour and the movement of pinned domain walls in ferromagnetism also involve the crossing of energy barriers.

§18.6

As an example of the calculation of the rate R of a thermally activated process, consider the two-dimensional motion of an atom adsorbed onto a solid surface (Fig. 16.2). Suppose that the atom is close to an adsorption site X, but that an empty site Y of lower energy exists nearby. The contours of potential energy show that of all routes from X to Y the lowest in energy is the one passing through the saddle point A. To discuss the rate of the process we imagine a large ensemble of such systems, each with its atom either to the left of or near the saddle point. (Once an atom is well to the right of its saddle point, we regard the process as having occurred, and remove the system concerned from the ensemble.) The rate R of transition from X to Y is then defined as the net flux of atoms across the line LL' to the right divided by the number of atoms in the ensemble.

The atoms, being thermally activated, are subject to random thermal forces, and this of course implies that they are also subject to damping as they move in the potential field. We shall assume that this damping is intermediate in strength, so that atoms approaching LL' from the left are not appreciably deviated by the thermal forces while in the immediate neighbourhood of LL', but that, on the other hand, atoms which fall back towards X do in effect reach

§ 16.5

The assumption of intermediate damping

Fig. 16.2. To illustrate thermal activation. In (*a*) we show potential energy ϕ as a function of position for a one-dimensional system. In (*b*) we show contours of potential energy for an atom moving in two dimensions. Here σ_x and σ_y are the rms displacements in x, y near the metastable state, and σ_ξ is the rms value of ξ near the activated state, which is a measure of the width of the 'window' through which the atom passes in travelling from X to Y via the activated state A.

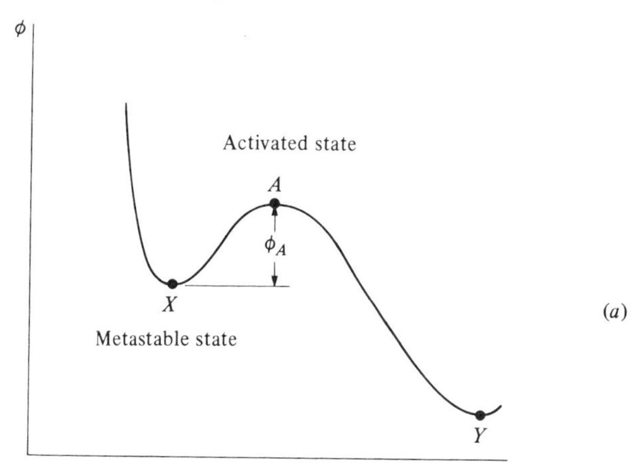

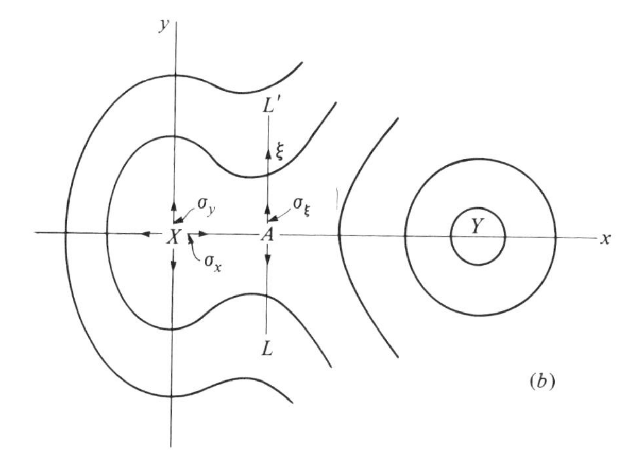

thermal equilibrium before they return to LL'. This means, firstly, that we may calculate the *net* flux from X to Y as the *actual* flux of atoms moving from left to right across LL' (since virtually all atoms crossing from left to right will continue towards Y and thus be removed from the ensemble, and the number returning will be negligible) and, secondly, that we may calculate this flux from left to right *as though* the distribution at A were in thermal equilibrium with the distribution near X (although in reality it is not, for there are very few atoms crossing LL' from right to left). This assumption of intermediate damping is often, but not always, a good approximation. (It tends to give the largest possible value for the rate R: if the damping were heavier we should have to think of the atoms *diffusing* past the activated state A, and it may be shown that this is a slower process than the simple ballistic one considered here, while if the damping were so light that the atoms crossing LL' from the left had not reached thermal equilibrium the number crossing would be *lower* than the equilibrium number adopted here. An example of a *diffusive* activated process appears in Chapter 18.)

§ 18.7

Using this assumption of intermediate damping we may write the flux density across LL' to the right as $\frac{1}{2}\bar{v}n$, where $\bar{v}$ is the mean value of $|v_x|$ for the Maxwell one-component velocity distribution, equal to $(2kT/\pi m)^{\frac{1}{2}}$, and n is the number of atoms per unit area in the *equilibrium* ensemble. The total flux from left to right is therefore

§ 5.2

$$\tfrac{1}{2}\bar{v}n_X\,\mathrm{e}^{-\phi_A/kT}\int_{LL'}\mathrm{e}^{-\delta\phi/kT}\,\mathrm{d}\xi, \qquad (16.27)$$

where we have used the Boltzmann distribution to write n as $n_X \exp(-\phi/kT)$, using ϕ_X as our zero of potential, $\delta\phi$ is $\phi-\phi_A$, and the integral is taken along the line LL'. At the same time, the total number of atoms in the ensemble, which will be concentrated near X, may be written as

$$n_X\int_A \mathrm{e}^{-\phi/kT}\,\mathrm{d}x\,\mathrm{d}y, \qquad (16.28)$$

where the integral is taken over the area of the plane near X. Now, for small values of ξ we may reasonably suppose that $\delta\phi$ varies quadratically with ξ along LL', so that the integral in (16.27) may be rewritten as $\sqrt{(2\pi)}\sigma_\xi$, where σ_ξ is the rms value of ξ, averaged over the Boltzmann distribution: σ_ξ is a measure of the width of the 'window' through which escaping atoms pass at A, and is temperature-dependent. In the same way the integral in (16.28) may be rewritten as $2\pi\sigma_x\sigma_y$, where σ_x and σ_y are the rms deviations in x and y for atoms oscillating in the potential minimum at X. In terms of these quantities the rate for the activated process, which is the ratio of expressions (16.27) and (16.28), may be written as

$$R=(1/2\pi)(kT/m)^{\frac{1}{2}}(\sigma_\xi/\sigma_x\sigma_y)\,\mathrm{e}^{-\phi_A/kT}. \qquad (16.29)$$

Now σ_x^2 is in fact equal to kT/α_x, where α_x is the force constant for vibration in the x direction at X, and we also recognise that $(1/2\pi)(\alpha_x/m)^{\frac{1}{2}}$ is just v_x, the corresponding vibration frequency. We may therefore conveniently express R in the form

$$R = \nu_x(\sigma_\xi/\sigma_y)\,\mathrm{e}^{-\phi_A/kT}. \tag{16.30}$$

The basic rate formula

This formula has a simple interpretation. The frequency ν_x is the number of times per second at which a given atom crosses the y axis at X while travelling to the right: we may regard this as an *attempt frequency*, the number of times per second that the atom tries to escape. The rest of the expression may be regarded as the probability of escaping at any given attempt. It contains two terms. The term σ_ξ/σ_y is the ratio of the widths of the 'windows' at A and at X through which the atom has to pass, and is independent of temperature. The final term is a temperature-dependent Boltzmann factor, which gives the ratio of the equilibrium number densities at A and X.

The calculation just described involves only the two-dimensional motion of a single particle in the classical limit. In more complicated examples of thermal excitation the configuration space will have more dimensions, and it is sometimes necessary to use a quantum rather than a classical treatment. The final results, however, look very similar to (16.30). For instance, this method of calculation may be applied to the problem of determining from first principles the rate of a chemical reaction in the gas phase, in which a set of molecules have to meet and form an activated complex, which then decays to some new set of molecules of lower energy. It may be shown that, when we have just one of each of the molecules required in some given volume, the reaction rate is given by

The absolute rate theory of chemical reactions

$$R = (kT/h)\left(\zeta^* \Big/ \prod_i \zeta_i\right)\mathrm{e}^{-\phi_A/kT}, \tag{16.31}$$

where the ζ_i are the partition functions for the separate reacting molecules and ζ^* is the part of the partition function for the activated complex corresponding to its bounded motions, that is, its motions *parallel* to the 'window' in configuration space, analogous to motion in the ξ direction in Fig. 16.2. This calculation is known to chemists as the *Eyring absolute rate theory* of chemical reactions, and you will find details of it in texts of chemical statistical mechanics. You may like to check that for simple decomposition of a single molecule in the classical limit (16.31) reduces to an expression equivalent to (16.30).

The temperature dependence of the rate R of a thermally activated process is concentrated in the Boltzmann factor which, since ϕ_A is in practice, as we shall see, considerably greater than kT, increases rapidly as the temperature is raised. The remainder of the expression has little temperature dependence, and may safely be treated as constant over a small temperature range. In this situation the activation energy may be determined experimentally from an *Arrhenius plot* of the experimental value of $\ln R$ against $1/T$. The result should be a straight line of slope $-\phi_A/k$. An example of such a plot is shown in Fig. 16.3.

The temperature dependence of the rate R

Two useful rules of thumb

Two useful facts about thermally activated processes may be deduced from the fact that the attempt frequency is typically an atomic or molecular vibration frequency, of order 10^{12} to 10^{13} s^{-1}, or thereabouts, while the ratio of the 'windows' is not usually very far from unity, linking this with the fact that if we observe the activated process and measure its rate, this in itself means that R probably lies between, say, 10^{-2} and 10^{2} s^{-1}. (Otherwise we should regard the process either as not occurring at all, or as explosive and virtually instantaneous.) It follows that $\exp(-\phi_A/kT)$ normally lies between 10^{-10} and 10^{-15}, and thus that ϕ_A/kT lies between about 23 and 35. Two remarkable conclusions follow. Firstly, at room temperature *most observed activation energies have the same order of magnitude*, about 10^{-19} J or rather less than 1 eV. Secondly, *if T increases by a factor of $\frac{1}{30}$* (that is, 10 K at room temperature) *the rate of the activated process will roughly double.* These simple rules of thumb apply to many common microscopic activated processes.

Fig. 16.3. Experimental Arrhenius plot for the chemical reaction $D + H_2 \rightarrow HD + H$ (after Ridley, Schulz & Le Roy, 1966). Such plots are often less linear than expected because mechanisms other than simple activation are involved. In this case the curvature at low temperatures arises because the system may pass *through* the potential energy barrier by quantum mechanical tunnelling as well as over it by thermal excitation. The solid curve is fitted to (16.31) with tunnelling correction, with ϕ_A as adjustable parameter: the fit is clearly very good. See problem 16.10.

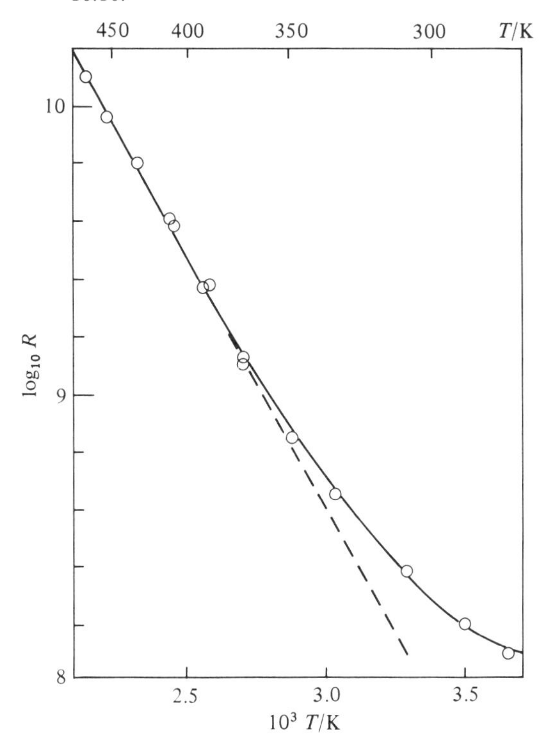

16.7 The dependence of scattering rates on occupation numbers

Generally speaking the problem of finding expressions for rates of scattering is a matter of detailed kinetics. We shall, however, consider briefly here the special question of how scattering rates for photons, electrons and other particles depend on their occupation numbers. There are two reasons for doing so. It will throw a useful light on the nature of thermal equilibrium in such systems. It will also provide a method for calculating the rate of approach to thermal equilibrium in certain important cases.

Einstein considered the problem of a quantum system (such as an atom) which is free to interact with the photons inside some enclosure (Fig. 16.4). Consider in particular two states 1 and 2 of the system separated by an energy $\Delta\varepsilon = \varepsilon_2 - \varepsilon_1$, and the corresponding photons of energy $h\nu = \Delta\varepsilon$. It is natural to assume that when the system is in the lower state the probability of excitation to the upper state by absorption of a photon is proportional to n, the number of photons present, but that when the system is in the upper state the probability of decay, with emission of a photon, is a property of the system alone and independent of the number of photons already present. Einstein showed that *these assumptions are inconsistent with the Boltzmann and Planck distributions*, as follows. We assume that the enclosure is in equilibrium at temperature T. Then the net rate at which upward transitions occur is given by

Einstein's analysis of the radiation probabilities

$$dp_2/dt = p_1\bar{R}_\uparrow - p_2\bar{R}_\downarrow, \tag{16.32}$$

where p_1 and p_2 are the Boltzmann probabilities for states 1 and 2, and $\bar{R}_\uparrow$ and $\bar{R}_\downarrow$ are the mean rates for the upward and downward transitions at temperature T. We deduce at once that, since $dp_2/dt = 0$ in equilibrium, $\bar{R}_\uparrow/\bar{R}_\downarrow = p_2/p_1 = \exp(-h\nu/kT)$. Now, according to Planck's distribution, the mean occupation number $\bar{n}$ for photons of frequency ν is $[\exp(h\nu/kT) - 1]^{-1}$. Using this to express $\exp(h\nu/kT)$ in terms of $\bar{n}$ we find that

§9.1.1

$$\bar{R}_\uparrow/\bar{R}_\downarrow = \bar{n}/(1+\bar{n}). \tag{16.33}$$

This is clearly inconsistent with the assumptions that $R_\uparrow \propto n$ and $R_\downarrow$ is independent of n. Einstein suggested instead that

$$\blacktriangleright \quad \begin{aligned} R_\uparrow &= R_0 n \\ R_\downarrow &= R_0(n+1) \end{aligned}, \quad \text{for absorption and emission of photons,} \tag{16.34}$$

where n is the number of photons present *before* the jump occurs. These hypotheses were later confirmed when quantum theory was applied to the radiation field of quantum systems by Dirac. Notice that the formula for $R_\downarrow$ contains two parts. There is a term R_0 which is independent of n, and represents the rate of *spontaneous emission*, which can occur when there are initially no photons present. There is also a term $R_0 n$. This represents the rate of *stimulated emission*, an extra emission process whose rate is proportional to the number of photons present. The constant R_0 is known as the *Einstein coefficient*.

One may reach the same conclusion in a slightly simpler way by considering the enclosure to be thermally isolated. Let us assume that $R_\uparrow = R_0 n = R_0 n_A$, where n_A is the number of photons present *before* the photon is absorbed (state A in the middle row of Fig. 16.4). Now, according to the principle of jump rate symmetry the probability for the reverse transition (from state B to state A in the diagram) should be the *same*. Thus $R_\downarrow = R_0 n_A = R_0(n_B + 1)$, where $n_B = n_A - 1$ is the number of photons present in state B *before* the new photon is emitted. This is consistent with (16.34).

Similar considerations apply when the energy required to excite the quantum system comes not from absorption of a photon, but from the scattering of a boson or fermion from a higher energy state to one of lower

Fig. 16.4. The effect of occupation number on scattering rates. The top box represents a small system with levels ε_1 and ε_2 interacting with photons of frequency ν at temperature T, as considered by Einstein. The second row shows two states of the same box which have the same energy: according to the principle of jump rate symmetry the jump rate from A to B should be equal to the reverse rate from B to A. The third row shows a similar situation in which the energy to excite the small system comes from inelastic scattering of a boson.

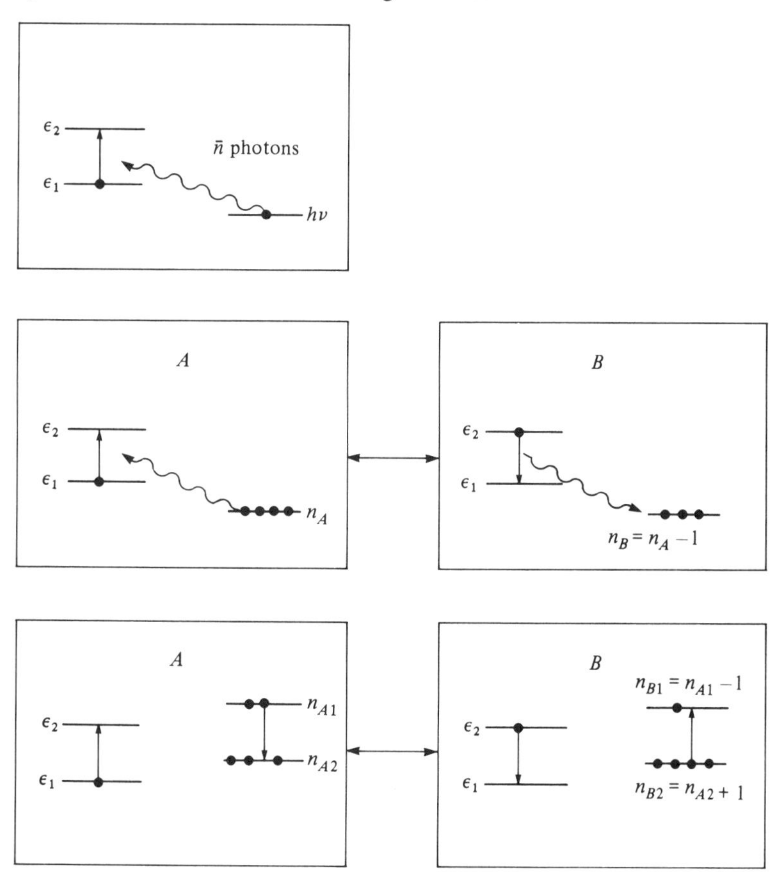

energy. In the case of the boson (bottom row of Fig. 16.4) it is natural to assume that the rate of the forward transition from A to B is proportional to n_{A1}, and that the rate of the reverse transition from B to A is proportional to n_{B2}. The principle of jump rate symmetry applied to the complete enclosure then forces us to conclude that the reverse transition rate is proportional not only to n_{B2} but also to $n_{A1}=n_{B1}+1$, and that the forward rate is proportional to $n_{B2}=n_{A2}+1$ as well as n_{A1}. Thus

$$\left.\begin{aligned} R_\uparrow &= R_0 n_1(n_2+1) \\ R_\downarrow &= R_0 n_2(n_1+1) \end{aligned}\right\} \text{ for boson scattering,} \tag{16.35}$$

where in each case n_1 and n_2 refer to the boson occupation numbers *before* the transition occurs, and R_0 is a constant. Again we observe an effect of stimulated emission: the jump rate is increased as the number of bosons in the *final* state increases. For fermions, on the other hand, the exclusion principle requires that the transition can only occur if the final fermion state is *empty*. Thus

$$\left.\begin{aligned} R_\uparrow &= R_0 n_1(1-n_2) \\ R_\downarrow &= R_0 n_2(1-n_1) \end{aligned}\right\} \text{ for fermion scattering.} \tag{16.36}$$

These results, too, are in accord with exact quantum theory.

Let us now see how such results can help us in understanding the approach to equilibrium. Consider, for instance, the approach to equilibrium of the electron distribution in a metal or semi-conductor under the influence of elastic scattering by lattice defects, and inelastic scattering by absorption and emission of phonons. For elastic scattering between states 1 and 2, if the corresponding occupation numbers are $\tilde{f}_1$ and $\tilde{f}_2$, we have a *net* rate of scattering into state 2 given by

Elastic scattering of electrons

$$\begin{aligned} \mathrm{d}\tilde{f}_2/\mathrm{d}t &= R_0[\tilde{f}_1(1-\tilde{f}_2)-\tilde{f}_2(1-\tilde{f}_1)] \\ &= R_0(\tilde{f}_1-\tilde{f}_2). \end{aligned} \tag{16.37}$$

In equilibrium $\tilde{f}_1$ and $\tilde{f}_2$ are both equal to the Fermi function, so $\mathrm{d}\tilde{f}_2/\mathrm{d}t=0$. When out of equilibrium we see that the net scattering rate is proportional to $\tilde{f}_1-\tilde{f}_2$. For phonon scattering, on the other hand, we have to take the phonon occupation into account (using the same rule as for photons). If an electron can be scattered from state 1 into state 2 with *absorption* of a phonon, we have, on combining rules (16.34) and (16.36),

$$\begin{aligned} \mathrm{d}\tilde{f}_2/\mathrm{d}t &= R_0[\tilde{f}_1(1-\tilde{f}_2)\tilde{n}_{\mathrm{ph}}-\tilde{f}_2(1-\tilde{f}_1)(1+\tilde{n}_{\mathrm{ph}})] \\ &= R_0[(\tilde{f}_1-\tilde{f}_2)\tilde{n}_{\mathrm{ph}}-\tilde{f}_2(1-\tilde{f}_1)], \end{aligned} \tag{16.38}$$

where $\tilde{n}_{\mathrm{ph}}$ is the phonon occupation number. If we substitute the Fermi and Planck distributions into (16.38) we find that $\mathrm{d}\tilde{f}_2/\mathrm{d}t=0$ in equilibrium, as expected. For a small departure from equilibrium in the electron distribution, represented by $g=\tilde{f}-f$, where f is the Fermi function, we find to first order in g that

Inelastic scattering of electrons by phonons

$$\mathrm{d}g_2/\mathrm{d}t = R_0[(g_1-g_2)n_{\mathrm{ph}}-g_2(1-f_1)+g_1 f_2], \tag{16.39}$$

if we assume that n_{ph} is undisturbed. This result is less intuitively obvious than (16.37). The term in n_{ph} represents the effect of absorption and stimulated emission. The remaining two terms are due to spontaneous emission. Using such results we may find how the non-equilibrium distribution changes as it approaches the Fermi distribution, which would be useful in calculating the electrical or thermal conductivity as a function of temperature, for instance, as we shall see in the next chapter.

Problems

Fluctuations in the Fermi function

16.1 Apply (16.1) to the fluctuations in occupation of a fermion state in contact with a reservoir, and show that the mean square fluctuation $\overline{(\Delta f)^2}$ is equal to $\bar{f}(1-\bar{f})$, where $\bar{f}$ is the mean occupation. [You may find it helpful to prove first the useful result $\overline{(\Delta x)^2} = \overline{x^2} - \bar{x}^2$.]

Fluctuations of electric dipole moment

16.2 Apply (16.3) to the fluctuations in p_z, the z component of the electric dipole moment $\mathbf{p}_0$ of a polar molecule in zero electric field, using classical statistics. Show first that, as in problem 15.1, $g(p_z)\,\mathrm{d}p_z$ is proportional to $\sin\theta\,\mathrm{d}\theta$, where θ is the angle between $\mathbf{p}_0$ and the z axis, and deduce that the mean square fluctuation in p_z is equal to $\frac{1}{3}p_0^2$.

Show also that the fluctuations in p_x, p_y and p_z are not statistically independent.

Accuracy of the Gaussian distribution of fluctuations in large systems

16.3 The complete expansion of $S(x)$ around the equilibrium point to be used in (16.9) is

$$S(x) = S_0 + \frac{1}{2!}(\partial^2 S/\partial x^2)\,\Delta x^2 + \frac{1}{3!}(\partial^3 S/\partial x^3)\,\Delta x^3 \ldots.$$

Consider a change in x with $\Delta x = \delta$ large enough to make $(\partial^2 S/\partial x^2)\delta^2$ of order Nk. This implies that the number of configurations *per particle* has changed by a small but not negligible factor. Thus, at the microscopic level, the new state is appreciably altered, but not unrecognisably different. In general we may expect the corresponding *change* in $\partial^2 S/\partial x^2$ to be neither much smaller nor much greater than $\partial^2 S/\partial x^2$ itself. Use this expectation to show that in a typical fluctuation the third term in the expansion is likely to be smaller than the second by a factor of order $N^{\frac{1}{2}}$. [It follows that in systems of everyday size the approximations used in (16.10) and (16.13) are likely to be extremely good.]

Fluctuations of V at fixed P for a system in thermal isolation

16.4 Show that for changes in V at fixed N in a thermally isolated system $\mathrm{d}A$ may be written as $(P_R - P)(T_R/T)\,\mathrm{d}V$. Deduce that the mean square volume fluctuation in a thermally isolated system subject to surroundings at fixed pressure is equal to $-kT(\partial V/\partial P)_{S,N}$, which may be found by measuring the *adiabatic* compressibility of the system.

16.5 Integrate the double fluctuation distribution (16.15) over E, and check that you recover the fluctuation distribution $f(V)$ at fixed P and T, and hence (6.14). To do so, first write the exponent of (16.15) as

$$-\tfrac{1}{2}\alpha_{EE}[\Delta E + (\alpha_{EV}/\alpha_{EE})\,\Delta V]^2/kT - \tfrac{1}{2}(\alpha_{VV} - \alpha_{EV}^2/\alpha_{EE})\,\Delta V^2/kT.$$

On integrating over ΔE, the first term reduces to a constant, and we find that $f(V) \propto \mathrm{e}^{-\frac{1}{2}(\alpha_{VV} - \alpha_{EV}^2/\alpha_{EE})\,\Delta V^2/kT}$.

To recover the usual form for $f(V)$ we need to show that the term in brackets is $-(\partial P/\partial V)_T = -T[\partial(P/T)/\partial V]_T$. Do this conveniently by writing $d(P/T)$ in terms of its partial derivatives, using E and V as variables.

16.6 We saw in § 9.3.1 that when a vacancy is created in a crystal the change in free energy may be written as $\delta F_c - kT \ln [(N+N_v)/N_v]$. Use the quasi-Boltzmann distribution to show that the rms fluctuation in N_v is close to $N_v^{\frac{1}{2}}$ when $\delta F_c \gg kT$.

Vacancy fluctuations

16.7 Show that when two white noise voltage sources are connected in series, the *squares* of the noise voltages (not their magnitudes) should be added. Confirm that this conclusion is consistent with the Nyquist formula applied to a pair of resistances R_1 and R_2 in series at temperature T, considered as a single unit. If the resistances are at different temperatures T_1 and T_2, what is the effective noise temperature of the combination?

The Johnson noise for resistors in series

16.8 A resistance R is connected across the terminals of an ideal capacitance C, the whole circuit being at temperature T. What is the power spectrum of the noise voltage across C? Deduce from it the mean square noise voltage across C, and comment on your result.

Thermal noise voltages apparent in a circuit

16.9 By an argument analogous to that of § 16.5, show that the power spectral density of the noise force F_N acting on the bob of a lightly damped pendulum in the direction in which it is free to swing is equal to $4\alpha kT$ at the pendulum frequency, where α is the drag coefficient for the damping force acting on the bob.

The mechanical equivalent of the Nyquist formula

16.10 Using the upper part of the Arrhenius plot of Fig. 16.3, deduce the activation energy for the reaction $D + H_2 \rightarrow HD + H$.

17

Transport properties

When we apply a temperature gradient ∇T to a block of copper, heat flows through the copper. For small temperature gradients the ratio of the heat flux density $\mathbf{H}$ to $-\nabla T$ is constant, and is known as the *thermal conductivity*, κ. In the same way a voltage gradient ∇V will drive an electric current density $\mathbf{J}$ through the copper, and the ratio of $\mathbf{J}$ to $-\nabla V$ is the *electrical conductivity*, σ (Fig. 17.4). Experiment shows that there are also cross-terms. It turns out that a temperature gradient may drive a small electric current and that a voltage gradient may drive a small heat flow: these are *thermoelectric effects*. There are many other such transport properties. For instance, the *diffusion coefficient* of one gas through another is a measure of the flux density of particles generated by a gradient of number density; and even the *viscosity* of a fluid may be treated as a measure of the flux density of transverse momentum generated by a gradient of transverse velocity in the fluid. In each of these cases the system itself reaches a steady state of statistical equilibrium, subject to the driving force applied. But we note that the assembly as a whole is not in true thermal equilibrium, because some quantity is being transported through the system, and the reservoirs to which it is attached are changing their states: in the case of thermal conduction, for instance, one reservoir is warming, and one reservoir is cooling, even if only very slowly. In fact, the study of transport processes is a special part of the general study of how systems *approach* equilibrium, on which we first embarked in the preceding chapter.

Problems 17.1, 17.2

The discussion of transport processes turns on how the constituent particles move through the system, how they are scattered, and how the entropy increases during this process. We shall begin the chapter by considering two elementary approaches which, though often useful, have serious limitations. We shall then study a general and more precise formalism, the *Boltzmann transport equation*, in the particular context of transport problems in metals. Finally we shall explore the behaviour of the entropy in the thermoelectric effects, and discuss some newer, general ideas, introduced by Onsager, which have become the basis of the growing subject of *irreversible thermodynamics*.

17.1 The mean free path method

The mean free path method assumes that the quantity transported is carried by the particles in the system (the molecules in a gas, the electrons in a metal or semi-conductor, and the phonons, or quanta of lattice vibration, in an insulating crystal, for instance) and that these particles move freely between one collision and the next. If we choose a particle at random at some instant in time, it will have travelled some distance p, known as its *free path*, since its last collision. The free path will have some statistical distribution, and there will be a *mean free path*, l. In its simplest form, the method assumes that a particle found at any given point belongs to an equilibrium distribution which is characteristic not of the point where it now is, *but of the point where it last collided*: for heat flow in a gas, for instance, we assume that the molecules crossing any plane of interest have a Maxwell–Boltzmann velocity distribution characteristic of the temperature and number density at the point of previous last collision.

Let us indeed, to illustrate the method, calculate the thermal conductivity of a gas in terms of its mean free path. We consider the net energy crossing unit area of a plane PP', normal to the temperature gradient in the gas, from the hot side to the cold side (Fig. 17.1). We segregate the molecules according to their speeds c and the angles θ which their velocities make with the normal to the plane. Then the heat flux density from hot to cold may be written down as

Thermal conductivity of a gas calculated in terms of l

$$H=\int_0^{\pi}\int_0^{\infty}\underbrace{[nf(c)\,\mathrm{d}c\,\tfrac{1}{2}\sin\theta\,\mathrm{d}\theta]}_{\text{number density}}\underbrace{(c\cos\theta)}_{\text{normal velocity}}\underbrace{[-c_{\mathrm{m}}l\cos\theta\,\mathrm{d}T/\mathrm{d}z]}_{\text{mean excess energy}}. \tag{17.1}$$

Here the first bracket represents the number density of molecules having speeds between c and $c+\mathrm{d}c$, and angles between θ and $\theta+\mathrm{d}\theta$. This is multiplied by their component of velocity normal to the plane, to yield a flux density of *particles* from hot to cold. This is converted to a heat flux density on multiplying by the final bracket, which represents the *excess* heat carried by each molecule, assuming that its energy is, on average, characteristic of a point a distance l away: here c_{m} is the heat capacity per molecule, and $\mathrm{d}T/\mathrm{d}z$ is the temperature gradient. The expression is to be integrated over all speeds c, and over values of θ from 0 to π, which allows for the heat deficits of molecules arriving from the cold side, as well as the energy surplusses of those arriving from the hot side. On performing the integrals we find that the thermal conductivity is given by

▶ $$\kappa=-H/(\mathrm{d}T/\mathrm{d}z)=\tfrac{1}{3}c_{\mathrm{m}}nl\bar{c}. \tag{17.2}$$

This result is only of use if we can make some estimate of l. For collisions in a gas we may make a *crude* estimate as follows. We take the molecules to be hard spheres of diameter σ. Then if any two molecules are

Crude estimate of l in a gas

moving so that their centres would pass within a distance σ of one another if undeflected, a collision will occur. To estimate l, we treat the gas as though all molecules other than the molecule of interest are at rest. Then we may treat the moving molecule as sweeping out an imaginary tube of cross-section $\pi\sigma^2$, the *collision cross-section*, such that if the centre of any other molecule lies within the tube, a collision will occur (Fig. 17.1). If the number density of molecules is n, the mean number of collisions for a path of length s will be $n\pi\sigma^2 s$, the mean number of collisions per unit path will be $n\pi\sigma^2$, and the mean free path will be given by

Fig. 17.1. Transport properties calculated by the mean free path method. The upper diagrams show heat transport in a gas in the z direction. We assume that molecules crossing the plane PP' bring excess heat corresponding to point S, distance l away, where the temperature excess is $-l\cos\theta(\mathrm{d}T/\mathrm{d}z)$. The unit sphere on the right shows that the fraction of molecules whose velocities have angles to the z axis between θ and $\theta+\mathrm{d}\theta$ is equal to $2\pi\sin\theta\,\mathrm{d}\theta/4\pi$, or $\frac{1}{2}\sin\theta\,\mathrm{d}\theta$. The lower diagram shows how l may be estimated, using a collision cross-section equal to $\pi\sigma^2$, where σ is the diameter of the molecule.

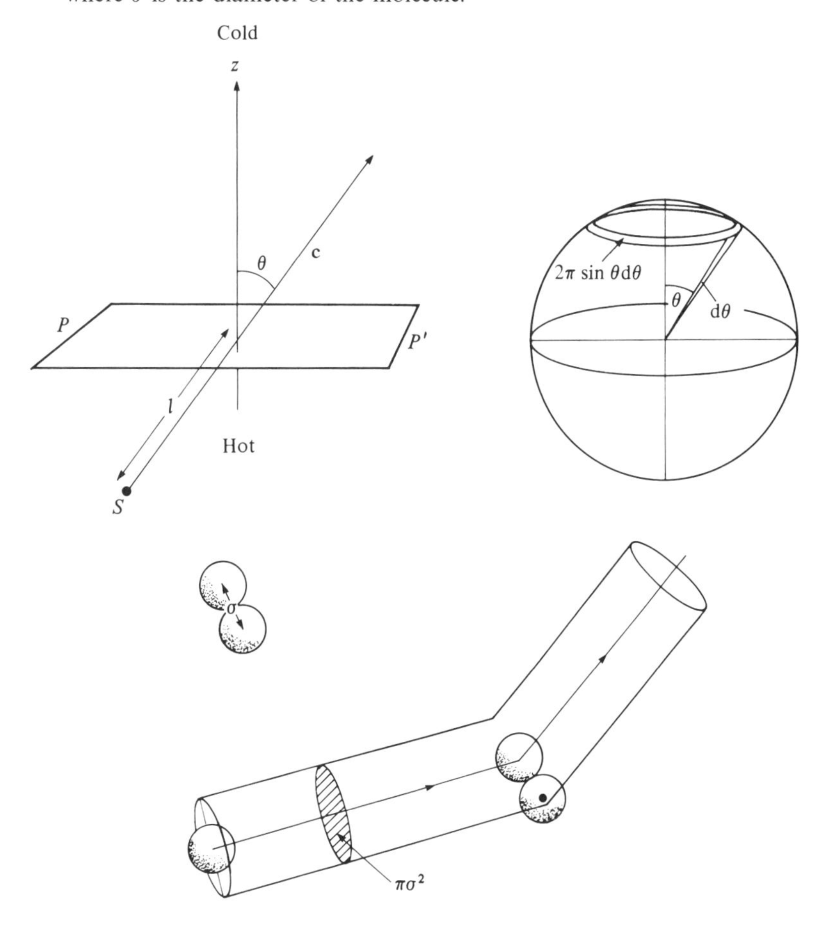

$$l = 1/n\pi\sigma^2. \tag{17.3}$$

According to this rather crude estimate, the conductivity should be given by

$$\kappa = c_m \bar{c}/3\pi\sigma^2. \tag{17.4}$$

The success of this formula may be judged by inspecting the experimental results for argon shown in Fig. 17.2. Our formula predicts first that the thermal conductivity κ should be *independent* of number density at fixed temperature. This perhaps surprising result comes about because at low densities the decrease in the number of particles available to carry the heat is exactly offset by the increase in the mean free path due to the reduced probability of collision. The diagram shows that this prediction is accurately confirmed, at ordinary pressures. (Our theory becomes inappropriate at very low and very high pressures, at low pressures because the free path is limited by the size of the apparatus, and at high pressures because the molecules are so close together that the free path concept breaks down.) However, when we come to consider the absolute value of κ and its temperature dependence, we find much less satisfactory agreement. The dashed plots in Fig. 17.2 are obtained from (17.4), using $\frac{3}{2}k$ for c_m, the Maxwell–Boltzmann value for $\bar{c}$ (which is proportional to $T^{\frac{1}{2}}$), and a fixed value of 3.41×10^{-10} m for σ, which we take to be equal to r_0 in Fig. 14.1. At *high* temperatures the effective scattering cross-section indicated by the measured value of κ is evidently

p. 154

Fig. 17.2. The thermal conductivity of argon (after Flowers & Mendoza, 1970). On the left, the solid line shows the measured values of κ at 313 K for a wide range of pressures (Waelbroeck & Zuckerbrodt, 1958, and Michels *et al.*, 1956). On the right the points show measurements of κ as a function of temperature at atmospheric pressure (Kannuluik & Carman, 1952) compared with a solution of the Boltzmann equation using a 6–12 potential to calculate the scattering term (solid line): the agreement is excellent. The dashed lines show the predictions of (17.4) with $\sigma = 3.41 \times 10^{-10}$ m: this simple mean free path model predicts correctly that κ is independent of density, and gives the right order of magnitude, but is clearly not a precise theory.

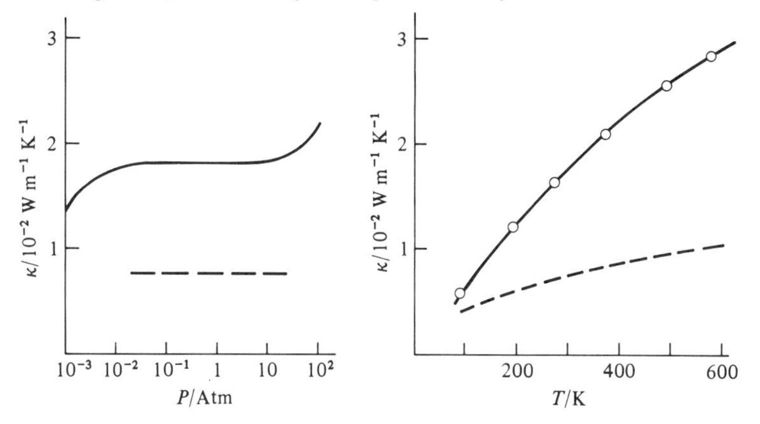

considerably smaller than our hard sphere value, corresponding to an unexpectedly long free path. As the temperature is lowered, however, the effective scattering cross-section appears to increase by a factor of more than 2. Moreover, if we make an equivalent analysis of the *viscosity* of argon, we find that the effective mean free path needed to explain the measured viscosity proves to be about 2.4 times smaller than the effective value needed to explain the thermal conductivity at the same density and temperature.

Problem 17.2

Defects of the mean free path method

It is not hard to see how such discrepancies arise. The method, as here presented, has three serious defects. First, in calculating l we ignored the motions of the other molecules. As Maxwell showed, this is only valid for very fast-moving molecules. For slow-moving molecules l actually tends to zero. This consideration reduces the effective value of l. Secondly, we assumed that molecules which have just collided have an *equilibrium* distribution corresponding to the local values of n and T. But in reality it takes more than one collision to establish equilibrium – the quantities transported tend to *persist*. Because of this persistence the effective mean free path for transport is usually considerably greater than the actual mean free path. This is the reason for the large apparent values of l at high temperature. It is also the reason why the effective value of l is greater for thermal conductivity than it is for viscosity: transport of energy turns out to be more persistent than transport of transverse momentum. Finally, we used a hard sphere model. This is reasonable for argon at high temperatures, where the molecules move so fast that the intermolecular attractions have little effect. But at lower temperatures the intermolecular attractions lead to substantial deflections – small-angle scatterings, in effect – when pairs of molecules pass at distances considerably greater than σ: these 'attractive collisions' explain the increase in effective scattering cross-section at low temperatures.

The mean free path method may be applied to transport problems of many types, and considerable efforts have been made to overcome the difficulties described. It is, however, not easy to do so in a systematic and straightforward way. It seems best to regard the method as an approximation which is successful in predicting the general pattern of transport behaviour in an understandable way, but is not appropriate as an *exact* treatment.

17.2 Relaxation time approximations

A rather different elementary approach to transport properties is to make what is known as a *relaxation time approximation*. The idea is that the steady state set up in a transport process is the result of two competing influences, the driving force applied to the system, which tries to increase the transport, and the scattering, which tries to decrease it. For instance, in a free electron metal containing a number density of electrons n, an electric field $\mathscr{E}$ will give an acceleration $-e\mathscr{E}/m_e$ to each electron. If there were no scattering, the current density would increase at a rate given by

$$\mathrm{d}\mathbf{J}/\mathrm{d}t = (ne^2/m_e)\mathscr{E}.$$

But the scattering tends to decrease the current. In the absence of any field we might expect the current density to decay exponentially with time, corresponding to the equation

$$\mathrm{d}\mathbf{J}/\mathrm{d}t = -\mathbf{J}/\tau_J,$$

where τ_J is a characteristic time for decay of the current by scattering, the *relaxation time for electric current density*. This is the relaxation time approximation for this phenomenon. If the approximation is a good one, we may combine these two competing effects in a single equation,

The connection between transport and relaxation time

$$\mathrm{d}\mathbf{J}/\mathrm{d}t = (ne^2/m_e)\mathscr{E} - \mathbf{J}/\tau_J. \tag{17.5}$$

In the steady state $\mathrm{d}J/\mathrm{d}t$ is zero, and we deduce that the electrical conductivity must be given by the formula

▶ $$\sigma = J/\mathscr{E} = ne^2\tau_J/m_e. \tag{17.6}$$

According to this argument, then, there is a direct connection between the transport property σ and the corresponding relaxation time τ_J. Similar arguments may be applied to other transport processes.

Physicists commonly make a relaxation time assumption of this sort, in discussions of low and high frequency conductivity in metals and semiconductors, for instance, and in analysing the high frequency behaviour of dielectric and paramagnetic substances. It is often simply assumed that a relaxation time exists, and may be found if required by considering the detailed kinetics of the relaxation mechanism. Note carefully that τ is not necessarily the same for the relaxation of transport currents of different quantities, nor is it necessarily the same as the *collision time*, the average time until its next collision for a particle chosen at random. In some cases these times are the same. For instance, in a free electron metal in which the dominant scattering mechanism is elastic scattering by small impurities, each scattering completely randomises the direction in which the electron is moving. In this situation the relaxation times for both the electric current and the heat current are equal to the scattering time. On the other hand, for a metal in which the dominant mechanism is small-angle inelastic scattering by low temperature phonons, the electric current relaxation is much slower than that of the heat current, and both relaxation times are longer than the scattering time, as we shall see in more detail later.

Problems 17.3, 17.4

The meaning of the relaxation time, and the validity of relaxation time approximations

The relaxation time approximation is more often valid than the simple mean free path model, but it is by no means always correct. In the case of an electric current, for instance, a given current density may be carried by many different electron distributions. Since the different distributions decay by different scattering processes, and, therefore, in general, at different rates, there can be no unique τ_J which applies to all cases. And, indeed, a current carrying distribution need not retain the same form as it decays: a current originally carried by a special small group of electrons may well become spread over a larger group as it decays. Since the rate of decay of current for the two groups may well be different, we have no reason even to suppose that an electric current will necessarily decay exponentially with time; and there are indeed important cases where it does not. Thus the relaxation time method, like the mean free path method, though often useful, is inappropriate as a general exact treatment of transport properties.

17.3 The Boltzmann transport equation

The Boltzmann transport equation is the equation of motion of the distribution function for a system of particles which are moving under the influence of external fields, and subject to scattering. It was developed by Boltzmann for studying the dynamics of ordinary gases, and has since been successfully applied to many other systems.

We shall develop the equation for the special case of electrons in a metal or semi-conductor. Adopting a semi-classical picture, we treat each electron as having a definite position **r** and crystal momentum **p**. As it moves in one of the allowed energy bands of the crystal, it obeys the following dynamical equations, which are derived in texts of solid state physics:

$$\varepsilon = \varepsilon_{\mathbf{p}} - e\phi(\mathbf{r}) \tag{17.7}$$

$$\mathbf{v}(\mathbf{p}) = \nabla_{\mathbf{p}}\varepsilon_{\mathbf{p}} \tag{17.8}$$

$$\mathrm{d}\mathbf{p}/\mathrm{d}t = \mathbf{F} = e\nabla\phi. \tag{17.9}$$

Here ε, the total energy of the electron, is the sum of the *band energy* $\varepsilon_{\mathbf{p}}$ ($p^2/2m$ for free electrons) and a potential energy $-e\phi(\mathbf{r})$, where ϕ is the part of the electrostatic potential due to the external field; **v** is the velocity of the electron, which is equal to the gradient of $\varepsilon_{\mathbf{p}}$ in momentum space; and **F** is the external force acting on the electron, equal to $e\nabla\phi$ if there is no magnetic field present. We suppose that there is some scattering mechanism present which causes electrons to jump locally from one value of crystal momentum to another, in accordance with the rate formulae discussed in the preceding chapter.

§16.7

In thermal equilibrium in the absence of external driving forces the occupation number for an electron state of crystal momentum **p** is given by the Fermi function

§12.3

$$f = \frac{1}{\mathrm{e}^{(\varepsilon-\mu)/kT}+1}, \tag{17.10}$$

and is independent of position. But in a general non-equilibrium state the corresponding occupation number $\tilde{f}$ will differ from f, and may depend on **r** as well as on **p**. We must therefore consider $\tilde{f}$ to be defined in a six-dimensional space, with three real space and three momentum space co-ordinates. Our aim is to find the equation of motion of $\tilde{f}(\mathbf{r}, \mathbf{p}, t)$.

Consider first a system without scattering. An electron originally at **r** with crystal momentum **p** will move in real space and in momentum space, according to (17.8) and (17.9). In the absence of scattering, the occupation number $\tilde{f}$ for the state in which the moving electron finds itself must remain constant, so that $\mathrm{d}\tilde{f}/\mathrm{d}t=0$ when the total derivative applies to a point moving with the electron in the six-dimensional space. Writing $\mathrm{d}\tilde{f}$ in terms of its partial derivatives we find that

$$\mathrm{d}\tilde{f}/\mathrm{d}t=\partial\tilde{f}/\partial t+\nabla\tilde{f}\cdot(\mathrm{d}\mathbf{r}/\mathrm{d}t)+\nabla_{\mathbf{p}}f\cdot(\mathrm{d}\mathbf{p}/\mathrm{d}t)=0,$$

where $\mathrm{d}\mathbf{r}/\mathrm{d}t$ is the real space velocity of the electron and $\mathrm{d}\mathbf{p}/\mathrm{d}t$ is its momentum space velocity. We may rearrange this result as an equation for $\partial\tilde{f}/\partial t$, the rate of change of occupation at fixed $\mathbf{r}$ and $\mathbf{p}$. Adding now an extra term to allow for the scattering, we find using (17.9) that

The Boltzmann transport equation

$$\blacktriangleright \quad \partial\tilde{f}(\mathbf{r},\mathbf{p},t)/\partial t=-\nabla\tilde{f}\cdot\mathbf{v}-\nabla_{\mathbf{p}}\tilde{f}\cdot\mathbf{F}+(\mathrm{d}\tilde{f}/\mathrm{d}t)_{\mathrm{scatt}}. \qquad (17.11)$$

This is the Boltzmann transport equation for electrons in a crystalline solid. Very similar equations apply to other systems of particles. The first term on the right is known as the *convection term*: it describes the effect on $\tilde{f}$ at a fixed point due to the arrival of particles from neighbouring points. The second term is the *force term*: it describes the effect on $\tilde{f}$ at a fixed momentum due to the momentum changes produced by the external force. The final term is the *scattering term*. Its details depend on the nature of the scattering, so we omit them for the moment. Its value for a particular momentum will depend on how the electrons are distributed in momentum at the point of interest.

When the Boltzmann equation is written in the form (17.11) the only explicit driving term is the external force, $\mathbf{F}$. But this is somewhat misleading, for we know that the true driving forces for transport of electrons and heat are the gradients of the electrochemical potential and the temperature, $\nabla\mu$ and ∇T. These gradients are concealed in the convection term. To make them explicit we rewrite $\tilde{f}$ as

§ 15.6

$$\tilde{f}=f+g. \qquad (17.12)$$

Here f represents the *local equilibrium occupation*, defined as being equal to a Fermi function (17.10) in which μ and T take appropriate *local* values. Clearly, f so defined is isotropic, but varies with position. The quantity g, on the other hand, represents the *departure from local equilibrium*. When we have a uniform electrical or heat current density, g is independent of position, and is antisymmetric in momentum space, an excess on one side of the Fermi surface being balanced by a deficit on the opposite side (Fig. 17.3). If we restrict ourselves to the *small current limit*, in which the driving forces $\mathbf{F}$, $\nabla\mu$ and ∇T are all small, g will be small also. On inserting (17.12) in (17.11) we find the following equation for g,

The linearised Boltzmann equation for g, the departure from local equilibrium

$$\partial g/\partial t=-\nabla f\cdot\mathbf{v}-\nabla_{\mathbf{p}}f\cdot\mathbf{F}+(\mathrm{d}g/\mathrm{d}t)_{\mathrm{scatt}},$$

where we have assumed that in the cases of interest f is independent of time and ∇g is negligible, and we have dropped $\nabla_{\mathbf{p}}g\cdot\mathbf{F}$ as a small quantity of second order. On evaluating the gradients of f we find that

$$\blacktriangleright \quad \partial g(\mathbf{r},\mathbf{p},t)/\partial t=f'\mathbf{v}\cdot[\nabla\mu+(\varepsilon-\mu)\nabla T/T]+(\mathrm{d}g/\mathrm{d}t)_{\mathrm{scatt}}, \qquad (17.13)$$

where $f'=(\partial f/\partial\varepsilon)_{\mu,T}$ and the term in $\mathbf{F}$ has cancelled with part of the term in

$\nabla \varepsilon$. This is the *linearised Boltzmann equation*. It is a linear equation of motion for g, the departure from local equilibrium, having $\nabla \mu$ and ∇T as explicit driving terms. If it can be solved, we may write down the transport currents in terms of g, and the transport problem is solved also.

17.4 Transport properties of metals

In-scattering and out-scattering

To apply (17.13) to the problem of the transport of charge and heat in metals by the electrons, we must first discuss the scattering term. We showed in the previous chapter how typical scattering rates depend on occupation number. For elastic scattering of electrons, for instance, (16.37) shows that the net rate of scattering into state 1 from state 2 is $R_0(g_2 - g_1)$, while (16.39) is a corresponding expression for inelastic scattering of electrons by phonons. When we use such expressions to write down $(\mathrm{d}g/\mathrm{d}t)_{\mathrm{scatt}}$ for a state of momentum $\mathbf{p}$ we find terms of two types. We have negative terms proportional to $g_{\mathbf{p}}$, which represent the rate of scattering out of state $\mathbf{p}$ due to its excess population: these are the *out-scattering terms*. We also find positive terms proportional to the excess population of other states $\mathbf{p}'$, which represent excess scattering into state $\mathbf{p}$: the *in-scattering terms*.

The difficulty of solving the linearised Boltzmann equation depends crucially on the importance of the in-scattering terms. When in-scattering is

Fig. 17.3. Solutions of the Boltzmann equation for a free electron metal plotted in momentum space for current flow (left) and heat flow (right). In the upper diagrams dots and open circles represent an excess and a deficit of electrons respectively. The lower diagrams represent cross-sections through the Fermi sphere. Of the scattering processes (a) is a wide-angle process, effective in reducing both types of current, (b) is a small-angle elastic process, ineffective in both cases, and (c) is a small-angle inelastic process, which is effective in reducing heat current but not electric current.

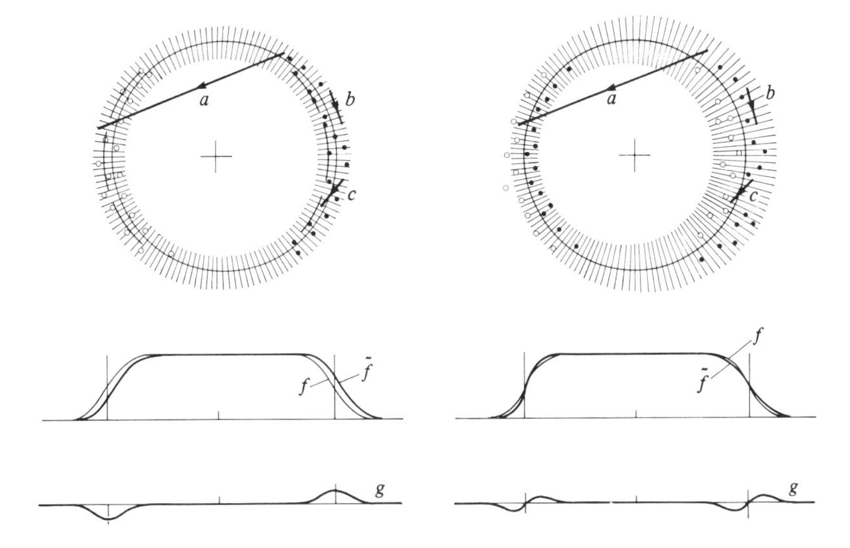

negligible the equations (17.13) constitute a set of *separate* equations for the various values of $g_{\mathbf{p}}$, which are easily solved. But when the in-scattering terms are important, the different equations are coupled together, and harder to solve. Generally speaking, when we have wide-angle scattering, the in-scattering is not important, because the contributions to it from opposite sides of the Fermi surface tend to cancel out. But for small-angle scattering (which occurs in low temperature phonon scattering, and in elastic scattering by dislocations and grain boundaries) $g_{\mathbf{p}'}$ for the important states $\mathbf{p}'$ is often not very different from $g_{\mathbf{p}}$. Scattering of this type may be relatively ineffective, because the in-scattering tends to cancel the effect of the out-scattering, and the relaxation time is often much longer than the scattering time for a single electron (Figs. 17.3 and 17.4). We shall leave the detailed analysis of such cases to more specialised texts.

Problem 17.5

Fig. 17.4. Electrical and thermal conductivities and Lorenz ratio $\kappa/\sigma T$ for a sample of copper measured by Berman & MacDonald (1952). Below 5 K the scattering is elastic scattering by impurities, as discussed in §17.4. At higher temperatures the relaxation times fall as phonon scattering increases. At 40 K small-angle phonon scattering is important, and the Lorenz ratio is considerably lower than the Wiedemann–Franz value. At room temperature, the phonon scattering is nearly isotropic, and the Wiedemann–Franz law is obeyed quite closely.

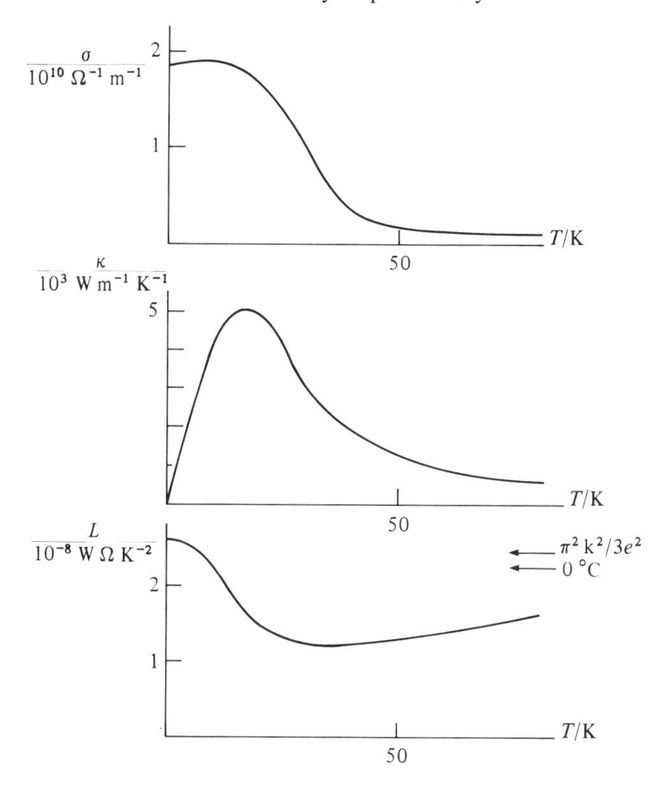

The case of isotropic elastic scattering by impurities

Instead, let us consider one important case which is quite simple, that of elastic scattering by small impurity atoms (as in an alloy), which is isotropic to a good approximation. Elastic scattering occurs only between states which lie on the same energy surface in momentum space. When it is also isotropic the scattering rate R_0 is the same for all pairs of states on the surface. Thus the total in-scattering rate is $R_0 \sum g_{\mathbf{p}'}$, the sum being taken over all states $\mathbf{p}'$ on the energy surface, and because $g_{\mathbf{p}'}$ is anti-symmetric in momentum space (as we shall confirm later) the in-scattering rate is zero, the contributions from opposite sides of the Fermi surface cancelling one another exactly. The out-scattering rate is $-\sum_{\mathbf{p}'} R_0 g_p$, which we shall write simply as $-g_p/\tau(\varepsilon)$, where $\tau(\varepsilon)$ is a relaxation time for the decay of g on the energy surface concerned. For uniform one-dimensional transport in the x direction in which the steady state has been established the linearised Boltzmann equation (17.13) now takes the form

$$0 = f' v_x[\mathrm{d}\mu/\mathrm{d}x + (\varepsilon - \mu)(\mathrm{d}T/\mathrm{d}x)/T] - g/\tau(\varepsilon),$$

The steady state departure from equilibrium

which has the simple solution

$$g(\mathbf{p}) = f' v_x \tau(\varepsilon)[\mathrm{d}\mu/\mathrm{d}x + (\varepsilon - \mu)(\mathrm{d}T/\mathrm{d}x)/T]. \tag{17.14}$$

The forms of $g(\mathbf{p})$ predicted by (17.14) which correspond to gradients in μ only and in T only are shown in Fig. 17.3. Because f' is sharply peaked near the Fermi surface, in both cases $g(\mathbf{p})$ is appreciable only in that region. As we anticipated, in both cases $g(\mathbf{p})$ is anti-symmetric in $\mathbf{p}$. Observe that, to a good approximation, a gradient in μ simply shifts the Fermi distribution sideways in momentum space, corresponding to a uniform flux of electrons. A gradient in T, on the other hand, broadens one side of the distribution but narrows the other, corresponding to a temperature difference between the groups of electrons travelling in opposite directions.

17.4.1 *The electrical conductivity*

Having determined $g(\mathbf{p})$ we may easily write down the transport terms. For the electric current density we have

$$J_x = -e \sum_{\text{states}} v_x g$$

$$= -e \int [\mathcal{N} \tau \overline{v_x^2}] f'(\varepsilon)[\mathrm{d}\mu/\mathrm{d}x + (\varepsilon - \mu)(\mathrm{d}T/\mathrm{d}x)/T]\, \mathrm{d}\varepsilon \tag{17.15}$$

where $\mathcal{N}$ is the density of states in energy, and the average $\overline{v_x^2}$ is taken over the energy surface in momentum space. We have already noted that $f'(\varepsilon)$ is sharply peaked near the Fermi surface $\varepsilon = \mu$. Generally speaking, the quantity $[\mathcal{N}\tau\overline{v_x^2}]$ is a more slowly varying function of ε, and we may replace it with sufficient accuracy by the first two terms of its Taylor expansion in powers of $\varepsilon - \mu$,

$$[\mathcal{N}\tau\overline{v_x^2}]_\varepsilon \simeq [\mathcal{N}\tau\overline{v_x^2}]_\mathrm{F} + [\mathcal{N}\tau\overline{v_x^2}]'_\mathrm{F}(\varepsilon - \mu). \tag{17.16}$$

This allows us to perform the integrals, and we find that

$$J_x \simeq e[\mathcal{N}\tau\overline{v_x^2}]_F(d\mu/dx) + e[\mathcal{N}\tau\overline{v_x^2}]'_F\tfrac{1}{3}\pi^2k^2T(dT/dx), \quad (17.17)$$

where we have used the results $\int_0^\infty f'(\varepsilon)\,d\varepsilon = -1$ and $\int_0^\infty (\varepsilon-\mu)^2 f'(\varepsilon)\,d\varepsilon = -\frac{1}{3}(\pi kT)^2$. Remembering that the effective voltage V is $-\mu/e$, we see that in the absence of a temperature gradient we may identify the electrical conductivity as

§ 15.6

Formula for σ valid for any band structure

$$\blacktriangleright \quad \sigma = -J_x/(dV/dx) = e^2[\mathcal{N}\tau\overline{v_x^2}]_F. \quad (17.18)$$

This expression applies to metals of any band structure, for elastic isotropic scattering. You may like to show, with the help of (12.9), that in the case of free electrons it reduces to (17.6), our result based on the relaxation time approximation, which is valid in this case because non-equilibrium distributions on all of the important energy surfaces relax at much the same rate.

§ 12.4

17.4.2 *The thermopower*

The term in dT/dx in our expression for the current density (17.17) represents a *thermoelectric effect*. The expression may be rewritten in the form

$$J_x = \sigma\{-dV/dx + (1/3e)\pi^2k^2T[\mathcal{N}\tau\overline{v_x^2}]'_F/[\mathcal{N}\tau\overline{v_x^2}]_F\,dT/dx\}, \quad (17.19)$$

where $V = -\mu/e$ is the effective voltage. On comparing this with an ordinary circuit equation we see that, when dT/dx is finite, a sample of metal having a temperature difference dT between its ends contains an internal *thermoelectric emf* $d\eta$ acting in the forward x direction, given by

$$d\eta = (1/3e)\,\pi^2k^2T[\mathcal{N}\tau\overline{v_x^2}]'_F/[\mathcal{N}\tau\overline{v_x^2}]_F\,dT.$$

The quantity $d\eta/dT$ is known as the *thermopower* of the metal at temperature T. It may be expressed in the form

Expression for the thermopower valid for any band structure

$$\blacktriangleright \quad d\eta/dT = (1/3e)\pi^2k^2T(d\ln\sigma/d\varepsilon)_F, \quad (17.20)$$

where σ is the formal expression for the electrical conductivity (17.18). We notice that the thermopower depends on how the quantities $\mathcal{N}$, τ and v vary with energy. It may have either sign. Notice, too, the interesting fact that for elastic scattering the thermopower depends only on what type of scattering happens to be dominant, and not on the number of scattering centres present, for scaling σ by a constant factor does not affect (17.20). (We should perhaps observe that in other circumstances different mechanisms may contribute to the thermopower. For instance, when electron–phonon scattering is dominant, the flow of phonons from hot to cold may drive electrons in the same direction. This *phonon drag effect* can produce large thermopowers in semi-conductors, and is sometimes the dominant mechanism in pure metals at low temperatures.) For the elastic scattering thermopower described by (17.20) one would expect, unless τ varies abnormally rapidly with ε, that $d\ln\sigma/d\varepsilon$ would be of order $1/\varepsilon_F$, suggesting a thermopower of order $k^2T/e\varepsilon_F$, typically a few $\mu V\,K^{-1}$ at room temperature. This is indeed what is seen in most metals.

To *observe* the thermopower, we need a circuit containing two

Thermocouples and circuit thermopower

different metals, known as a *thermocouple* (Fig. 17.5). In a circuit made of a single metal there is no net thermal emf, for the integral $\oint d\eta$ taken around the complete circuit is zero. But if we have a circuit containing two dissimilar metals A and B, with a hot junction at temperature T_2 and a cold junction at temperature T_1, a net circuit emf, known as the *Seebeck emf* η_S, appears, and may be measured with a suitable voltmeter. The net Seebeck emf driving conventional current anti-clockwise around the circuit is given by

$$\eta_S = \int_{T_1}^{T_2} [(d\eta/dT)_A - (d\eta/dT)_B]\, dT. \tag{17.21}$$

If T_1 is held fixed, a thermocouple provides a convenient small thermometer measuring T_2, the temperature of the hot junction. In this context the rate of change of η_S with T_2 is known as the *circuit thermopower*, and is given by

$$(\delta\eta_S/\partial T_2)_{T_1} = [(d\eta/dT)_A - (d\eta/dT)_B]_{T=T_2}. \tag{17.22}$$

The circuit thermopower is a measure of the sensitivity of the thermocouple as a thermometer.

17.4.3 *Thermal conductivity and the Wiedemann–Franz law*

We may also use (17.14) to calculate the x component of energy flux density $U_x = \sum \varepsilon v_x g$. On expressing this as an integral over the electron density of states we find that it may be written as

$$U_x = J_x V + \int_0^\infty [\mathcal{N}\tau\overline{v_x^2}]_\varepsilon f'(\varepsilon)[(\varepsilon-\mu)\, d\mu/dx + (\varepsilon-\mu)^2 (dT/dx)/T)]\, d\varepsilon, \tag{17.23}$$

where V is the local voltage, $-\mu/e$. We have separated the first term from the

Fig. 17.5. The thermoelectric effects. The circuit on the left is open, and the Seebeck emf, which drives conventional current anti-clockwise, produces a measurable voltage at the open terminals. Conventional current is circulating anti-clockwise around the circuit in the centre: the heat flows shown are the reversible heat flows associated with the circulation of unit charge. η_S, Π and σ may have either sign: the diagrams correspond to the conventional positive signs. The right-hand diagram shows the configuration of reservoirs of energy and electrons used in the Onsager treatment of thermoelectric effects in wire A.

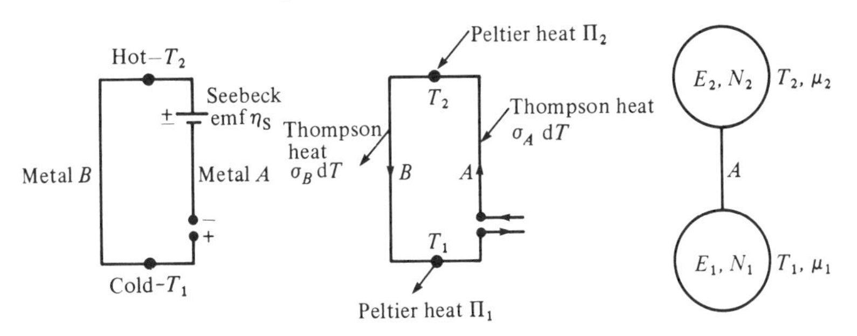

rest of the expression because it is conventionally regarded as a flow of *electrical* energy: it represents the flow of charge multiplied by the local voltage of that charge. It is the second term which is conventionally regarded as the flow of *heat*. Just as in (17.15) we discovered that the electric current J_x contains a small term in $\mathrm{d}T/\mathrm{d}x$, corresponding to the thermoelectric emf, so here we find that the *heat* flow includes a small thermoelectric term driven by the effective electric field, in addition to the usual term in $\mathrm{d}T/\mathrm{d}x$. We shall return to this term in the next section. Here we for the moment ignore the small term in $\mathrm{d}\mu/\mathrm{d}x$, and deduce that the heat flux density H_x may be written to a good approximation as

The meaning of heat flow in the presence of an electric current

$$H_x = U_x - J_x V \simeq [\mathcal{N}\tau\overline{v_x^2}]_{\mathrm{F}} \int_0^\infty f'(\varepsilon)(\varepsilon-\mu)^2 \,\mathrm{d}\varepsilon\, (\mathrm{d}T/\mathrm{d}x)/T.$$

We deduce on performing the integral that the thermal conductivity κ is given by

Expression for the thermal conductivity of a metal

▶ $$\kappa = [\mathcal{N}\tau\overline{v_x^2}]_{\mathrm{F}}\, \tfrac{1}{3}\pi^2 k^2 T. \tag{17.24}$$

By combining (17.18) and (17.24) we may calculate the ratio of the thermal and electrical conductivities, and we find that

▶ $$\kappa/\sigma T = \pi^2 k^2/3e^2. \tag{17.25}$$

The Wiedemann–Franz law

Thus, for the elastic isotropic scattering model considered here, we find that the *Lorenz ratio* $\kappa/\sigma T$ is independent of the band structure of the metal, and is equal to a simple combination of universal constants. This result is the well-known *Wiedemann–Franz law*. It holds quite well for phonon scattering at room temperature, too, and was discovered empirically many years ago. (It is worth noticing, however, that the law fails for small-angle inelastic phonon scattering. As Fig. 17.3 shows, a small-angle inelastic process has little effect on an electric current, but may be effective in reducing a thermal current. When such scattering processes are dominant the electrical conductivity becomes abnormally large compared with the thermal conductivity. This is why the Lorenz ratio is abnormally small in some pure metals at certain temperatures, as Fig. 17.4 shows.)

17.5 Thermoelectric heat flows and the Thomson relations

So far we have considered the approach to equilibrium as a matter of detailed kinetics, based on particular models. It turns out, however, that there are certain considerations which apply generally to systems near equilibrium, and to transport processes. The study of these general considerations is the field of *non-equilibrium thermodynamics*. For the sake of concreteness we shall discuss them in the particular context of the thermoelectric effect in metals.

We have seen already how an emf may develop within a metal which is subject to a temperature gradient, and how this emf may be described in terms of the thermopower, $\mathrm{d}\eta/\mathrm{d}T$. Suppose now that we have a mixed circuit with a temperature difference between the junctions (Fig. 17.5), which contains

§ 17.4.2

a Seebeck emf, represented by the battery symbol. If we allow charge to flow slowly around this circuit the Seebeck emf does electrical work on the charge. Where does this work come from? This problem was considered by Thomson (Lord Kelvin). It had already been discovered that when charge Q flows around such a circuit heat equal to $\Pi_2 Q$ is absorbed at the hot junction and heat equal to $\Pi_2 Q$ is released at the cold junction. These are the *Peltier heats* and Π is the *Peltier* coefficient. The Peltier heat flow is *reversible*: if the current is inverted the heat flows are reversed. But experiment shows that the Peltier heat alone cannot account for the work done by the thermal emf, as Thomson soon realised. He therefore suggested that there must be a second reversible heat flow present, now known as the *Thomson heat*. His idea was that the electricity is behaving like a thermodynamic fluid which is being carried around a thermal cycle. If the Peltier heat is like a latent heat involved in changing the state of the fluid when it passes from metal A to metal B, the Thomson heat is like a heat capacity: he suggested that when charge Q flows up through a temperature difference $\mathrm{d}T$, a heat $\sigma Q\,\mathrm{d}T$ has to be supplied. The quantity σ is the *Thomson coefficient*. The existence of this second reversible heat flow associated with the circulation of charge was subsequently confirmed experimentally.

Peltier heat and Thomson heat

Thomson observed that the thermoelectric circuit is an example of a heat engine, and he asked himself what might be deduced by applying the usual thermodynamics of reversible heat engines to the circuit. The calculation is simple. When a charge Q circulates slowly around the circuit work equal to $\eta_S Q$ may be extracted, where η_S is the Seebeck emf for the circuit. If we equate this with the Peltier and Thomson heats supplied we find that

$$\eta_S = \Pi_2 - \Pi_1 + \int_{T_1}^{T_2} (\sigma_A - \sigma_B)\,\mathrm{d}T. \tag{17.26}$$

On the other hand, the total entropy supplied should be zero – it corresponds to the total entropy change of the electric fluid when it passes once around the circuit. Thus

$$0 = \Pi_2/T_2 - \Pi_1/T_1 + \int_{T_1}^{T_2} (\sigma_A - \sigma_B)\,\mathrm{d}T/T. \tag{17.27}$$

These equations imply that Π and σ are connected in a simple way with η_S. If we differentiate both equations with respect to T_2, holding T_1 fixed, we find that

$$\mathrm{d}\eta_S/\mathrm{d}T_2 = \mathrm{d}\Pi_2/\mathrm{d}T_2 + (\sigma_A - \sigma_B)_{T_2},$$

$$0 = \mathrm{d}\Pi_2/\mathrm{d}T_2 - \Pi_2/T_2 + (\sigma_A - \sigma_B)_{T_2}.$$

By subtraction we deduce that

The Thomson relations

$$\Pi_2/T_2 = \mathrm{d}\eta_S/\mathrm{d}T_2, \quad \text{and} \tag{17.28}$$

▶

$$[(\sigma_B - \sigma_A)/T]_{T_2} = \mathrm{d}^2\eta_S/\mathrm{d}T_2^2, \tag{17.29}$$

the second relation being obtained by observing that $(\sigma_B - \sigma_A)/T$ is equal to $(\mathrm{d}/\mathrm{d}T_2)(\Pi_2/T_2)$. These results are the *Thomson relations*, and they are confirmed by experiment.

Thomson's analysis was therefore very successful. And yet, as he himself realised, it was incomplete. He had treated the circulation of the electric fluid as though it were a completely reversible process, associated with the Peltier and Thomson heats only. In fact, of course, there are other heat changes occurring in the system. In particular, there is an irreversible flow of heat by ordinary thermal conduction from hot to cold and, when current is allowed to flow, there is ordinary irreversible resistive Joule heating. We cannot eliminate both of these irreversible effects simultaneously: if we do the experiment fast or make the wires thin to eliminate the thermal conduction, the resistive heating will be large, but if we eliminate the resistive heating by using thick wires and letting the charge circulate slowly, the entropy increase due to thermal conduction will be large. Thus in any thermoelectric effect we can never say that the system *as a whole* is behaving reversibly, and it is far from obvious that we are entitled to separate the reversible Peltier and Thomson heats from the various irreversible heat flows which occur simultaneously. These doubts are reinforced when we recall our analysis of the thermopower using the Boltzmann equation. This shows clearly that the thermopower is *not* an equilibrium property. It is driven by scattering processes which are exactly the same as those responsible for the irreversible thermal conduction and Joule heating. Moreover, though Thomson said that σ is *like* an electronic heat capacity, the two quantities are certainly not numerically equal: although in some metals σ has the same order of magnitude as the electronic heat capacity per unit charge, this is not always so, and it may in fact have either sign. Why, then, did Thomson's treatment work so well?

Incompleteness of Thomson's argument

§ 17.4.2

17.6 Non-equilibrium thermodynamics applied to thermoelectricity

A more legitimate and quite different approach to the problem of the thermoelectric effects was suggested by Onsager. Imagine a single wire which links two large reservoirs of energy and electrons (Fig. 17.5), in which the temperatures are T_1 and T_2, and the electrochemical potentials are μ_1 and μ_2, as shown. We expect that a steady state will be set up in which energy and electrons will flow along the wire under the influence of the driving terms $T_2 - T_1$ and $\mu_2 - \mu_1$. Onsager suggests, however, that we define the driving forces concerned in such problems by using a particular formal prescription. As the energy and particles flow, the joint entropy of the reservoirs will be increasing. Onsager suggests that for each variable x_α which is changing as a system approaches equilibrium we should define a corresponding *thermodynamic driving force*, $F_\alpha = \partial S/\partial x_\alpha$. In the thermoelectric case we choose as our variables x_α the energy E_1 and number of electrons N_1 in the lower reservoir. The corresponding driving forces are

Thermodynamic driving forces

$$F_E = (\partial S/\partial E_1)_{N_1} = 1/T_1 - 1/T_2$$
$$F_N = (\partial S/\partial N_1)_{E_1} = \mu_2/T_2 - \mu_1/T_1. \quad (17.30)$$

Note that these 'thermodynamic' forces are related to, but not the same as, the gradients dT/dx and $d\mu/dx$ which appeared in our treatment of thermoelectricity using the Boltzmann equation. Now for given values of F_E and F_N we expect the wire to settle down to some definite steady state, and if F_E and F_N are not too large, we expect that the rates of change of E_1 and N_1 will depend linearly on the forces F_E and F_N. (This assumption may be confirmed by examining the Boltzmann equation which, as we saw earlier, may be made linear when the forces and responses are sufficiently small.) Thus in general we expect to be able to write $\dot{x}_\alpha = \sum L_{\alpha\beta} F_\beta$, and in the case of the thermoelectric effect we have

The Onsager equations

$$dE_1/dt = L_{EE}F_E + L_{EN}F_N \quad (17.31)$$
$$dN_1/dt = L_{NE}F_E + L_{NN}F_N, \quad (17.32)$$

provided that the driving forces are not too large. These are the *Onsager equations* for the system. The coefficients $L_{\alpha\beta}$ are known as the *Onsager coefficients*, and it is the off-diagonal coefficients L_{EN} and L_{NE} which are responsible for the thermoelectric effects. Note that these equations are not tied to any particular model, and that the Onsager coefficients are supposed to take all mechanisms into account.

It is not difficult to use the Onsager equations to find the thermopower and the Peltier and Thomson coefficients directly. But this approach, though it would eventually confirm the Thomson relations, would give us little insight into why Thomson's calculation succeeded, so we leave it as an exercise. Instead we start by considering the behaviour of the *entropy*. What did Thomson assume about this? He knew perfectly well that one cannot simultaneously eliminate thermal conduction and resistive heating, with their attendant irreversible generation of entropy, from a thermoelectric circuit. What he in fact supposed is that when we superimpose thermal conduction and electric current, as we do in a thermoelectric circuit, there is no *new* entropy generating process – that the entropy generation is the sum of the usual heat conduction term, proportional to the square of the temperature gradient, and the usual Joule heating term, proportional to the square of the current. In other words, he assumed that the heat flows associated with thermal conduction and Joule heating continue to deliver net entropy to the reservoirs in the usual way, but that there is no *extra* net entropy flux out of the circuit associated with the Peltier and Thomson heats: it is this assumption which is expressed by (17.27). Let us see whether this agrees with what the Onsager equations tell us. In general, we may write the rate of entropy production in an Onsager process as

Problem 17.7

Condition for the validity of the Thomson calculation

$$dS/dt = \sum_\alpha (\partial S/\partial x_\alpha)\, dx_\alpha/dt = \sum_{\alpha,\beta} F_\alpha L_{\alpha\beta} F_\beta. \quad (17.33)$$

In the thermoelectric case this rate of entropy production is equal to $L_{EE}F_E^2 + (L_{EN}+L_{NE})F_EF_N + L_{NN}F_N^2$. To compare this with Thomson's assumption it is helpful to use F_E and $\dot{N}_1$ rather than F_E and F_N as the variables describing the state of the system. With the help of (17.32) we therefore rewrite the rate of entropy production in the wire as

$$\mathrm{d}S/\mathrm{d}t = (L_{EE} - L_{EN}L_{NE}/L_{NN})F_E^2 + \dot{N}_1^2/L_{NN} + [(L_{EN} - L_{NE})/L_{NN}]F_E\dot{N}_1. \tag{17.34}$$

If no current flows we have $\dot{N}_1 = 0$, so we must identify the first term as the entropy production due to ordinary thermal conduction driven by the temperature gradient expressed by F_E. In the absence of a temperature gradient, on the other hand, we have $F_E = 0$, so we must identify the second term as the ordinary entropy production due to resistive heating, which is proportional to the square of the current. Thomson assumed that these were the *only* entropy production processes present. What (17.34) shows us is that *Thomson's assumption is only valid if* $L_{EN} = L_{NE}$. If this is not true there is a cross-term in the rate of entropy production, the last term of (17.34). It represents an interference term between the two processes, a situation in which the total entropy generation due to the two processes together is *not* just equal to the sum of the entropy generations for each process separately. Evidently Thomson's treatment of thermoelectricity, and Equation (17.27) in particular, can only be justified if we can prove that $L_{EN} = L_{NE}$, and we shall return to this question in the final section of the chapter.

First, however, let us assume that L_{EN} and L_{NE} *are* equal, and see how the Peltier and Thomson heats should be understood. We have not only generation of entropy *in* the wire, but also flux of entropy *along* the wire. For instance, at the bottom end of the wire the entropy flux is equal to the entropy entering the lower reservoir. Since for the reservoir $T_1\,\mathrm{d}S_1 = \mathrm{d}E_1 - \mu_1\,\mathrm{d}N_1$, the downward entropy flux Φ_S at the bottom of the wire may be written as

Entrained entropy and its relation to Π, σ and $\mathrm{d}\eta_S/\mathrm{d}T$

$$\begin{aligned}\Phi_S &= (\mathrm{d}E_1/\mathrm{d}t - \mu_1\,\mathrm{d}N_1/\mathrm{d}t)/T_1\\ &= [(L_{EE} - \mu L_{NE})/T]F_E + [(L_{EN} - \mu L_{NN})/T]F_N\\ &= (1/T)[(L_{EE} - L_{EN}L_{NE}/L_{NN})F_E + (L_{EN}/L_{NN} - \mu)\dot{N}_1],\end{aligned} \tag{17.35}$$

where we have again chosen to use F_E and $\dot{N}_1$ as the independent variables, with the help of (17.32). Here the first term represents the flow of entropy associated in the usual way with heat conduction in a temperature gradient. The second term represents *entrained entropy*, extra entropy carried along with the electric current. In fact the entrained entropy per unit charge S^* may be written as

▶ $$S^* = -(L_{EN}/L_{NN} - \mu)/eT. \tag{17.36}$$

Note carefully that the *entrained* entropy S^* depends on the nature of the scattering processes which determine L_{EN} and L_{NN}: it is certainly not the same as the *equilibrium* entropy per unit charge of the electron fluid.

Now if the entrained entropy falls as the charge entraining it passes through some region of the circuit, we have a mechanism which delivers excess entropy to that region, in addition to the usual heat flow mechanism and the local entropy generation due to Joule heating. It is this extra entropy leaving the circuit in the form of heat which accounts for the Peltier and Thomson heats. Positive Peltier heat, for instance, arises if $S_B^* > S_A^*$ so that entropy is released by the charge when it moves from metal B to metal A at the lower junction, and we see by writing down the entropy released that

$$\Pi/T = S_B^* - S_A^*. \tag{17.37}$$

In the same way entropy is released by charge which is flowing towards lower temperatures if S^* increases with temperature, corresponding to a positive Thomson heat. Thus for either metal

$$\sigma/T = \mathrm{d}S^*/\mathrm{d}T. \tag{17.38}$$

We also find that S^* is related to the thermopower. On rewriting F_E and F_N in terms of $T_2 - T_1$ and $V_1 - V_2$ (remembering that the measured voltage V is equal to $-\mu/e$) we find from (17.32) that, in the case where these differences are small, the electric current flowing in the direction of increasing temperature may be written as

$$I = (e^2 L_{NN}/T)[V_1 - V_2 - (T_2 - T_1)S^*], \tag{17.39}$$

if we again assume that $L_{EN} = L_{NE}$. This is the equation for a circuit element of conductance $e^2 L_{NN}/T$ which contains an emf acting in the direction of increasing temperature equal to $-(T_2 - T_1)S^*$. Thus, for either metal, we may identify the thermopower as

$$\mathrm{d}\eta/\mathrm{d}T = -S^*. \tag{17.40}$$

Equations (17.37), (17.38) and (17.40) are consistent with the Thomson relations (17.28) and (17.29), as expected.

17.7* Onsager's proof of the reciprocal relations

We have seen that Thomson's line of argument for the thermoelectric effect is valid *provided* that we can assume that $L_{EN} = L_{NE}$. Now it is quite easy to show using (17.15) and (17.23) that this requirement is indeed satisfied for the particular model of elastic isotropic scattering which we considered with the help of the Boltzmann equation in § 17.4. Onsager, however, was interested in the question of whether a relation of the form

The Onsager reciprocal relations

$$L_{\alpha\beta} = L_{\beta\alpha}, \tag{17.41}$$

known as a *reciprocal relation*, would be *universally* true, and was able to find a general proof of this result which applies under many circumstances.

Onsager's proof of the reciprocal relations

Onsager's proof uses an ingenious argument based on fluctuation theory. We note that when we first connect a wire between two reservoirs, as in Fig. 17.5, there will be a short transient behaviour during which the steady state flow is established, but that if the reservoirs are large the time required to establish the steady state is very small compared with the time required for $T_2 - T_1$ and $\mu_2 - \mu_1$ to decay to zero. Onsager's idea is that, instead of deliberately setting up two reservoirs for which $T_2 - T_1$ and $\mu_2 - \mu_1$ are non-zero, we should take a system of two reservoirs connected by a wire and watch its fluctuations over a *very* long period. Very occasionally large fluctuations will arise in which $T_2 - T_1$ and $\mu_2 - \mu_1$ take the values of interest, and we argue that once the transient is past it is of no significance whether the non-zero values were first established artificially or by natural fluctuation: in both cases we expect the average behaviour of the subsequent decay to be described by the Onsager equations. Thus Onsager suggests that we watch the fluctuating states of the system, and pick out from them the particular cases in which the variables of interest ($x_1, x_2, \ldots, x_n$ in the general case) happen to take the values of interest. We then use this subset of states, the *Onsager ensemble*, as the representative starting states for our statistical description of the subsequent decay. So long as we measure the behaviour of the system over a time τ which is large compared with the transient decay time, but short compared with the decay towards equilibrium of the variables $x_1, x_2, \ldots, x_n$ themselves, we may safely assume that the behaviour of the Onsager ensemble will not differ significantly from that of an ensemble corresponding to experiments in which we prepare the initial values $x_1, x_2, \ldots, x_n$ artificially. Thus we expect to be able to make the identification

$$\overline{[x_\alpha(t+\tau) - x_\alpha(t)]}/\tau = \sum_\beta L_{\alpha\beta} F_\beta, \tag{17.42}$$

where the left-hand side is an average over Onsager's ensemble (t being the time at which the required configuration $x_1, x_2, \ldots, x_n$ arose by chance) and the right-hand side is the expression for the mean value of $\mathrm{d}x_\alpha/\mathrm{d}t$ according to the Onsager equations.

In order to pick out a particular coefficient $L_{\alpha\beta}$ on the right of this equation Onsager used a mathematical device. We multiply both sides of the equation by one particular variable x_γ, and then average both sides of the equation over all possible starting configurations $(x_1, x_2, \ldots, x_n)$ according to their fluctuation probabilities. On the left-hand side this means that we now average over *all* possible fluctuations according to their probabilities instead of just the particular fluctuations originally picked out by Onsager: in other words, we take a time average for a very long time over all fluctuations. On the right-hand side we insert the fluctuation probability density ρ_c explicitly by writing

§ 16.4

$$\rho_c = \mathrm{e}^{S_c/k} \bigg/ \int \mathrm{e}^{S_c/k}\, \mathrm{d}x_1\, \mathrm{d}x_2 \ldots \mathrm{d}x_n, \tag{17.43}$$

where the suffix c indicates that the quantity is to be calculated for a particular configuration. On the right-hand side we now have

$$\int x_\gamma \sum_\beta L_{\alpha\beta} (\partial S_c/\partial x_\beta) \rho_c\, \mathrm{d}x_1\, \mathrm{d}x_2 \ldots \mathrm{d}x_n = k \sum_\beta L_{\alpha\beta} \int x_\gamma (\partial \rho_c/\partial x_\beta)\, \mathrm{d}x_1\, \mathrm{d}x_2 \ldots \mathrm{d}x_n.$$

On performing the integrals with respect to x we discover that this expression reduces to $-k\sum_\beta L_{\alpha\beta}\delta_{\gamma\beta}=-kL_{\alpha\gamma}$, because ρ_c tends to zero for both positive and negative values of x. We thus obtain an important relation between the Onsager coefficient $L_{\alpha\gamma}$ and a particular time average over all fluctuations,

$$\overline{[x_\gamma(t)x_\alpha(t+\tau)-x_\gamma(t)x_\alpha(t)]}/\tau=-kL_{\alpha\gamma}. \tag{17.44}$$

§ 19.2

The requirement of microscopic time-reversal symmetry

Onsager now makes a crucial step. As we shall see in the final chapter, many systems have internal *microscopic time-reversal symmetry*, which means that if we reversed the velocity of every particle at some point in time, the whole system would retrace its path, as though we were running backwards a ciné film of the system's earlier history. This has the important consequence that, once the system has *aged*, so that all memory of its starting state has been lost and it is fluctuating around equilibrium, *the fluctuations themselves have time-reversal symmetry*. Thus we may change the sign of τ in the above argument without changing the average. (A little thought shows that this apparently paradoxical statement is not unreasonable. Any configuration in which the variables $x_1, x_2, \ldots, x_n$ differ appreciably from their equilibrium values is relatively unlikely, and the fact is that in the overwhelming majority of cases the system not only *will* be closer to equilibrium at the future time $t+\tau$, but also *was* closer to equilibrium at the past time $t-\tau$. In other words, it is very much more likely that the unlikely configuration was reached by climbing from some much more probable configuration than by falling from some much less probable configuration.) On this assumption we may rewrite the left-hand side of (17.44) as

$$\overline{[x_\gamma(t)x_\alpha(t-\tau)-x_\gamma(t)x_\alpha(t)]}/\tau=\overline{[x_\alpha(t)x_\gamma(t+\tau)-x_\alpha(t)x_\gamma(t)]}/\tau,$$

where in the second expression we have in the first term shifted the time origin backwards so as to increase both arguments by τ, which does not alter the time average. But according to (17.44) the final expression is just $-kL_{\gamma\alpha}$, so we find that $L_{\alpha\gamma}=L_{\gamma\alpha}$ for any system having internal microscopic time-reversal symmetry. This completes the proof of the reciprocal relations. These relations are of great general importance. They not only justify Thomson's analysis of thermoelectricity, but provide also the fundamental ground rule for the discussion of coupled non-equilibrium fluxes of many types, on which the subject of non-equilibrium thermodynamics is largely based.

The reciprocal relations fail in a magnetic field

We close this chapter with two further comments on the reciprocal relations. Note first that situations do sometimes arise when the system does *not* have internal microscopic time-reversal symmetry. The example of most practical importance is that of charged particles moving in a fixed external magnetic field: the reciprocal relations need modification if they are to be applied to the Hall effect, for instance. Secondly, when the reciprocal relations do apply, we note that $L_{\alpha\beta}$ is a real symmetric matrix, with real eigenvalues.

The upper limit on the thermoelectric coefficients

Using (17.33) we see that, since $\mathrm{d}S/\mathrm{d}t$ cannot be negative for any set of forces F_α, these eigenvalues cannot be negative either. This implies some further relations between the coefficients $L_{\alpha\beta}$. In the case of the thermopower, for instance, we easily find that

$$L_{EN}^2 \leqslant L_{NN} L_{EE}. \tag{17.45}$$

Thus for given values of the ordinary electric and thermal transport coefficients there is a definite upper limit on the strength of the thermoelectric coefficients.

Problems

17.1 When gas A diffuses through gas B in response to a gradient of number density n for gas A in the z direction, the ratio of the particle flux density in the z direction to $-\mathrm{d}n/\mathrm{d}z$ is known as the *diffusion coefficient*, D. Use an equation analogous to (17.1) to show that $D = \frac{1}{3} n l \bar{c}$.

Diffusion coefficient in a gas

17.2 When the x velocity of a gas varies in the z direction, a viscous drag force P in the x direction acts across unit area of the x–y plane between the upper and lower bodies of gas. The ratio of P to $\mathrm{d}v_x/\mathrm{d}z$ is known as the *viscosity*, η. Recognising that any mechanism which transfers momentum from one system to another is equivalent to a force, identify P as the flux density of x momentum in the z direction. Use an equation analogous to (17.1) to show that $\eta = \frac{1}{3} \rho l \bar{c}$, where ρ is the density of the gas.

Viscosity of a gas

17.3 Use the connection between conductivity and relaxation time and the data of Fig. 17.4 to find τ_J at $T=0$ for the sample of copper concerned. If the scattering is isotropic impurity scattering for which the relaxation time is equal to the scattering time, what is the electronic mean free path? Use (17.5) to show that at angular frequency ω the complex conductivity $\sigma(\omega)$ is equal to $\sigma(0)/(1+\omega\tau_J)$. At roughly what frequency would you expect Ohm's law to fail in this sample of copper? [Take n for copper to be $8.5 \times 10^{28}\ \mathrm{m}^{-3}$. With the help of (12.10) you should be able to show that the corresponding velocity of free electrons at the Fermi surface is $1.6 \times 10^6\ \mathrm{ms}^{-1}$.]

The failure of Ohm's law at high frequencies

17.4 A dielectric liquid consists of polar molecules which are free to rotate. The atoms of the liquid polarise very rapidly when a field is applied, but the molecules take an appreciable time to orient themselves. If τ is a suitable relaxation time for the molecular orientation, obtain *Debye's equation* for the complex permittivity ε at angular frequency ω

Debye's equation for $\varepsilon(\omega)$ in polar dielectrics

$$\varepsilon(\omega) = \varepsilon(\infty) - [\varepsilon(\infty) - \varepsilon(0)]/(1 + \mathrm{i}\omega\tau).$$

Show that $\mathrm{Im}\ \varepsilon(\omega)$, which is a measure of the dissipation in the dielectric, has a broad peak near the frequency $\omega = 1/\tau$. Explain physically why this dissipation occurs. How would you attempt to justify the relaxation assumption in this case?

17.5 Show how the in-scattering term in the Boltzmann equation may make a relaxation mechanism ineffective, using the following model. Assume no driving terms and elastic scattering, so that (17.13) becomes $\partial g_{\mathbf{p}}/\partial t = \sum_{\mathbf{p}'} R_{\mathbf{p},\mathbf{p}'}(g_{\mathbf{p}'} - g_{\mathbf{p}})$. Assume that we have a transport current for which $g_{\mathbf{p}}$ on a given energy surface has the form $g_0 \cos\theta$, where θ is the angle between $\mathbf{p}$ and the current direction (Fig. 17.3). Show first that for isotropic scattering, in which $R_{\mathbf{p},\mathbf{p}'} = R_0$, $g_{\mathbf{p}}$ decays with relaxation time equal to the scattering time, given by $1/\tau_s = \sum_{\mathbf{p}'} R_{\mathbf{p},\mathbf{p}'}$. Then show that if $R_{\mathbf{p},\mathbf{p}'}$ has the form of a simplified scattering function $R(\delta\theta)$, which is symmetric in $\delta\theta = \theta_{\mathbf{p}} - \theta_{\mathbf{p}'}$, and is only appreciable when $\delta\theta$ is small, the relaxation time is given by

$1/\tau = \sum_{\mathbf{p}'} R_{\mathbf{p},\mathbf{p}'} \frac{1}{2}(\delta\theta)^2$. Deduce that τ is of order τ_s/α^2, where α is a typical scattering angle. [Hint: expand $g_{\mathbf{p}'}$ in powers of $\delta\theta$ to second order.]

Behaviour of a typical thermocouple

17.6 Near room temperature the circuit thermopower of a thermocouple may usually be written quite accurately as $\alpha + \beta t$, where t is the temperature of the hot junction expressed in °C. For a copper–lead thermocouple $\alpha = 2.8\ \mu\mathrm{V\,K}^{-1}$, $\beta = 0.012\ \mu\mathrm{V\,K}^{-2}$, while for a constantan–lead thermocouple $\alpha = -38.1\ \mu\mathrm{V\,K}^{-1}$, $\beta = 0.09\ \mu\mathrm{V\,K}^{-2}$. Compare these results with the prediction of (17.20) that the thermopower is proportional to T for elastic impurity scattering with a weak energy dependence. For a copper–constantan thermocouple whose cold junction is kept at 0 °C, find α, β, the hot junction temperature at which η_S is maximised, and, at 100 °C, the Peltier coefficient and the difference between the two Thomson coefficients.

17.7 Calculate directly from the Onsager equations the thermopower, Peltier heat and Thomson heat in terms of the Onsager coefficients and μ, and hence confirm the Thomson relations between them. [The following route is suggested: Use (17.32) to write the electric current in terms of δT and $\delta\mu$ for a short length of wire, and deduce the thermopower. Then write the energy flux $\dot{E}$ in terms of F_E and $\dot{N}$, and show that an electric current delivers to the short length of wire an energy $\delta(L_{EN}/L_{NN})$ per particle, in addition to any energy delivered by the temperature gradients. Subtract from this the Joule heat, in the form current multiplied by appropriate voltage, allowing for the thermal emf in the wire. What is left represents Peltier or Thomson heat.]

18

Phase transitions

A *phase transition* is a process in which a thermodynamic system changes, over a negligible range of temperature, pressure or some other intensive variable, from one state into another which has different properties. The melting of ice to form water at the melting temperature, the evaporation of water to form water vapour at the vapour pressure, the superconducting–normal transition at the critical field, and the appearance of ferromagnetism below the Curie temperature are all phase transitions. Such transitions do not all follow precisely the same pattern, but do nevertheless have important features in common. The emphasis in this chapter will be on these common features, and on the theoretical models which can account for them. To illustrate our discussion we shall concentrate particularly, but not exclusively, on two transitions, the liquid–vapour transition and the ferromagnetic–paramagnetic transition. In the first section we shall survey the phenomena which occur in phase transitions. This will lead to a set of questions which we shall set out to answer in the rest of the chapter.

18.1 Phenomena

The best-known examples of phase transitions are the liquid–vapour transition (evaporation), the solid–liquid transition (melting) and the solid–vapour transition (sublimation). We may see evidence of all of these transitions if we examine in detail the equation of state of a simple substance. Fig. 18.1 shows the equation of state for argon, using a logarithmic plot so that a wide range of states may be shown conveniently on a single diagram, by giving the *P–V isotherms* (P as a function of V at fixed T). The diagram also shows the corresponding *phase diagram*, which exhibits the regions of existence of the solid, liquid and vapour phases in the P–T plane, P and T being the intensive variables needed to describe the state of each phase in this case.

The equation of state and the phase diagram

Let us examine these isotherms, starting with the hottest. On the 600 K isotherm and at large volume argon is an ideal gas, and, as PV is constant, the isotherm on this logarithmic plot is a straight line of slope -1. At smaller volumes, since we are above the Boyle temperature, the gas is harder to

§ 14.4

The solid, liquid and vapour isotherms of argon

compress than an ideal gas, and the pressure rises above the straight line. On the 200 K isotherm, on the other hand, we are below the Boyle temperature, and we see that the intermolecular attractions at first make the gas *easier* to compress than an ideal gas until, at the highest densities, the hard cores dominate the situation, and the system becomes very hard to compress. We have passed on this isotherm from a gaseous state to a denser essentially liquid state without seeing a phase transition. The 150 K isotherm passes through the *critical point* C, and is known as the critical isotherm. At point C the intermolecular attractions have become so strong that the system is on the point of collapsing and the *compressibility is infinite*, though beyond that point the hard cores again ensure that the system becomes hard to compress. At still lower temperatures separate phases appear. Where the 100 K isotherm enters the shaded region at point V the pressure has reached the *liquid vapour pressure*. At this point drops of liquid argon condense from the vapour. As we reduce the volume more and more of the vapour condenses, the pressure remaining constant and equal to the vapour pressure, until at L the system has been converted entirely into liquid. The shaded region is the *mixed phase region* for liquid and vapour. Within the mixed phase region a sample of uniform density would be *unstable*: in any small density fluctuation the intermolecular attractions in the regions of high density would tend to increase the density of those regions at the expense of the low density regions, and the system would break up into a high density liquid phase and a low density vapour phase. Continuing along the 100 K isotherm we see that the liquid phase is hard to compress, but if we continue to reduce the volume we notice that a second condensation phenomenon occurs: the liquid is converted into solid when the pressure reaches the *melting pressure*, along the horizontal line L–S. The solid is even harder to compress than the liquid. Finally, on the 80 K isotherm we see that the vapour changes directly into solid along the horizontal line V–S, when the pressure reaches the *sublimation pressure*.

The critical point

Liquefaction: the mixed phase region

Sublimation

Turning to the phase diagram, which is plotted in terms of the intensive variables P and T, we see that each of the mixed phase regions is represented by a single *co-existence line*. The value of the chemical potential must be the same for any two phases in equilibrium, and is therefore the same for the two phases involved at any point on a co-existence line. It follows that if the solid–liquid and liquid–vapour co-existence lines meet on the phase diagram, such a point must *also* lie on the solid–vapour co-existence line: it is known as a *triple point*. The triple point pressure is marked TP on the P–V diagram.

§11.2

The triple point

We must note carefully, however, that a given phase may sometimes persist beyond the point at which a transition to another phase should occur. On the 100 K isotherm, for instance, in clean conditions it is possible to compress the vapour beyond point V without condensation occurring. The vapour is then said to be *supercooled*, and $\mu_V > \mu_L$. The supercooled region of the P–V relation is shown dotted. If a drop of liquid, or even a charged particle,

Metastable states

is inserted into a supercooled vapour, condensation occurs at once. In the same way, a clean liquid may be superheated without boiling; in that case $\mu_L > \mu_V$. (Solids, by contrast, may only rarely be superheated without melting.) Superheated liquid and supercooled vapour are examples of *metastable states*, states which appear to be internally stable, but are thermodynamically unstable because another state of lower chemical potential exists.

One can draw similar diagrams for other types of phase transition. For instance, in a magnetic system the extensive variable corresponding to the volume is the magnetic moment, and the intensive variable corresponding to the pressure is the applied magnetic field. Fig. 18.3(*b*) shows the *magnetic equation of state* of an ideal 'soft' ferromagnetic, the isotherms being now the plot of magnetic moment per ion $\bar{m}$ against applied magnetic field $\mathscr{B}_E$ for various temperatures. (In accordance with the usual magnetic convention, I have plotted the intensive variable $\mathscr{B}_E$ horizontally, so the phase transition

Ferromagnetic isotherms and phase diagram

Fig. 18.1. Isotherms for 1 mole of argon (after Flowers & Mendoza, 1970) with the corresponding phase diagram (facing page). Note that P and V are plotted on logarithmic scales. Mixed phase regions are shaded. C is the critical point, T_c being 150 K. TP is the triple point.

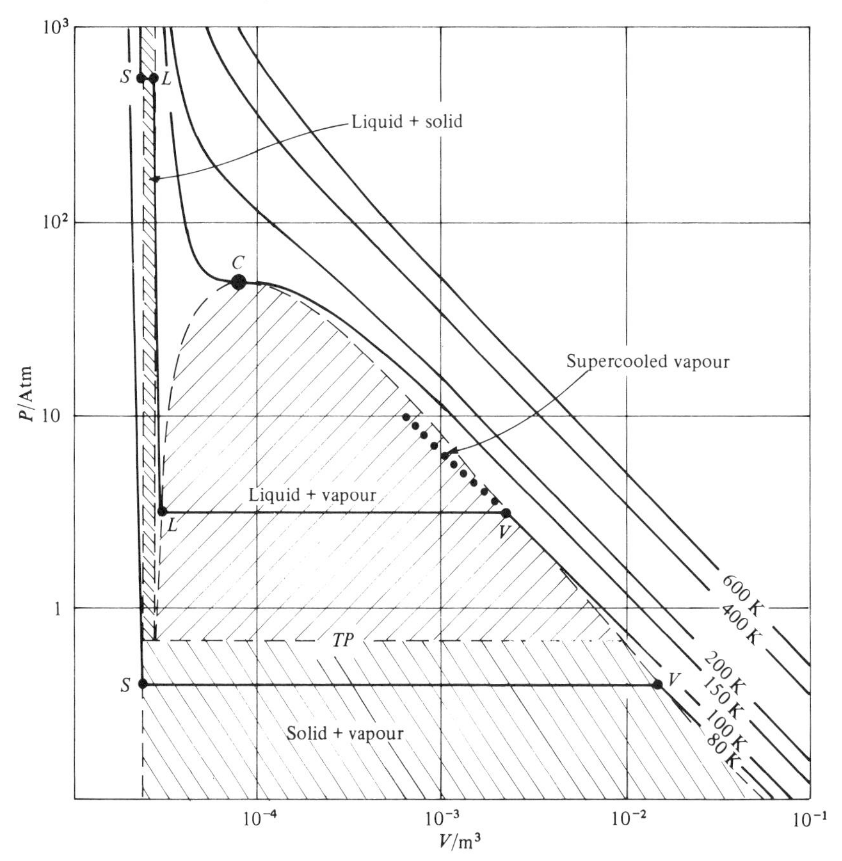

appears as a *vertical* section on the magnetic isotherm.) The corresponding phase diagram is also shown. At the high temperature T_3 the system behaves much as an ideal paramagnet would. We see, however, that as the temperature is lowered internal interactions which tend to align the magnetic dipoles parallel with each other make the substance easier to magnetise, until at $T_2 = T_c$, the *Curie temperature*, the magnetisability becomes infinite in zero field. At lower temperatures the interactions are able to maintain a finite residual magnetisation even when the applied field is zero, and we see a phase transition when the applied field is varied. At T_1, for instance, as we increase the field we find that when $\mathscr{B}_E$ passes through zero the residual magnetisation changes sign, and there is a phase change from the spin down to the spin up state. In this phase change the system passes through a mixed phase region, in which *domains* of spin up magnetisation grow at the expense of other domains of spin down magnetisation.

In general, when we cross a co-existence line on the phase diagram we see a *first-order transition* such as the liquid–vapour transition, in which the system passes through a two-phase region, the two phases being separated in space (Table 18.1). Each phase is internally stable, and some range of

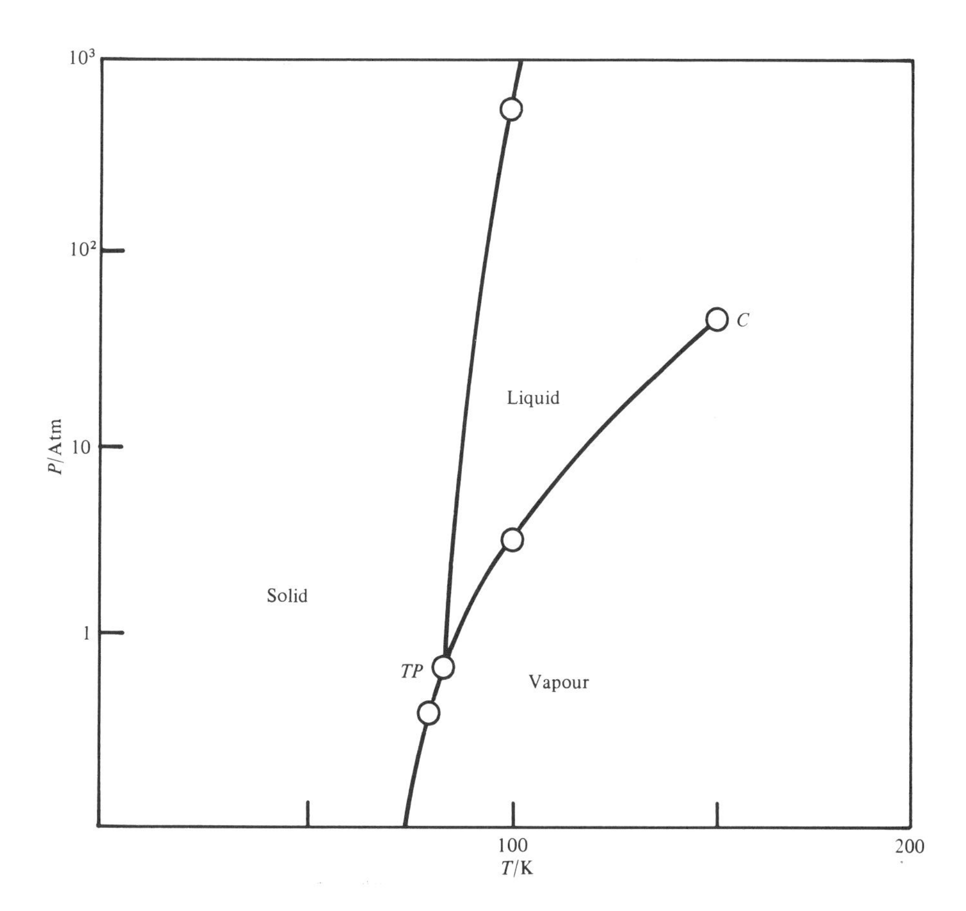

metastable states can usually be found on the 'wrong' side of the line. The phases are substantially different. In particular some of their extensive functions of state take different values. Thus there is typically a substantial change in entropy (corresponding to a latent heat), or change of magnetisation, or change of enthalpy, or change of volume at a first-order phase transition. If, on the other hand, we take a route on the phase diagram which passes through a critical point, the situation is different, and we say that a *higher-order* or *critical point transition* occurs. There is then no separation into two spatially distinct phases. At the critical point the two phases are *identical*, so there can be no supercooling or superheating, and the whole system passes into the new state imperceptibly and uniformly. There is no discontinuity in any function of state, although, as we have seen, *derivatives* of the functions of state such as the compressibility, magnetisability and heat capacity may become infinite at the critical point. A critical point transition may often be regarded as the *onset of an ordering process*: when we cool a ferromagnet in zero field through the Curie temperature magnetisation begins to appear spontaneously in the individual domains; and when we cool a superconductor through the critical temperature in zero field, the density of 'superelectrons' begins to rise from zero.

First-order and higher-order transitions

In what follows we shall try to understand how such phenomena arise. I have already hinted that phase transitions involve *instabilities*. The

Table 18.1. *Examples of phase transitions*

First-order transitions	Latent heat:	Other discontinuities in:
Liquid–vapour	yes	density, viscosity, etc.
Liquid–solid	yes	crystalline order, rigidity, etc.
White tin–grey tin	yes	lattice type, etc.
Type I superconductor – normal metal, in finite magnetic field	yes	conductivity, magnetic moment, etc.
Ferromagnetic reversal, as function of magnetic field	no, by symmetry	magnetisation direction
Critical point transitions	Order parameter:	Anomaly in:
Liquid–vapour at P_c	density	compressibility, heat capacity
Ferromagnetic–paramagnetic, in zero magnetic field	magnetisation	susceptibility, heat capacity
Superconductor – normal, in zero magnetic field	density of superelectrons	heat capacity
Order–disorder in alloys	occupation of sublattice	heat capacity

complete statistical treatment of such instabilities is difficult, as we shall discover. Nevertheless, a simple type of theory known as *mean field theory* gives a useful and in some circumstances quite accurate account of most phase transitions. For ferromagnetism, mean field theory takes a form which is particularly clear and natural, and we shall therefore start, in §§ 18.2 and 18.3, by developing it for that case. We shall find that the mean field theory successfully predicts the existence of the Curie temperature, at which a critical point transition from the paramagnetic to the ferromagnetic state occurs, and by analysing at lower temperatures the magnetic equation of state which the theory predicts we may understand why it is that a first-order magnetic transition occurs when the applied field is varied. In § 18.4 we show how the vapour–liquid transition may be understood in very similar terms. In §§ 18.5, 18.6 and 18.7 we explore the mechanisms of first-order mixed phase transitions, and discover what limits the stability of supercooled and superheated states. In § 18.8 we extend our study of first-order transitions to systems of mixed composition, and discuss Gibbs' *phase rule*. Finally, in §§ 18.9 and 18.10 we explore the region of *critical point phenomena*, for which the mean field theory largely fails, and new theoretical techniques have to be developed.

Plan of the chapter

18.2 The Weiss model of ferromagnetism in zero applied field

Ferromagnetics are metals or salts which below a critical temperature known as the *Curie temperature* T_c may have a permanent magnetic moment in zero field. T_c may be quite high: in iron, for instance, it is 1043 K, and at all ordinary temperatures iron and many iron alloys are ferromagnetic.

In 1907 Weiss proposed a very simple model of the ferromagnetic. He suggested that it should be regarded simply as a paramagnetic material containing permanent magnetic dipoles which are coupled together by an *internal field* $\mathscr{B}_{\text{int}}$, generated by the dipoles themselves. If $\mathscr{B}_{\text{int}}$ lies in the z direction, Weiss took as his model that

$$\mathscr{B}_{\text{int}} = \alpha \bar{m}, \tag{18.1}$$

The Weiss internal field model

where $\bar{m}$ is the mean z component of magnetic moment, averaged over the dipoles in the sample. Thus $\mathscr{B}_{\text{int}}$ acts in such a direction as to *increase* the magnetic moment which generates it. Weiss here introduced into his model a feature common to many phase transitions. We commonly find that the high temperature state is *disordered*: the *order parameter* – the mean magnetic moment $\bar{m}$ in this case – is zero. But as the temperature is lowered the disordered state becomes *unstable*. This comes about because the system has, in the language of circuit theory, *positive feedback*: if $\bar{m}$ departs from the value zero it generates an internal field which tends to maintain the departure. At low temperatures this feedback effect is strong enough to make the disordered state unstable, and the system settles to a new stable state with, in this case, a finite magnetisation.

It must not be imagined that the Weiss internal field $\mathscr{B}_{\text{int}}$ is magnetic in origin. (Indeed, I shall assume for simplicity that we are dealing with a long rod of material parallel to the direction of magnetisation, so that the *magnetic* field $\mathscr{B}_M$ due to the magnetisation is zero.) The internal field is, in fact, due to a much stronger but short-range effect known as the Heisenberg *exchange interaction.* This is a quantum effect in which the spins of electrons on neighbouring atoms whose wave functions overlap may have a lower Coulomb energy when parallel than when anti-parallel. In writing the exchange field as $\alpha\bar{m}$ we are making a *mean field assumption*: we are assuming that $\bar{m}$ for the immediate neighbours, which actually generate the exchange field acting on a given spin, is the same as $\bar{m}$ for the whole sample.

The exchange interaction

The mean field assumption

Weiss further assumed that the mean magnetic moment $\bar{m}$ is given in the usual way by the Brillouin function

§ 15.3

$$\bar{m} = m_0 B_J(m_0\mathscr{B}_{\text{int}}/kT), \tag{18.2}$$

where I have written the quantity $gJ\mu_B$ as m_0. A simple and illuminating method of finding the state of the ferromagnetic according to the Weiss theory is to plot this relation between $\bar{m}$ and $\mathscr{B}_{\text{int}}$ on the same diagram as (18.1), which tells us the value of $\mathscr{B}_{\text{int}}$ generated by a given $\bar{m}$ (Fig. 18.2). Where the two relations intersect we have a *self-consistent state* in which the value of $\bar{m}$ present generates an internal field $\mathscr{B}_{\text{int}}$ just large enough to maintain itself. The diagram is drawn for the case $J = S = \frac{1}{2}$, and we use the dimensionless variables $x = m_0\mathscr{B}_{\text{int}}/kT$ and $y = \bar{m}/m_0$. Thus Equations (18.1) and (18.2) become

The self-consistent state

$$y = Tx/T_c \tag{18.3}$$

and

$$y = \tanh x, \tag{18.4}$$

where

The Curie temperature and spontaneous magnetisation

▶ $$T_c = \alpha m_0^2/k. \tag{18.5}$$

We have here introduced T_c simply as a convenient parameter, but it is in fact the predicted Curie temperature. To see this we must examine the nature of the ferromagnetic instability predicted by the Weiss theory. The slope of (18.3) at the origin is T/T_c, while the slope of (18.4) is unity. Thus if $T > T_c$ (18.3) will cut (18.4) at the origin only, in a single self-consistent state. For $T > T_c$ this state of zero magnetisation is stable. But if $T < T_c$ there are *three* intersections, corresponding to three possible self-consistent states. Of these, the disordered state U with zero magnetisation is now clearly *unstable*: if, in the course of thermal fluctuations, $\bar{m}$ increases from zero we see by inspecting the diagram that it generates a value of $\mathscr{B}_{\text{int}}$ which is *greater* than what is needed to maintain $\bar{m}$, so $\bar{m}$ will tend to increase. This generates a still greater internal field, and the process of increase continues until the stable state S_+ is reached. In the same way, if $\bar{m}$ happens to decrease from zero it will generate an internal field which makes it continue to decrease until it reaches the stable state S_-. Thus the two intersections S_+ and S_- represent stable states having *spontaneous magnetisation.* We see that there will be spontaneous magnetisation whenever $T < T_c$.

Fig. 18.2. The Weiss model of ferromagnetism for the case $S=J=\frac{1}{2}$. The top diagram shows the self-consistency of the ferromagnetic state: curve (*a*) is the magnetisation produced by a given internal field and line (*b*) is the internal field produced by a given magnetisation for $T<T_c$. The self-consistent states S_+ and S_- are stable, but U is unstable. The lower graphs show the predicted magnetisation (compared with data for nickel obtained by Weiss & Forrer, 1926), the exchange energy $\bar{\varepsilon}$ and the magnetic heat capacity C_M as functions of temperature.

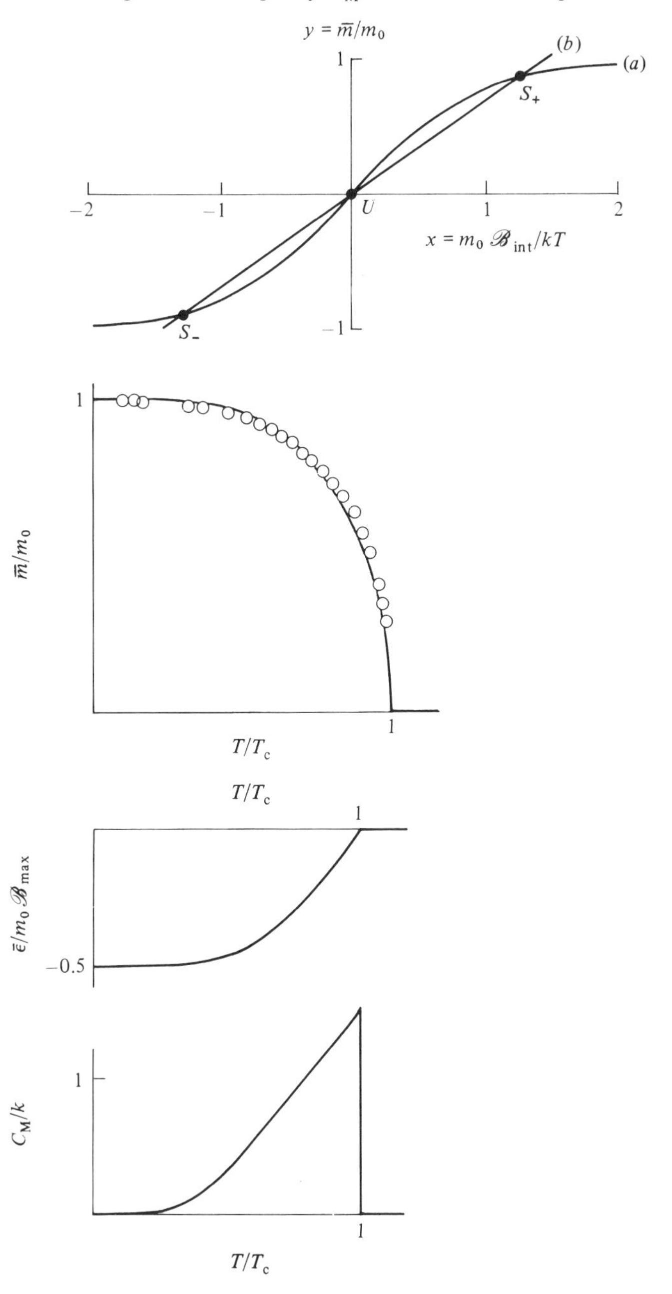

Clearly if we reduce the temperature in zero applied field we have, according to the Weiss model, a critical point transition when $T = T_c$. Let us explore how the system behaves as we reduce T beyond this point. First we note that, by finding how the position of intersection varies as we decrease the slope of (18.3) (or, equivalently, by solving the equation $y = \tanh(T_c y/T)$ numerically) we may find how the order parameter $\bar{m}$ increases as the temperature is lowered. The result is compared with some experimental data in Fig. 18.2. The agreement is quite good (but far from perfect at low temperatures and near T_c, for reasons which we shall discuss later). Notice how the order parameter rises with infinite slope at T_c. At low temperatures $\bar{m}$ saturates, approaching its maximum value m_0, since the internal field aligns the individual moments more and more completely as T approaches zero.

$M(T)$, $E(T)$ and $C_M(T)$ below T_c

§§ 18.3, 18.10

We may also investigate how the energy and heat capacity of the system vary with temperature. In zero applied field the only internal energy is the *exchange energy* of the dipoles in the internal field. In the Weiss model the exchange forces do work $\alpha\bar{m}\,d\bar{m}$ on each dipole when the magnetisation changes. By integrating this expression we deduce that the exchange energy must be $-\frac{1}{2}\alpha\bar{m}^2$ per ion. This energy and the corresponding *magnetic heat capacity* per dipole $C_M = -\alpha\bar{m}(d\bar{m}/dT)$ are also shown as functions of temperature in the diagram. The energy of the ordered state is lower than that of the disordered state. The heat capacity represents the thermal energy needed to break up the order as the temperature is raised. According to the Weiss theory C_M increases steadily with temperature up to T_c, where it drops discontinuously to zero, because there is then no further order to break up. As we shall see later this does *not* correspond to what is actually observed near T_c.

§ 18.9

18.3 The Weiss model in an applied field

If there is present an applied field in addition to the internal field we have an *effective field* $\mathscr{B}_{eff}$ given by

$$\mathscr{B}_{eff} = \mathscr{B}_E + \mathscr{B}_{int} = \mathscr{B}_E + \alpha\bar{m}, \tag{18.6}$$

and we must use this effective field in place of $\mathscr{B}_{int}$ in (18.2). Thus using (18.6) to substitute for $\mathscr{B}_{eff}$ in (18.2) we find the equation

▶ $$\mathscr{B}_E = (kT/m_0)B^{-1}(\bar{m}/m_0) - \alpha\bar{m}, \tag{18.7}$$

The magnetic equation of state

where B^{-1} stands for the inverse Brillouin function. This equation giving the intensive variable $\mathscr{B}_E$ in terms of the magnetisation m and the temperature T is the *magnetic equation of state*. For the special case $J = S = \frac{1}{2}$ it takes the form

$$\mathscr{B}_E/\mathscr{B}_{max} = (T/T_c)\tanh^{-1}(\bar{m}/m_0) - \bar{m}/m_0. \tag{18.8}$$

Some isotherms for this equation are shown in Fig. 18.3(*a*). (In accordance with the usual convention in ferromagnetism the applied field $\mathscr{B}_E$ is plotted horizontally and the order parameter vertically, so on these isotherms a first-order change of state will appear as a *vertical* line.)

Note first that for temperature T_3, well above the Curie point, there is no phase transition as we vary the applied field $\mathscr{B}_E$. In fact the magnetisation curve is similar to that for a paramagnetic (Fig. 9.4(*a*)). However, the internal field has the effect of enhancing the magnetisability, especially at low fields. For small values of $\bar{m}$ we may approximate $\tanh^{-1}(\bar{m}/m_0)$ as $\bar{m}/m_0$, and we find

p. 95

Fig. 18.3. Isotherms of a ferromagnetic in an external field $\mathscr{B}_E$, for temperatures $T_1 < T_c$, $T_2 = T_c$, $T_3 > T_c$. (*a*) shows the predictions of the Weiss theory, showing hysteresis typical of a very 'hard' ferromagnetic below the Curie temperature. (*b*) shows the corresponding ideal equilibrium isotherms, with metastable states shown dotted: there is a first-order phase transition at $\mathscr{B}_E = 0$, corresponding to a very 'soft' ferromagnetic. (*c*) shows the corresponding magnetic phase diagram. (*d*) shows typical observed hysteresis intermediate in character between the 'hard' and 'soft' limits.

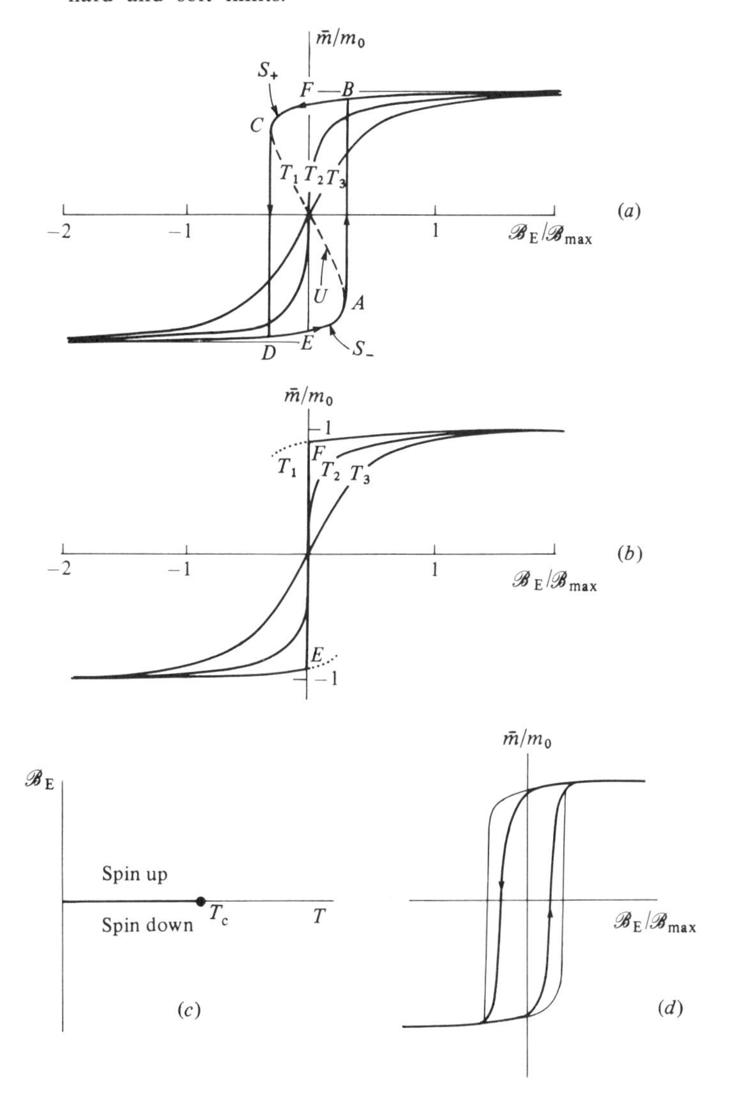

from (18.8) that the magnetisability α_M is given by

$$\alpha_M = \bar{m}/\mathscr{B}_E = m_0^2/k(T - T_c). \tag{18.9}$$

The Curie–Weiss law

This is the *Curie–Weiss expression* for the low field magnetisability of a ferromagnet. It is identical with the Curie expression for a paramagnet except that T is replaced by $T - T_c$. Thus the magnetisability is increased by the internal field, and diverges at T_c. This is apparent on the isotherm for $T_2 = T_c$, the critical isotherm, which has a vertical tangent at the origin.

§9.2.1

When we consider the isotherm for T_1, which lies below T_c, we find that there are, for small fields, *three* states S_-, U and S_+, equivalent to the three states of Fig. 18.2. The central state U is unstable since $\bar{m}$ *decreases* as $\mathscr{B}_E$ is increased: the unstable region is shown dashed in the diagram. The states S_- and S_+ are stable. In such a situation we would expect that on increasing $\mathscr{B}_E$ from a negative value the system would pass through points D and E and continue to the *limit of local stability* at A. The system would then become unstable, and $\bar{m}$ would increase rapidly until the system came to equilibrium at point B; this behaviour is represented by the vertical line AB. On decreasing $\mathscr{B}_E$ from this point we would expect $\bar{m}$ to decrease steadily through F to the limit of local stability at C. The system would then jump to D. Irreversible behaviour of this type is known as *hysteresis*. Certain ferromagnets known as *hard ferromagnets* do indeed trace out in the $\mathscr{B}_E - \bar{m}$ plane a *hysteresis loop* of approximately this form when we successively increase and decrease $\mathscr{B}_E$. We shall examine the mechanism of hysteresis in more detail later.

Hysteresis

§18.5

The hysteretic behaviour just described does not correspond, however, to the ideal first-order phase transition from spin down to spin up, which is *reversible*. In the hysteretic behaviour the system has, for some values of the applied field, two stable states S_- and S_+. But a thermodynamic system has only a *single* state of true thermal equilibrium, for given constraints and temperature – the state of minimum availability which corresponds to the Boltzmann distribution. It follows that one of the two states S_- and S_+ must be *metastable*. This means that it is *locally* stable; stable, that is, against *small* fluctuations, but *thermodynamically* unstable in the sense that another state of lower availability exists, so that if we wait long enough a large fluctuation will occur which takes the system into the state of lower availability, from which it is extremely unlikely to return – this is, in fact, an example of a thermally activated process. We shall examine the mechanism of the ideal first-order phase transition later. For the moment we simply note that it must occur at the point where $A_- = A_+$. For a mean field theory such as the Weiss theory this corresponds, as we shall see later, to *the point at which the applied field cuts off equal areas in the loops of the Weiss isotherm* (Fig. 18.3(*a*)). This, as one might have anticipated by appealing to symmetry, occurs at zero applied field. Thus, if we increase $\mathscr{B}_E$ very slowly from negative values, the system will start in state S_-, passing through D and reaching state E, at which the applied field is zero. At this point A_+ falls below A_-. As soon as we are past this point the system,

The ideal reversible first-order phase transition

§16.6
§18.5

The equal area rule

helped by thermal activation, will jump to S_+ at F and will then stay in state S_+. On reducing $\mathscr{B}_E$, the system, again helped by thermal activation, will jump from F to E and then stay in state S_-. Thus for sufficiently slow changes the *ideal equilibrium isotherm*, shown in Fig. 18.3(*b*), should show a vertical and reversible jump between states E and F. This vertical jump corresponds to a first-order phase transition in which the system makes a finite jump between its down spin and up spin states. Certain *soft ferromagnetics* do indeed have very narrow hysteresis loops which approximate to this ideal, reversible behaviour.

The ferromagnetic phase diagram

The corresponding ideal phase diagram is also shown in the diagram. Note that in ferromagnetism the transition line is parallel to the T axis. Thus the first-order transition occurs when we vary $\mathscr{B}_E$, and not on varying T. Moreover, although there is a substantial change in *magnetisation* at the transition it is obvious by symmetry that there is no change in *entropy*: we have here an example of a first-order transition, in which a substantial change of state occurs, but without latent heat.

We must end this section by discussing briefly the success and validity of the Weiss model. It accounts quite well for the saturation magnetisation and heat capacity of insulating and rare-earth ferromagnets in the mid-range of temperatures, and for their behaviour in the paramagnetic state well above T_c. It has some obvious defects, however. It does not apply, without considerable modification, to the common transition metal ferromagnetics for which the model of localised magnetic moments at definite points in the lattice is incorrect: in transition metals the magnetism is associated with the spin of d electrons which are free to move through the metal. Also, by oversimplifying the exchange interaction, it ignores the low energy excitations known as *spin waves*, waves of magnetic precession which can propagate through the spin lattice, rather as phonons propagate through a crystal lattice. By ignoring spin waves the Weiss theory underestimates the magnetic heat capacity and overestimates the magnetisation at low temperatures. For our present purposes, however, the most interesting defect of the Weiss theory is its failure, which it shares with all simple internal field theories, to account for the detailed behaviour near the critical point. We shall return to this question in the last section of the chapter.

Defects of the Weiss model

§18.10

18.4 van der Waals' model of the liquid–vapour transition

We saw in Chapter 14 that one of the few straightforward theories of the equation of state for a liquid–vapour system is the modified hard core model. For the sake of algebraic simplicity let us take this model in the rather crude van der Waals approximation, for which

§14.10

$$P = \frac{NkT}{V - Nb} - \frac{N^2 a}{V^2}. \tag{18.10}$$

(An analysis similar to what follows could equally well be applied to the more precise modified hard core equation of state.) This equation is closely analogous to the magnetic equation of state for the Weiss model (18.7). The applied pressure P represents an 'external field'. The hard core pressure $NkT/(V-Nb)$ is analogous to the paramagnetic magnetisation field. The effective pressure due to intermolecular attractions N^2a/V^2 is the term which in van der Waals' theory plays the role of the 'internal field', and it makes the fluid unstable at certain densities. Note that this term depends only on the volume V, which plays a role similar to that of the mean moment $\bar{m}$ in the ferromagnet.

§ 18.3

Some of the van der Waals isotherms $P(V)$ predicted by (18.10) are shown in Fig. 18.4, and may be compared with the magnetic isotherms of Fig. 18.3(a). There is in this case no single point through which all the isotherms pass, but it remains true that we have a critical point. For large T all states are stable. On the critical isotherm $T=T_c$ the isothermal compressibility $-(\partial V/\partial P)_T/V$ becomes infinite at one point, the critical point C. For temperatures below T_c and pressures below P_c we have two apparently stable states L (or liquid) and V (or vapour), with an unstable state between them. I leave it as an exercise in algebra to show that for the van der Waals' equation of state (18.10) T_c, P_c and the critical volume V_c are given by the relations

The critical point

Problem 18.3

$$T_c = 8a/27kb,$$
$$P_c = a/27b^2, \qquad (18.11)$$
$$V_c = 3Nb.$$

As in our consideration of the Weiss model, we know that there can be only one true equilibrium state for a given external pressure, whereas the van der Waals isotherms show two locally stable states for $P<P_c$ and $T<T_c$. The construction for finding the pressure at which the transition occurs is similar to the one which we used in the magnetic case: the availabilities A_L and A_V are equal, and the transition occurs, at the pressure which cuts off equal areas in the loops of the van der Waals isotherm. The resulting ideal equilibrium isotherms are shown in Fig. 18.4, and may be compared with the data for argon plotted in Fig. 18.1. The basis for the equal area rule and the mechanism by which the first-order transition occurs are described in the next two sections.

The equal area rule and the ideal equilibrium isotherm

The van der Waals' theory was important historically because it successfully predicted the existence of the critical point, and, as the diagram shows, it gave a good qualitative description of the equilibrium isotherms for the liquid–vapour transition. But, unsurprisingly, in view of the crudeness of the van der Waals' approximations, its quantitative predictions are poor. As with the Weiss model, its predictions of the details of the behaviour near the critical point are qualitatively as well as quantitatively incorrect (Fig. 18.11), and we shall explore the reasons for this failure later.

§ 14.10

§ 18.10

18.5 The mixed phase region

Both the Weiss model of ferromagnetism and van der Waals' model of the liquid–vapour transition have isotherms which below T_c have more than

Fig. 18.4. van der Waals' isotherms. The upper diagram is plotted on linear scales, and shows the equal area construction: the vapour pressure corresponds to the horizontal line which makes the two shaded areas equal. Note the negative pressures, which are discussed in problem 18.4. The lower diagram shows the corresponding ideal equilibrium isotherms on a logarithmic plot. Isotherms are labelled with the values of T/T_c, chosen to correspond with those of Fig. 18.1, with which the lower diagram should be compared.

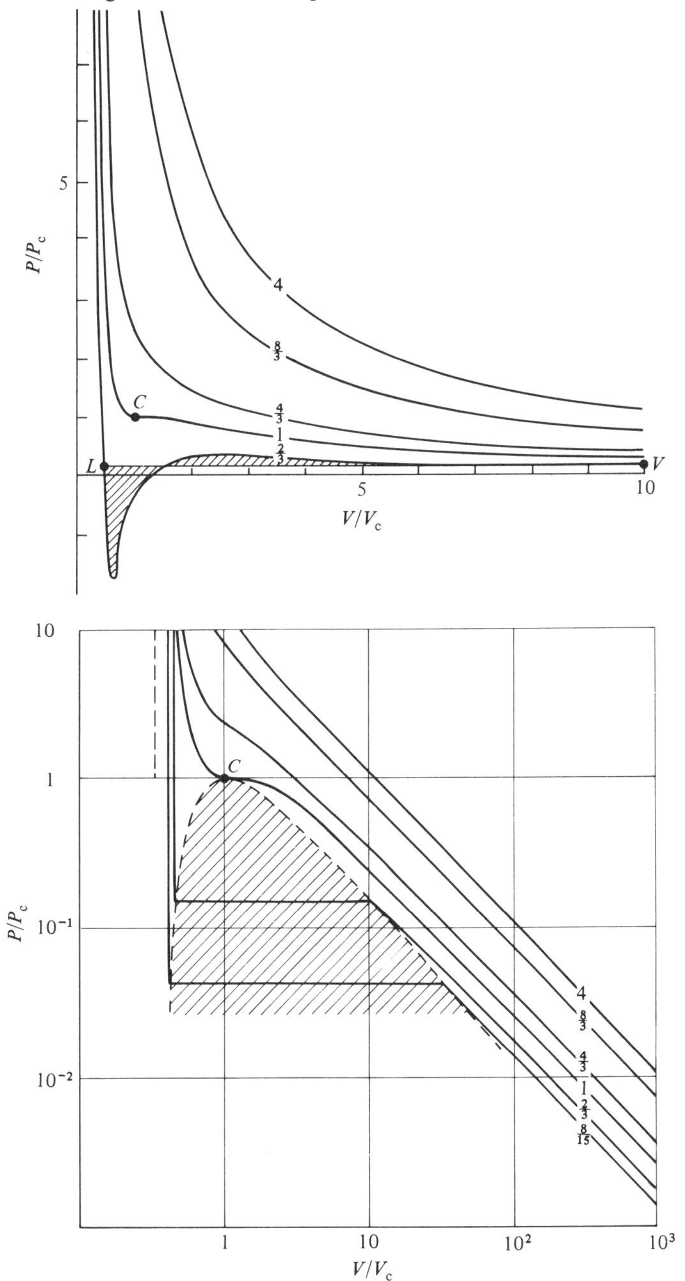

one stable state. Up to this point I have simply mentioned the equal area rule which allows us to determine the point at which the thermodynamic transition between these states occurs, without justifying it or discussing how the change of state takes place. We must now examine in more detail how the new phase appears at a first-order phase transition.

We begin by justifying the equal area rule. Let us concentrate on the liquid–vapour case and start by finding how the availability $A = E - T_R S + P_R V$ varies when we constrain the system at different non-equilibrium volumes on the van der Waals isotherm, assuming that we have a fixed external pressure P_R equal to the pressure (the vapour pressure) at which the transition occurs. We also assume that the system is in thermal equilibrium with the reservoir so that $T = T_R$. Under these circumstances the change in A associated with a change of volume is just the work which we have to do in changing the constraint reversibly, and $dA = (P_R - P)\, dV$. If we take the vapour state V as a reference point we have

Availability differences as areas cut off the isotherm

§7.5

$$A = A_V + \int_V (P_R - P)\, dV, \tag{18.12}$$

with the integral taken along the isotherm. In Fig. 18.5, for instance, we see that in state S the availability A_S is equal to A_V plus the shaded area.

We may use this method of calculation to find $A(V)$ for any volume. However, in passing from the vapour branch to the liquid branch a difficulty arises. In integrating (18.12) along the van der Waals isotherm we pass through a region in which the van der Waals state of uniform density is thermodynamically unstable. Strictly speaking, the entropy, and hence the availability also, of an unstable state has no meaning, so it is unclear in what sense the relation $dA = (P_R - P)\, dV$ holds in this region, if it holds at all. We get round this difficulty as follows. In the *stable* regions the availability of a van der Waals gas may be written as $A_h - N^2a/V$, where A_h is the availability of the corresponding hard core gas. (In the van der Waals theory we assume that at a given density and temperature the entropy is determined by the hard core configurations, and is the same for both gases; thus the availabilities differ only by the potential energy $-N^2a/V$ of the attractive forces.) But the hard core gas is stable at *all* densities, so A_h is always well defined. Thus for the unstable parts of the van der Waals isotherm we may simply *define* A as $A_h - N^2a/V$. Using (18.10) we see at once that, with this definition, dA remains equal to $(P_R - P)\, dV$. We may therefore safely use an integration along the unstable and metastable parts of the van der Waals isotherm to link the values of the availabilities on the liquid and vapour branches.

The problem of thermodynamically unstable states

Fig. 18.5 shows a plot of $A(V)$ calculated in this way. We see that the stable liquid and vapour states correspond to minima of the availability, and that A_L is equal to A_V when the external pressure line cuts off equal areas in the loops of the van der Waals isotherm. This is therefore the external pressure at which the ideal first-order transition should occur: we have verified the equal

The equal area rule

area rule. An essentially similar argument applies to the Weiss model of ferromagnetism, if we define the availability for the unstable regions as the availability A_p of the equivalent paramagnetic system, with no internal field, plus the exchange energy.

The plot of $A(V)$ is a considerable help to our understanding of the phase transition. Note first that $A(V)$ has a large maximum corresponding to the unstable state U of the van der Waals isotherm. Since this state is unstable, we must not take this value of A as physically meaningful. Note, however, that if this maximum were really present we should always, in practice, observe hysteresis and never the ideal equilibrium isotherm. The reason is that the probability of a fluctuation is proportional, as we have seen, to $\exp(-A(V)/kT)$. In a large system the availability barrier corresponding to

§ 16.4

Fig. 18.5. Calculation of availability from the van der Waals' isotherm. The upper diagram shows how A is calculated: A_S is equal to A_V plus the shaded area. P_R has been chosen to cut off equal areas on the isotherms, and this makes A_L equal to A_V. The lower diagram shows $A(V)$. The horizontal line between L and V refers to the mixed phase region, and corresponds to the horizontal section of the ideal equilibrium isotherm shown in the upper diagram.

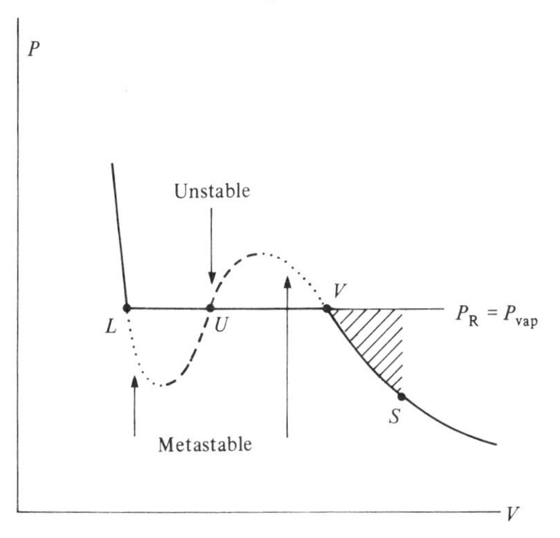

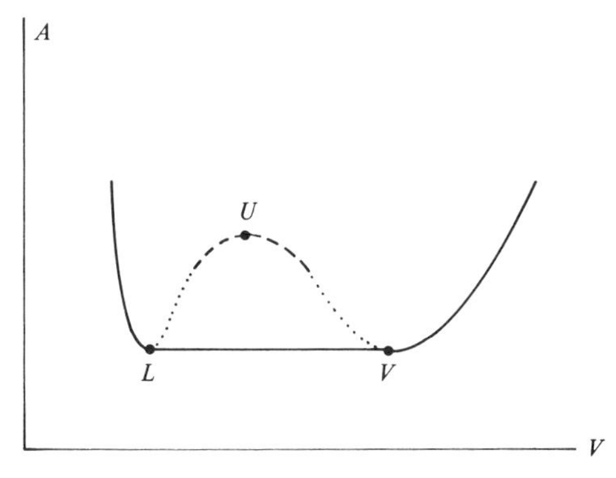

the maximum would be of order $10^{23}kT$. We could therefore confidently say that, although in *principle* an ideal phase transition should take place, according to the equal area rule, at the point where the two minima were at the same level, in *practice* this would never be observed, because the fluctuation required to carry the system over the large maximum in $A(V)$ would be so improbable that we could safely say it would never occur. The real situation is very different. We know that the unstable state actually breaks up into a *mixture* of phases. A mixture having a fraction f_L of the molecules in the liquid state L, and a fraction $f_V = 1 - f_L$ in the vapour state V has a total availability A equal to $A_L f_L + A_V f_V = A_L = A_V$, independent of f_L. At the same time the total volume V is equal to $V_L f_L + V_V f_V$, which may lie anywhere between V_L and V_V. Thus, as the diagram shows, a state containing a mixture of L and V corresponds to a horizontal branch of the plot of $A(V)$ between L and V which lies *below* the van der Waals curve. *This mixed phase state is the preferred state in this range of volumes.* It follows that the phase transition from vapour to liquid occurs not by an extremely unlikely fluctuation in which the system contracts from vapour to liquid at uniform density (corresponding to the van der Waals maximum) but by the *nucleation* of a liquid droplet within the vapour in the course of a local density fluctuation, and subsequent passage through a mixed phase region.

The mixed phase region

Somewhat similar considerations apply to the first-order transitions in ferromagnetics. As Weiss himself realised, the ordinary ferromagnet below the Curie temperature in zero field consists of a mixed phase region containing *domains*, small regions of up spin and down spin magnetisation, analogous to the regions of liquid and vapour present at the fluid transition. You will find details of the physics of magnetic domains in texts of solid state physics. Briefly, the domains are separated by *Bloch walls*, narrow regions within which the spin direction rotates from spin down to spin up. Like the liquid–vapour interface, the Bloch wall has a positive free energy. When we apply an up field to a ferromagnet, if the Bloch walls can move freely through the solid the spin up domains will grow rapidly at the expense of the spin down domains, and we see an almost ideal reversible first-order phase transition, passing through a mixed domain region (Fig. 18.3(*b*)). This is what is thought to occur in the 'soft' ferromagnets which are easily magnetised and have little hysteresis. However, in a ferromagnetic (unlike the fluid) the domain walls are not always free to move. Crystalline defects and inclusions may pin the Bloch walls in position, producing a large availability barrier to magnetisation. In very 'hard' ferromagnets this pinning is so strong that magnetisation can only occur by taking each domain to the limit of local stability as shown in Fig. 18.3(*a*). In practice, most ferromagnets lie between these two limits, the degree of hysteresis being determined by how firmly the Bloch walls are pinned (Fig. 18.3(*d*)).

Ferromagnetic domains

18.6 The nucleation barrier and metastability

In the vapour–liquid transition we see that on the scale of Fig. 18.5 there appears to be no availability barrier to the phase transition if it occurs by way of a mixed phase region. Why, then, do metastable supercooled vapour states exist? The reason is that the process of nucleation may itself present a small but significant availability barrier. Frequently there are present condensation nuclei such as dust particles, solid surfaces or even charged particles which encourage condensation and from which droplets can grow, in which case there is no barrier. In very clean conditions, however, it is not easy for a droplet to form. One can look at this problem from two points of view, which are closely related, as we shall see. First, we observe that a liquid–vapour surface of area $\mathscr{A}$ has a positive free energy equal to $\Gamma\mathscr{A}$, where Γ is the surface tension, which we have ignored so far in calculating the availability. Thus the work done in creating a small drop may be considerably greater than kT, and we can see that the drop may therefore be unlikely to form. Alternatively, it is known that the vapour pressure is *greater* over a curved surface than it is over bulk liquid. Thus, if a small drop forms by thermal fluctuation at the bulk vapour pressure, we can see that it will tend to evaporate again.

We may link these two points of view by imagining that we have a supercooled vapour subject to a reservoir pressure P_R which is *greater* than the bulk vapour pressure P_{vap}, and calculating the change in availability ΔA which occurs when a droplet of liquid of radius r is formed in the vapour. In fact,

$$\Delta A = (a_L - a_V)n + 4\pi r^2 \Gamma. \tag{18.13}$$

The first term on the right represents the bulk contribution: n is the number of molecules in the droplet, equal to $4\pi r^3/3v_L$ if v_L is the molecular volume in the liquid, a_L is the bulk availability per molecule inside the droplet and a_V is the bulk availability per molecule in the vapour. The second term is the surface free energy. In finding $a_L - a_V$ we must remember that a_L and a_V depend on the pressure and density of the medium. In fact, writing $a = e - T_R s + P_R v$ (where e, s and v are the bulk energy, entropy and volume per molecule) we find that, since $T = T_R$ everywhere in this system,

$$\mathrm{d}a = (P_R - P)\,\mathrm{d}v + v\,\mathrm{d}P_R, \tag{18.14}$$

if we allow for the possibility that P_R may change. Now we know that $a_L = a_V$ when $P = P_R = P_{vap}$. In the case of interest, however, we have $P_V = P_R > P_{vap}$ for the vapour, while $P_L = P_R + 2\Gamma/r > P_V$ for the liquid, since the surface tension generates an excess pressure $2\Gamma/r$ inside the droplet. In moving from a state in which $P_V = P_R = P_{vap}$ to one in which $P_V = P_R > P_{vap}$ for the vapour, the change in a_V is given by

$$\delta a_V = \int_{P_{vap}}^{P_V} v_V \, dP_V \quad (\text{since } P_V = P_R)$$
$$= \int_{P_{vap}}^{P_V} kT \, dP_V/P_V \quad (\text{assuming } P_V v_V = kT)$$
$$= kT \ln (P_V/P_{vap}). \tag{18.15}$$

Taking the usual case in which the liquid is much denser than the vapour we find that, since both dv_L and v_L are small, the corresponding change δa_L for the liquid is negligible when compared with δa_V. Thus we may write $a_L - a_V = \delta a_L - \delta a_V \simeq -kT \ln (P_V/P_{vap})$, and inserting this result into (18.13) we find that

▶ $$\Delta A = -kT \ln (P_V/P_{vap}) \, 4\pi r^3/3v_L + 4\pi r^2 \Gamma. \tag{18.16}$$

So long as $P_V > P_{vap}$ and the vapour is supersaturated, this expression rises to a maximum at the *critical radius* r_c, and then falls (Fig. 18.6). A little algebra shows that

The critical droplet

▶ $$r_c = 2\Gamma v_L/kT\lambda, \tag{18.17}$$

where λ stands for $\ln (P_V/P_{vap})$. Thus as the overpressure increases the critical radius slowly decreases. For water at room temperature and an external pressure P_V equal to $2.6P_{vap}$, for instance, r_c is about 10^{-9} m, and the critical droplet contains about 200 molecules. This is a typical critical droplet size for vapours well below T_c at realistic overpressures. The corresponding availability barrier ΔA_c is given by

▶ $$\Delta A_c = 4\pi r_c^2 \Gamma/3. \tag{18.18}$$

For water at room temperature this is 3.8×10^{-14} J. Since this is about $100kT$, we see that the availability barrier to nucleation is still extremely effective, even

Fig. 18.6. The availability barrier to drop nucleation, based on data for water at room temperature with $P_V = 2.6P_{vap}$. The critical droplet has a radius of just over 10^{-9} m and contains about 200 molecules. The availability barrier is about $100kT$, and is therefore very effective in preventing nucleation.

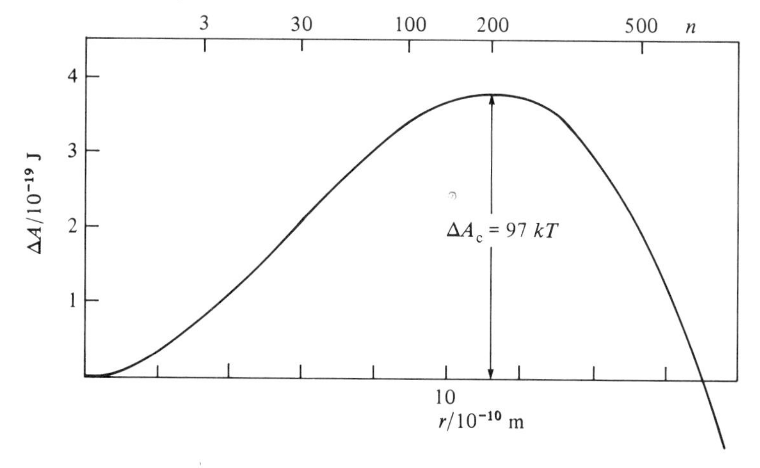

for such a small critical droplet. This makes it clear why the vapour so often persists in a metastable supercooled state.

Vapour pressure over a curved surface

It is worth observing at this point that for any radius r_c we may choose an external pressure which satisfies (18.17). At this external pressure the drop is of critical size, and is therefore in equilibrium with the vapour. We may then interpret P_V as the equilibrium vapour pressure over its curved surface. On rearranging (18.17) we find that

$$P_V = P_{vap}\, e^{2\Gamma v_L/rkT}. \tag{18.19}$$

This is *Kelvin's formula* for the vapour pressure over a surface whose radius of curvature is r. It reminds us that the equilibrium of the critical droplet is unstable: a smaller droplet has a larger vapour pressure and so tends to evaporate, while a larger droplet has a smaller vapour pressure, so that vapour tends to condense onto it, and it grows.

18.7* The rate of phase nucleation

In this section we shall discuss the rate at which droplets of supercritical size will form in a volume V of clean, supersaturated vapour, as an illustration of the general problem of determining the rate of phase nucleation. This will lead to an estimate of the degree of supersaturation likely to be observed in practice. The first step is to discuss the *equilibrium* concentration of droplets of various sizes. It is reasonable to treat *small* droplets as being in thermal equilibrium with the vapour. Droplets near the critical size, however, if they exist, are certainly not in thermal equilibrium. We shall see shortly how to find their concentration. For the moment we simply extrapolate the equilibrium distribution through the critical size, and so determine the hypothetical equilibrium number N_n of droplets containing n molecules for arbitrary n (Fig. 18.7). We do this in the following way. We note that when a small droplet is formed it is free to move around throughout the volume, and that this freedom substantially increases its entropy. When in the previous section we calculated the change in availability ΔA due to the formation of the droplet we tacitly ignored the entropy associated with its centre of mass motion. The total change in availability including this term should be written as

$$\Delta A_T = \Delta A + kT \ln\left[(N_n/V)(nmkT/2\pi\hbar^2)^{-\frac{3}{2}}\right], \tag{18.20}$$

§ 12.6

where the extra term is the centre of mass contribution to the chemical potential of the droplet considered as a particle of mass nm, using (12.25), and N_n is the number of such droplets present in volume V. When the system is in equilibrium the total availability is maximised, so on forming an extra droplet the first-order change in the total availability should be zero. Thus we may deduce from (18.20) that

The hypothetical equilibrium concentration of droplets

$$N_n = V(nmkT/2\pi\hbar^2)^{\frac{3}{2}}\, e^{-\Delta A/kT}, \tag{18.21}$$

and this shows us how the hypothetical equilibrium population of droplets of a given size varies with n.

We wish to find the rate at which drops of greater than the critical size are nucleated. This is a non-equilibrium thermally activated process. However, we cannot use our earlier calculation of the rates of such processes, for two reasons. Firstly, our earlier calculation required us to find the widths of the 'windows' in configuration space through which the system has to pass in its activated state, and the configuration space of the critical droplet has a very complicated structure in, typically, 600 dimensions. This is too difficult to handle. Secondly, our earlier calculation assumed, as we noted, a simple ballistic motion near the activated state which was relatively lightly damped. Droplet growth, by contrast, is the result of the random arrival and evaporation of individual molecules, and obeys a diffusive equation. But we may easily set up a simple kinetic model of this process. We imagine that we create a dynamic equilibrium in the vapour by continually removing from the system any large drops which are nucleated and returning their constituent molecules as uncondensed vapour. There will then be a steady net rate R at which droplets of size n are growing to size $n+1$: in the steady state R will be independent of n, and is simply the rate of nucleation which we wish to find. Now for any given droplet the rates of arrival and evaporation will of course show statistical fluctuations, with corresponding fluctuations in the size of the droplet. There is nevertheless for droplets of a given size a *mean* rate of arrival of molecules ν, due to bombardment by the vapour, which depends on the surface area of the droplet and on the density and temperature of the vapour, and is easily estimated using gas kinetic theory. There is also a definite mean rate at which molecules leave the surface of the droplet by evaporation. Since a droplet of critical size is in unstable equilibrium with the vapour, we know that

§ 16.6

Fig. 18.7. Populations of droplets near the critical size. The upper curve is the hypothetical equilibrium population. The lower curve is the population during continuous nucleation, according to (18.24). For $n \ll n_c$ the two curves coincide.

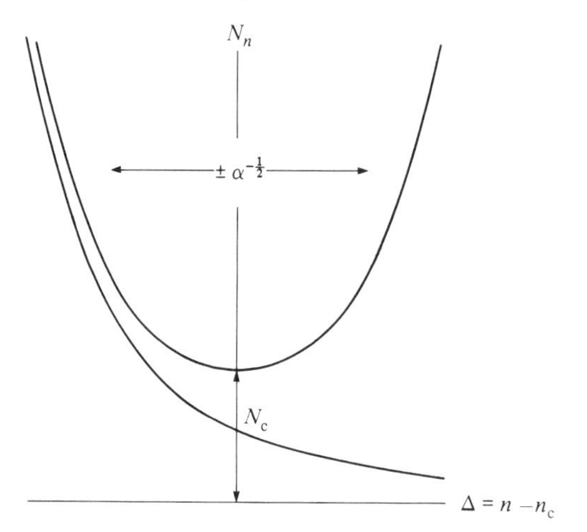

for such a droplet the rate of evaporation must be equal to v. For a smaller droplet, however, the mean rate of evaporation must be slightly greater than v, while for a larger droplet it must be slightly less than v. We therefore adopt a model in which the mean rate of evaporation is written as $v(1-\alpha\Delta)$, where $\Delta = n - n_c$, and the constant α is to be determined later. For this model the rate of nucleation R may be written as

$$R = \tilde{N}_n v - \tilde{N}_{n+1} v(1-\alpha\Delta), \tag{18.22}$$

where $\tilde{N}_n$ represents the number of droplets of size n in the steady state. If $(\tilde{N}_{n+1} - \tilde{N}_n)/\tilde{N}_n$ and $\alpha\Delta$ are small we may rewrite this relation as a differential equation,

$$\mathrm{d}\tilde{N}/\mathrm{d}\Delta - \alpha\Delta\tilde{N} = -R/v. \tag{18.23}$$

On integrating this equation we find, treating v as a constant, that

$$\tilde{N}(\Delta) = (R/v)\, \mathrm{e}^{\frac{1}{2}\alpha\Delta^2} \int_{\Delta}^{\infty} \mathrm{e}^{-\frac{1}{2}\alpha s^2}\, \mathrm{d}s, \tag{18.24}$$

where we have chosen the constant of integration so that $\tilde{N} \to 0$ for large n because we are removing the large drops as they are formed. The form of this result is shown in Fig. 18.7. To obtain the nucleation rate R from this formula we note that for negative Δ the integral rapidly approaches the value $(2\pi/\alpha)^{\frac{1}{2}}$, while $\tilde{N}$ rapidly approaches the equilibrium population N, so that

$$R = N\, \mathrm{e}^{-\frac{1}{2}\alpha\Delta^2} v(\alpha/2\pi)^{\frac{1}{2}}. \tag{18.25}$$

This result is expressed in terms of the unknown constant α. We may relate α to the availability by using the model again to calculate the hypothetical equilibrium distribution N in terms of α. In thermal equilibrium $R = 0$, and on integrating (18.23) for this case we find that

$$N = N_c\, \mathrm{e}^{\frac{1}{2}\alpha\Delta^2}. \tag{18.26}$$

On comparing this result with (18.21) we see that α must be identified with $-(\mathrm{d}^2 A/\mathrm{d}n^2)/kT$ at the critical size, and using (18.16) we find after a little algebra that

$$\alpha = 2\Delta A_c / 3kTn_c^2. \tag{18.27}$$

The quantity $\alpha^{-\frac{1}{2}}$ may be regarded as the width of the availability barrier at temperature T. Using the data quoted earlier for the critical water droplet we find that $\alpha^{-\frac{1}{2}} = \pm 9$ molecules. Using (18.27) and (18.21) we may now rewrite (18.25) as

The rate of droplet nucleation

$$R = [v(\alpha/2\pi)^{\frac{1}{2}}][V(n_c mkT/2\pi\hbar^2)^{\frac{3}{2}}]\, \mathrm{e}^{-\Delta A_c/kT}. \tag{18.28}$$

§16.6

This is our final result for the nucleation rate, and it is interesting to compare it with our earlier result (16.30) for other types of thermal activation. The factor $v(\alpha/2\pi)^{\frac{1}{2}}$ in the first bracket represents a diffusion rate: it is a measure of the rate at which droplets would pass through the effective width of the availability barrier if the process were driven by diffusion only. We may regard it as the effective *attempt frequency* in this case. For a typical water droplet it is about $10^7\ \mathrm{s}^{-1}$. The factor in the second bracket is a phase space factor proportional to the volume available. For a critical water droplet in a volume of 1 m^3 it is about 10^{34}. The final exponential term is analogous to the Boltzmann factor of (16.30).

Using this result we may obtain an estimate of the degree of supersaturation likely to be observable in a clean vapour. Let us first find the external pressure which would generate one nucleation per second in a volume of 1 m^3. Using the values for the two brackets just quoted we find from (18.28) that $\Delta A_c/kT$ has to be 97. Using (18.18) and (18.17) we deduce that the radius of the critical droplet must be 1.1×10^{-9} m, and that the external pressure P_V is $2.6P_{vap}$ – the values which we used in constructing Fig. 18.6. It turns out, however, that this conclusion is remarkably insensitive to the precise values of the brackets and the rate of nucleation chosen: an increase in the required nucleation rate by a factor of 10^4, for instance, increases the necessary value of P_V only by 10%. We may therefore regard our value of P_V as a rather precise estimate of the maximum observable degree of supersaturation: once P_V exceeds $2.5P_{vap}$ by even a small amount the rate of nucleation becomes explosively fast, while if it falls a little short of this value nucleation becomes very improbable. Calculations of this type have been generally successful in explaining the degree of supersaturation actually observed in clean vapours. Note that since the result is insensitive to the values of the two brackets in (18.28) the limiting value of $\Delta A_c/kT$ will be much the same in all cases, and it follows from (18.18) and (18.17) that λ^2 is approximately proportional to $\Gamma^3 v_L^2/(kT)^3$. The law of corresponding states shows us that this quantity will be roughly the same for different substances at the same value of T/T_c. However, since Γ increases quite quickly as temperature falls, the maximum observable ratio P_V/P_{vap} will always increase as the temperature is reduced. In water at low temperatures it may be as high as 4. This is the explanation of the familiar 'vapour trails' which form behind high-flying jet aircraft: the water vapour in the upper atmosphere is often substantially supersaturated, and condenses rapidly onto the condensation nuclei emitted by the aircraft engine.

The maximum degree of supersaturation

§ 14.7

Similar calculations may be applied to the nucleation of bubbles in superheated liquids, with similar results.

18.8 Phase transitions in systems of two or more species

A number of phase transitions of technical importance occur in systems of mixed composition. Fig. 18.8, for instance, shows a simple but common type of phase diagram for binary mixtures of two substances A and B, which might be metals or noble gases. Strictly speaking, this phase diagram should be plotted in three dimensions, with *pressure*, *temperature* and *composition* as the intensive variables. The diagram shown is plotted at atmospheric pressure as a function of temperature and composition, the pressure dependence being suppressed. There are four phases: vapour and liquid (which both show, in this case, the complete range of possible compositions) with solid A and solid B. To interpret such a diagram we draw a

Binary phase diagram

Fig. 18.8. Binary mixture of two species A and B: the top diagram shows a simple type of phase diagram in the composition–temperature plane at atmospheric pressure, the middle diagram shows the entropy of mixing, and the lower diagram shows the Gibbs function in the solid state at temperature T_1, on a simple model. The limits of solid solubility correspond to the points (c) and (d) where the Gibbs function curve has a common tangent.

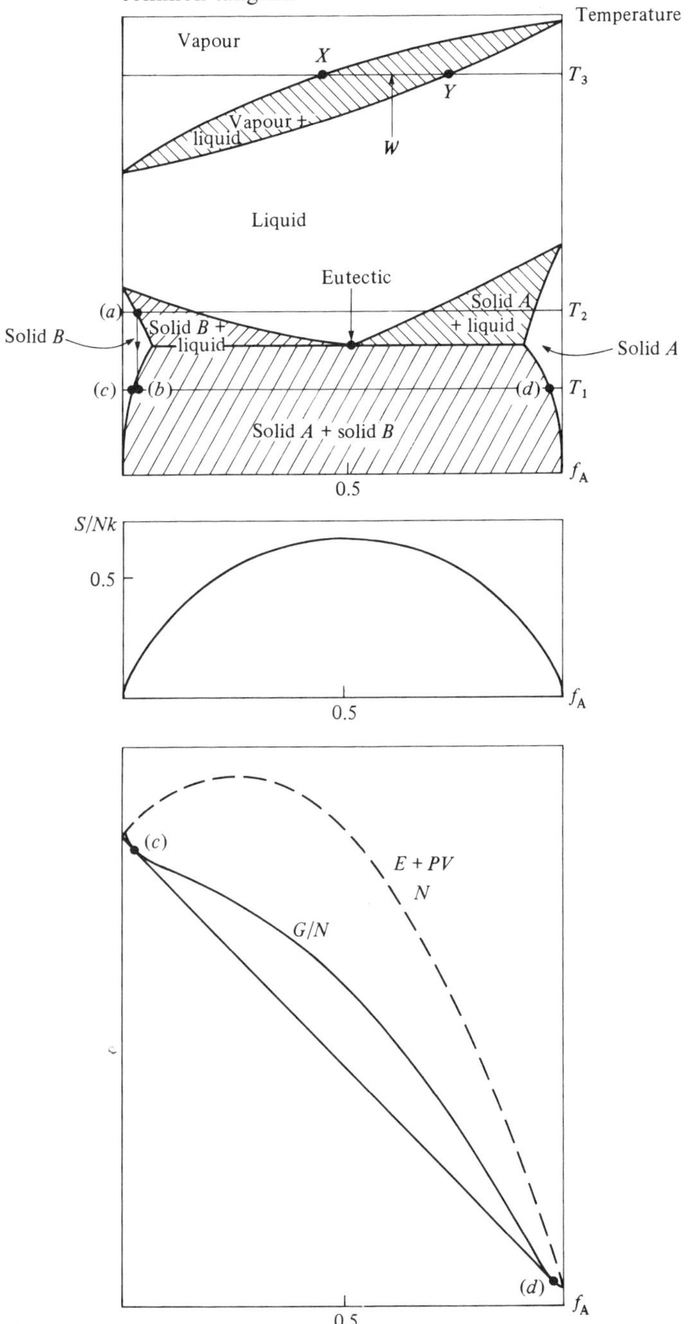

horizontal line to represent the temperature of interest. If, on this line, the overall composition falls in a white area, we have a single phase of this composition. If it falls in a shaded region we have a mixture of two phases whose compositions lie at the edge of the mixed phase region. For instance, at temperature T_3 a system of overall composition W will consist of a mixture of saturated vapour at X, and liquid at its boiling point at Y: in this case the liquid is richer than the vapour in component A. (One can use such a diagram to understand how a *distillation column* may be used to separate two mutually soluble liquids. If liquid of composition Y is boiled, vapour of composition X is produced, and rises up the column, which is a cooled vertical tube containing solid material of large surface area. As the vapour rises, it is cooled and becomes supersaturated, and liquid consequently condenses onto the solid surface and drips slowly back into the boiler. Once dynamic equilibrium has been established the rising vapour will be approximately in equilibrium with the falling liquid at each level in the column, but since the temperature falls as we move up the column the composition of both vapour and liquid will move to the left on the diagram until, at a sufficient height, we reach a region in which both rising vapour and falling liquid consist of pure B. At this point the pure B may be tapped off.)

Action of a distillation column

At the lower temperature T_2 we may have solid B, liquid or solid A, depending on the overall composition. Note in particular that if, at any composition, the liquid is cooled to the point at which solid appears, the solid is always *purer* than the liquid, and the liquid therefore becomes *less* pure, moving towards the *eutectic composition* shown on the diagram, as the temperature falls. At the eutectic point the remainder of the liquid solidifies at fixed temperature as a *eutectic mixture* of solid A and solid B. In metallurgical processes it is important to remember that unless we are at a eutectic point the solid metal appearing when a molten alloy is cooled does not have the same composition as the melt, and will not have *uniform* composition either, because the composition of the melt changes as solidification proceeds. At the eutectic point the character of the solid metal appearing changes completely: it now contains a mixture of crystallites of two different phases. The eutectic point is a *triple-point* at which three phases co-exist.

Solidification and the eutectic point

At still lower temperatures the only possible phase equilibrium is between solid A and solid B. We see that the range of mutual solid solubility decreases as the temperature falls. But we must remember that in the solid state phase equilibrium can only be established by diffusion through the solid. This is a relatively slow process. If, for instance, metal of composition (a) appears from the melt at T_2, and is then removed and cooled quickly, or *quenched*, to temperature T_1, it will arrive at a state of uniform composition (b). Strictly speaking, this state is unstable, and if kept long enough the alloy will separate into a mixture of crystallites having compositions (c) and (d). In practice, however, a solid having a composition such as (b) will persist for a considerable length of time.

Solid solubility and metastable quenched states

Each of the mixed phase regions in this diagram may be understood theoretically in the same sort of way. We know that a system held at fixed temperature and pressure comes to equilibrium when the *Gibbs function* $G = E + PV - TS$ is minimised. Let us consider, for instance, the mixture of solids at temperature T_1. We shall suppose that solid A and solid B have the same type of lattice structure and that in this lattice each atom interacts only with its $\mathcal{N}$ nearest neighbours. In this particular system the phase diagram shows that there is a wide *solubility gap*: the ranges of solubility of A in B and B in A are narrow. This comes about because the A–A bond and the B–B bond are stronger than the A–B bond. In fact, writing the bond energies as E_{AA}, E_{BB} and E_{AB}, and assuming that the atoms are arranged at random (which is not strictly true, but provides a convenient model) the lattice energy will be equal to

§7.4

$$E = -\tfrac{1}{2}N\mathcal{N}(E_{AA}f_A^2 + 2E_{AB}f_Af_B + E_{BB}f_B^2), \tag{18.29}$$

where f_A is the proportion of A atoms, $f_B = 1 - f_A$ is the proportion of B atoms and N is the number of lattice sites. The factor of $\frac{1}{2}$ is included to avoid counting each bond twice. (We ignore, for simplicity, forms of lattice energy other than bond energy.) The second term in the Gibbs function, PV, depends on the volume. It is usually a good approximation to write $V = N(f_Av_A + f_Bv_B)$, where v_A and v_B are the atomic volumes in the pure A and pure B lattices. If in place of f_A we use as variable $s = 1 - 2f_A$ (so that $s = -1$ for pure A; and $s = +1$ for pure B) we may rewrite the first two terms of G in the form

$$E + PV = N(E_2s^2 + E_1s + E_0), \tag{18.30}$$

where

$$E_2 = (\mathcal{N}/8)(2E_{AB} - E_{AA} - E_{BB}),$$
$$E_1 = (P/2)(v_B - v_A) + (\mathcal{N}/4)(E_{AA} - E_{BB}),$$
$$E_0 = (P/2)(v_B + v_A) - (\mathcal{N}/8)(E_{AA} + 2E_{AB} + E_{BB}).$$

This parabolic function is plotted as the dashed line in Fig. 18.8. In the case of interest E_2 is negative, which favours pure substances rather than mixtures, and E_1 is also negative, which means that $E + PV$ is lower for pure A than it is for pure B, so that pure A has the higher melting point.

The remaining term in the Gibbs function, $-TS$, involves the entropy. For simplicity let us ignore all forms of entropy apart from the *entropy of mixing*, given by

§12.8

$$S = k\ln\left(\frac{N!}{N_A!\,N_B!}\right), \tag{18.31}$$

where N_A and N_B are the numbers of A and B atoms, and we again assume that the atoms are distributed at random over the N sites. Using Stirling's approximation, this entropy may be rewritten as

§2.3

$$S = -Nk\left[\frac{1+s}{2}\ln\left(\frac{1+s}{2}\right) + \frac{1-s}{2}\ln\frac{(1-s)}{2}\right]. \tag{18.32}$$

This symmetrical function has a *maximum* at $s = 0$, and infinite slope at $s = \pm 1$:

it is shown in Fig. 18.8. When we subtract TS from $E+PV$ to form the Gibbs function the resulting relation $G(s)$ takes the form at low temperatures of a sinuous curve, also shown in the diagram. We may interpret the mixed phase region in terms of the sinuous curve as follows. Consider the tangent, shown in the diagram, which touches the curve at points (*c*) and (*d*). It is clear that a mixture containing fractions x of (*c*) and $1-x$ of (*d*) will have a mean composition equal to $f_c x+f_d(1-x)=f_c+(f_d-f_c)x$, and a total Gibbs function equal to $G_c+(G_d-G_c)x$. In other words, a mixture of (*c*) and (*d*) has a Gibbs function and composition corresponding to a point lying on the tangent between (*c*) and (*d*). This mixed phase branch of $G(f_A)$ lies *below* the corresponding single phase curve. We may therefore deduce that the single phase state is thermodynamically unstable, and will break up into a mixture of (*c*) and (*d*). These points therefore represent the *limits of solid solubility* of A in B and of B in A respectively. As the temperature rises the term in TS gets larger and the points of contact (*c*) and (*d*) approach one another, the ranges of solid solubility increasing with temperature. Such calculations, or improved versions of them which take proper account of other forms of energy and entropy, are successful in predicting the shapes of such binary phase diagrams.

Many materials have much more complicated phase diagrams, because more than one possible type of lattice structure is involved. Fig. 18.9, for instance, shows the phase diagram for copper–magnesium alloys. There are four possible solid phases, corresponding to copper, magnesium and the intermetallic structures $MgCu_2$ and Mg_2Cu. Each of the four has a different type of lattice structure. Consequently we have to consider four different curves of $G(f_{Mg})$ for the solid phases as well as curves for the liquid and vapour phases. How these might look is shown schematically in the same figure for a low temperature. (The curves shown are not based on a theoretical model, and should not be taken too seriously.) Once again we find the states of minimum Gibbs function by drawing tangents to pairs of curves, the tangents, shown dashed, representing the mixed phase regions. Note that in this case the curves for Mg_2Cu and magnesium show negligible ranges of solid solubility.

At higher temperatures the Gibbs functions for the liquid and vapour phases move downward with respect to those for the solid phases (because the liquid and vapour phases have a greater configurational entropy). When the liquid curve touches one of the dashed lines we reach a eutectic point, and it will be clear how the liquid region successively swallows up the various solid phases as the temperature rises. Eventually the vapour curve will overtake the liquid curve and the liquid phase will disappear, leaving the vapour phase as the only stable phase at the highest temperatures.

When we have more than two components present the phase diagrams become still more complicated. I shall not pursue the question of phase diagrams for more complex systems, but it may be useful to discuss briefly the question of the *number of degrees of freedom* in phase equilibrium. In a system of a single component, for instance, we know that for a single phase

The phase rule

we have two degrees of freedom: we can vary both temperature and pressure. If *two* phases are to co-exist, however, we have only one degree of freedom: at a given pressure there is only one temperature at which water and steam can be in equilibrium together. At the triple point we have no degrees of freedom. Gibbs was able to generalise such ideas and include cases in which the system contained many different components, and cases in which chemical reactions

Fig. 18.9. Phase diagram for Mg–Cu alloys at atmospheric pressure (upper diagram) with mixed phase regions shaded. The lower diagram shows a schematic plot of G as a function of composition for each of the six phases at about 450°C. The dashed lines correspond to mixed phase regions. As the temperature is raised, the liquid curve moves down with respect to the solid curves.

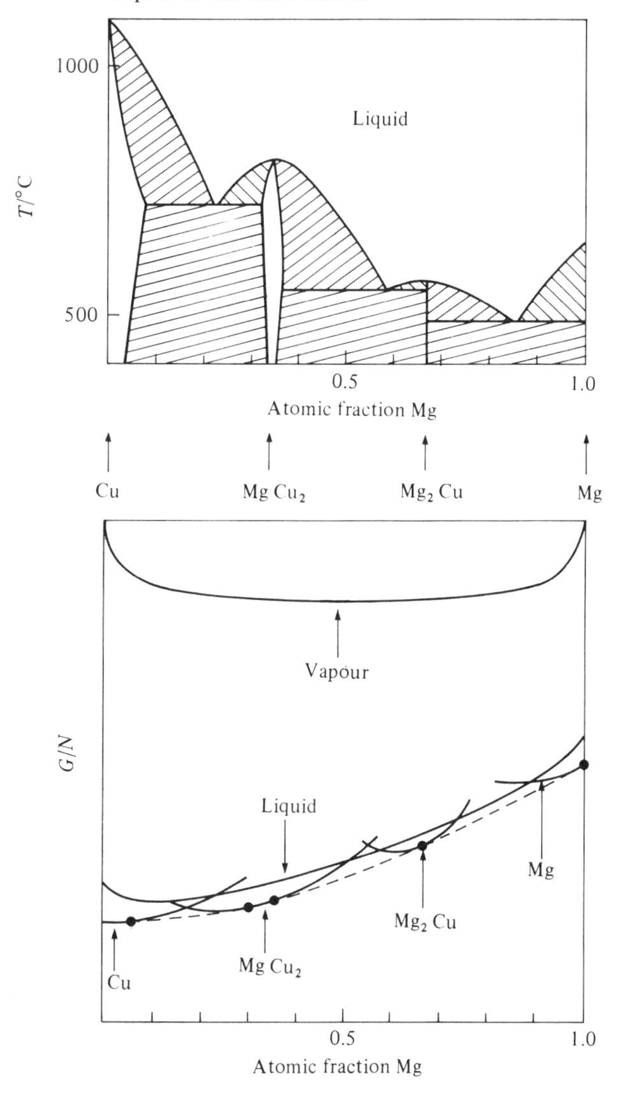

occurred between the components. His argument is very simple. We are not interested in the *quantity* of each phase present, but only in whether equilibrium is possible between the phases. We first ask what is the number of independent variables N_v required to specify the states of the *separate* phases in the absence of chemical interaction. If we are not interested in the quantity present we find that

$$N_v = I + \phi(c-1). \qquad (18.33)$$

Here I is the number of *intensive* variables, such as temperature, pressure and magnetic field, which we intend to vary. (We have some choice here: if we choose to vary, for instance, the applied electric field, this becomes an intensive parameter which has to be specified, but if it is understood that the electric field is zero, it is not a variable requiring specification.) It is understood that in equilibrium the intensive variables are the same in each phase. The variables ϕ and c stand for the number of phases and the number of components, respectively. For c components it takes $c-1$ variables to fix the *composition* of each phase, so $\phi(c-1)$ is the number of variables needed to fix the compositions of *all* the phases. Once these N_v quantities are fixed, the state of each phase, including the chemical potentials of each component in each phase, are fixed. We next observe that if the system is to be in phase equilibrium we must satisfy a number of conditions which restrict the chemical potentials. §11.2 We have first the requirement that the chemical potential of each component must be the same in each phase. There will be $\phi-1$ separate relations involved in specifying this for a single component, and $c(\phi-1)$ equations required to specify it for *all* components. In addition, if there are r independent chemical reactions which can occur between the components, there are a further r relations required to specify the *chemical* equilibrium. Thus the total number of restrictive conditions N_c is given by

$$N_c = c(\phi-1) + r.$$

Assuming that these restrictive conditions are independent, we may use each of them to eliminate one of the independent variables. This leaves f independent degrees of freedom, where $f = N_v - N_c$, or

$$\blacktriangleright \qquad f = I - \phi + c - r. \qquad (18.34)$$

This result is Gibbs' *phase rule.* To see how it works in a couple of cases consider first ordinary phase equilibrium in a pure substance in which we choose to vary P and T. Here $I=2$, $c=1$ and $r=0$. We confirm at once that for one phase $f=2$, for two phases $f=1$, and for three phases $f=0$, as we noted earlier. For a mixture of water, hydrogen and oxygen, on the other hand, if we choose to vary P and T, we have $I=2$, $c=3$ and $r=1$. In this case we can have up to *four* phases in mutual equilibrium – we could for instance, at a low temperature and pressure, have solid water, solid hydrogen and solid oxygen all in thermal equilibrium with vapour of an appropriate composition. Indeed, at low temperatures we could argue that the chemical reaction is too slow to be significant, and set $r=0$ for quasi-equilibrium measured over a short time. We could then have *five* phases such as solid water, solid hydrogen and solid oxygen, vapour and liquid (mainly hydrogen) all in mutual equilibrium at some definite pressure and temperature near the triple point of hydrogen.

18.9 Critical exponents and the scaling hypothesis

Theories such as the Weiss theory of ferromagnetism and the van der Waals theory of the liquid–vapour transition are known as *mean field theories*: they assume that the internal field acting at each point in the domain is the same, and equal to the mean internal field for the whole domain. Mean field theories often work well in regions away from the critical point. But it has long been known that the mean field predictions normally fail as the critical point is approached. Fig. 18.10 shows typical behaviour in a ferromagnetic. The dashed lines show the mean field predictions and the solid lines what is actually observed. Two particular points are worth noting. Firstly, the actual critical temperature T_c is always *lower* than the critical temperature predicted by the mean field theory T_{MF}. Secondly, although the mean magnetisation does drop to zero at T_c, the heat capacity does not: for $T>T_c$ there is a finite magnetic heat capacity even though there appears to be no magnetic order, a fact which seems entirely mysterious in the terms of mean field theory.

It is also worth noting that very close to the critical point the forms of C_M, $\bar{m}$ and $d\mathscr{B}_E/d\bar{m}$ as functions of the reduced temperature difference $t=(T-T_c)/T_c$ and the form of the critical isotherm are all different from what mean field theory predicts. The heat capacity has at T_c a cusp (or in some materials a divergence) rather than the simple discontinuity expected. The other quantities behave like simple powers of t near T_c, but the powers involved differ from the mean field predictions. For instance, we have seen that according to the Weiss theory the magnetisability $d\bar{m}/d\mathscr{B}_E$ should be proportional to t^{-1} just above the critical temperature. But experiment shows that in practice

Critical exponents

§ 18.3

$$(d\bar{m}/d\mathscr{B}_E)_{\mathscr{B}_E=0} \sim t^{-\gamma} \tag{18.35}$$

for t small and positive, where γ is typically about $\frac{4}{3}$. Powers such as γ which express the functional behaviour of quantities very near the critical point are known as *critical exponents*. Table 18.2 gives the definitions of the critical exponents associated with the four quantities plotted in Fig. 18.10, with a number of measured values, and the predictions of mean field theory. (In cases of doubt, if y is a function of t we define the critical exponent of y by writing

Problem 18.7

$$\text{exponent} = Lt_{t\to 0}(d \ln y/d \ln t). \tag{18.36}$$

Thus where the heat capacity has a cusp or a logarithmic divergence at $t=0$ the corresponding critical exponent is entered as zero.)

Note carefully several points about this table. Firstly, except for the doubtful case of the ^{4}He liquid–vapour transition, the exponents for the observed heat capacities and magnetisabilities are the same above and below

T_c, within experimental error. (This must be interpreted carefully: although the *exponents* are the same above and below T_c, the proportionality constants may not be, so that the heat capacity cusp, for instance, is not necessarily symmetrical at T_c.) Secondly, I have listed three types of critical point other than the ferromagnetic Curie point, with the analogous critical exponents (see

Fig. 18.10. Departures from mean field theory near the critical point. Dashed lines: mean field predictions. Solid lines: typical observations, described by the *critical exponents* given in Table 18.2. T_{MF} is the critical temperature predicted by mean field theory.

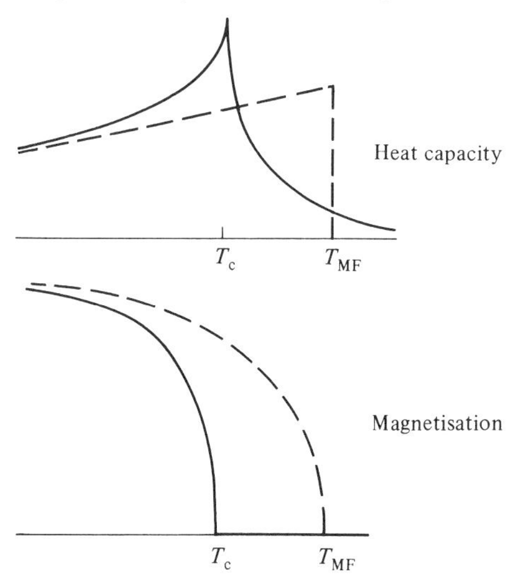

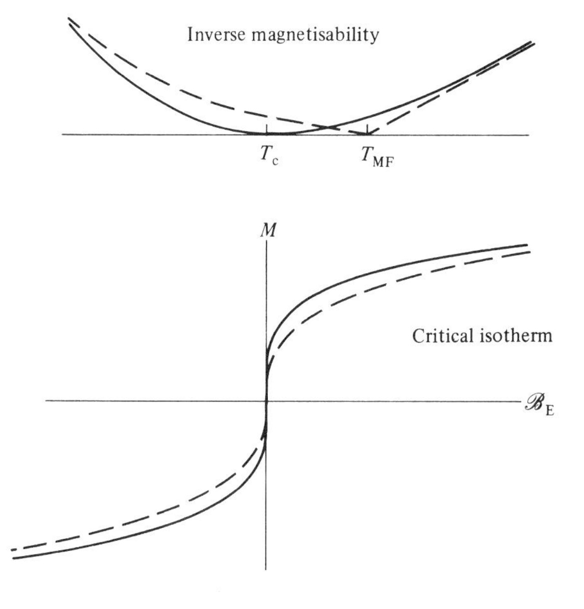

Universality

Fig. 18.11). The values of the critical exponents for the different types of critical point agree much more closely together than they do, as a group, with the mean field predictions. Thirdly, taking this point further, there is some suggestion that *there are groups of systems whose exponents are identical or almost identical* within experimental error: the ferromagnets quoted, the anti-ferromagnet $RbMnF_3$ and possibly the superfluid all agree quite well, while the anti-ferromagnet FeF_2 and the liquid–vapour transitions form a separate group. There seems to be some evidence of *universality*, universal sets of critical exponents which apply to certain critical points in systems of different types, involving quite different physics.

Scaling laws

A considerable step towards understanding the critical exponents was taken by Widom, Domb and others in 1965, who suggested that near a critical point the parameters obey laws of the type known as *scaling laws*. The magnetic equation of state according to the Weiss theory provides a good example of what this means. If we combine (18.4) and (18.6) we find the magnetic equation of state

$$x_E = \tanh^{-1} y - (T_c/T)y. \tag{18.37}$$

This is not, as it stands, a scaling law. But if for small y we expand $\tanh^{-1} y$ as $y + \frac{1}{3}y^3$ to third order in y we have, as an approximation valid near the critical point,

Table 18.2. *Values of critical exponents. The exponents α', γ' and β refer to $T < T_c$, while α, γ refer to $T > T_c$. Most of the experimental values have an accuracy of order ± 0.02, except where indicated by the significant figures quoted. For the liquid–vapour transition in ^4He α' differs from α. In all other cases $\alpha = \alpha'$ and $\gamma = \gamma'$ within experimental error (values from Ma 1976, Stanley 1971 and Wallace & Zia 1978)*

Exponent		α'	α	β	γ'	γ	δ
Quantity		Heat capacity in zero field		Spontaneous magnetisation	Magnetisability in zero field		Field on critical isotherm
Form near the critical point		$(-t)^{-\alpha'}$	$t^{-\alpha}$	$(-t)^{\beta}$	$(-t)^{-\gamma'}$	$t^{-\gamma}$	$y^{\delta}\,\mathrm{sgn}\,(y)$
Experimental values							
Ferromagnets:	Fe	0 (cusp)		0.34	1.33		
	Ni	0 (cusp)		0.33	1.32		4.20
	$YFeO_3$			0.35	1.33		
Anti-ferromagnets:	FeF_2	0.11					
	$RbMnF_3$	0 (cusp)		0.32	1.40		
Liquid–vapour:	CO_2			0.345	1.20		4.2
	Xe	0.08		0.34	1.203		4.4
	^4He	0.127	0.159	0.355	1.170		
Superfluid:	^4He	0 (log)		0.33			
Theoretical values							
Mean field theories		0		$\frac{1}{2}$	1		3
Two-dimensional Ising model: Onsager solution		0 (log)		$\frac{1}{8}$	$\frac{7}{4}$		15

$$x_E = ty + \tfrac{1}{3}y^3. \tag{18.38}$$

This equation has a *scaling property* in x, t and y. For any value of μ it may be rewritten in the form

$$(\mu^3 x_E) = (\mu^2 t)(\mu y) + \tfrac{1}{3}(\mu y)^3,$$

or

$$x'_E = t'y' + \tfrac{1}{3}y'^3, \tag{18.39}$$

where $x'_E = \mu^3 x_E$, $y' = \mu y$ and $t' = \mu^2 t$. In other words, if we replace x_E, μ and t by the *scaled* values x'_E, y' and t', the equation still holds, for any value of μ. Scaling laws arise very commonly in physics. In particular they apply whenever the physical description is not changed in any essential way when the scale of the system is changed: for instance, because the pattern of waves around a moving ship is simply scaled up when we scale up the ship and adjust its speed appropriately, the formula giving the wavemaking resistance in terms of the speed and length of the ship obeys an obvious scaling law.

Problem 18.7

Fig. 18.11. The co-existence curve, showing the densities of liquids and vapours in mutual equilibrium as a function of temperature (after Guggenheim, 1945). Eight different fluids are plotted using reduced variables, and agree very well, showing the success of the law of corresponding states (§ 14.7). In terms of critical exponents, this plot is analogous to the plot of spontaneous magnetisation against temperature for a ferromagnetic. The solid curve is a fit to a cubic equation, corresponding to $\beta = \frac{1}{3}$ at T_c. The dashed curve shows the van der Waals' prediction near T_c, for which $\beta = \frac{1}{2}$.

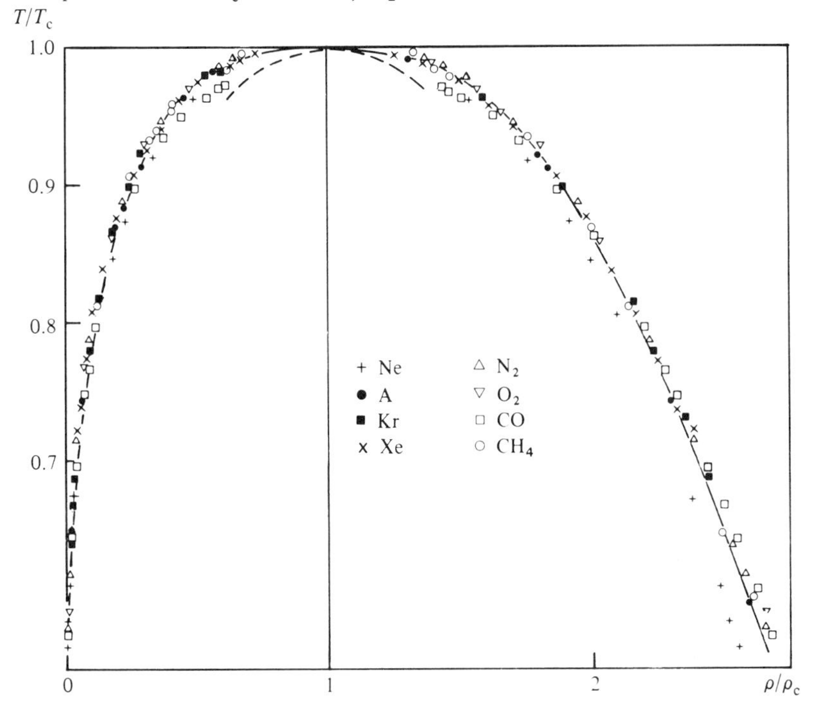

The scaling hypothesis

The suggestion made was that near a critical point the free energy may always be split into two parts, a regular part F_R which shows no singularity at the critical point, and a singular part $F_S = f(t, \mathscr{B}_E)$ which sufficiently near the critical point obeys the scaling rule

$$\lambda F_S = f(\lambda^n t, \lambda^m \mathscr{B}_E), \tag{18.40}$$

§ 18.10

for any λ. Let us treat this for the moment as an empirical suggestion. (We shall explore the reasoning behind it later.) It is then a simple matter to show that the critical exponents must be related to n and m. For instance, if we differentiate twice with respect to T, we find that the singular part of the heat capacity is given by

$$C_{MS} = -T_c(\partial^2 F_S/\partial T^2)_{\mathscr{B}_E=0} = -(1/T_c) f_{(1,1)}(\lambda^n t, 0)\lambda^{2n-1}, \tag{18.41}$$

where $f_{(1,1)}$ indicates the derivative of f obtained by differentiating twice with respect to its first argument. Since the value of λ is arbitrary, let us now, when we vary T, choose λ so that $\lambda^n|t|$ remains constant. Then, whether t is positive or negative, we find that $f_{(1,1)}$ remains constant, and thus

$$C_{MS} \propto \lambda^{2n-1} \propto |t|^{1/n-2}, \tag{18.42}$$

Problem 18.8

and we deduce that $\alpha' = \alpha = 2 - 1/n$. We may proceed using similar methods to find how $\bar{m}(t)$, $(\partial \bar{m}/\partial \mathscr{B}_E)_t$ and $\mathscr{B}_E(\bar{m})$ behave near the critical point. The results are

$$\begin{aligned} \alpha' = \alpha &= 2 - 1/n, \\ \beta &= (1-m)/n, \\ \gamma' = \gamma &= (2m-1)/n, \\ \delta &= m/(1-m). \end{aligned} \tag{18.43}$$

Eliminating m and n between these equations gives the following relations between critical exponents:

Relations between critical exponents

$$\begin{aligned} &\alpha' = \alpha; \quad \gamma' = \gamma, \\ &\alpha + 2\beta + \gamma = 2, \\ &\alpha + \beta(\delta + 1) = 2, \\ &\gamma(\delta + 1) = (2 - \alpha)(\delta - 1), \\ &\beta\delta = \beta + \gamma. \end{aligned} \tag{18.44}$$

Inspection of Table 18.2 shows that, with the possible exception of the liquid–vapour transition in ^{4}He, *the experimentally observed critical exponents do obey these rules within experimental error.* The theoretical values quoted also obey them. It seems then, that the scaling hypothesis is successful and that there are only two independent exponents for any given system.

Further evidence for the scaling hypothesis comes from the magnetic equation of state. Using the scaling rule (18.40) we find that

$$\bar{m}_S = -(\partial F_S/\partial \mathscr{B}_E)_T = f_{(2)}(\lambda^n t, \lambda^m \mathscr{B}_E)\lambda^{m-1}. \tag{18.45}$$

Let us now choose to vary λ so that $\lambda^n|t|$ remains constant, as before. Then

$$\lambda^{1-m}\bar{m}_S = f_{(2)}(\pm C, \lambda^m \mathscr{B}_E),$$

or

$$|t|^{(m-1)/n}\bar{m}_S = \mathscr{M}_{+,-}(|t|^{-m/n}\mathscr{B}_E), \quad \text{and}$$

▶ $$\bar{m}_S/|t|^{\beta} = \mathscr{M}_{+,-}(\mathscr{B}_E/|t|^{\beta\delta}), \tag{18.46}$$

where $\mathscr{M}_{+,-}$ is a definite function which is independent of t. This relation makes the remarkable prediction that if we plot the *reduced* magnetisation $m = \bar{m}/|t|^{\beta}$ against the *reduced* field $b = \mathscr{B}_E/|t|^{\beta\delta}$ *all isotherms are identical.* Experiment confirms that this is so, as Fig. 18.12 shows. (There are two branches of the function $\mathscr{M}(b)$, $\mathscr{M}_+$ for $t>0$ and $\mathscr{M}_-$ for $t<0$.) It seems, then, that the scaling hypothesis has strong experimental support.

18.10* Theories of critical point behaviour

The mean field theories fail near the critical point because they assume that the internal field is related to the *mean* degree of order. This is a good approximation if the fluctuations in order are small, or if the interaction producing the internal field is a long-range one which averages over the states of many sites. (This is true in superconductivity, in which the range of interaction is several hundred atomic spacings; and superconductivity *does* obey the mean field predictions quite well close to the critical point, as Fig. 15.5 shows for the heat capacity.) In most phase transitions, however, the real interaction has a very short range and the fluctuations become very large near T_c. Consider, for instance, a ferromagnetic just above T_c in zero field. Suppose

p. 193

Fig. 18.12. The magnetic scaling law near T_c. We plot reduced magnetisation $m = \bar{m}/|t|^{\beta}$ against reduced field $b = \mathscr{B}_E/|t|^{\beta\delta}$ for the ferromagnet $CrBr_3$ (after Ho & Litster, 1969: only a selection of the original data is shown). The plot shows that data from 18 different temperatures in the range $-0.11 < t < 0.067$ all fit on the *same* universal curve (which has separate branches for $T<T_c$ and $T>T_c$). For $CrBr_3$, $\beta = 0.368$, $\beta\delta = 1.583$ and $T_c = 32.844$ K. (This figure corresponds to a very small region very near the origin of Fig. 18.3(*b*).)

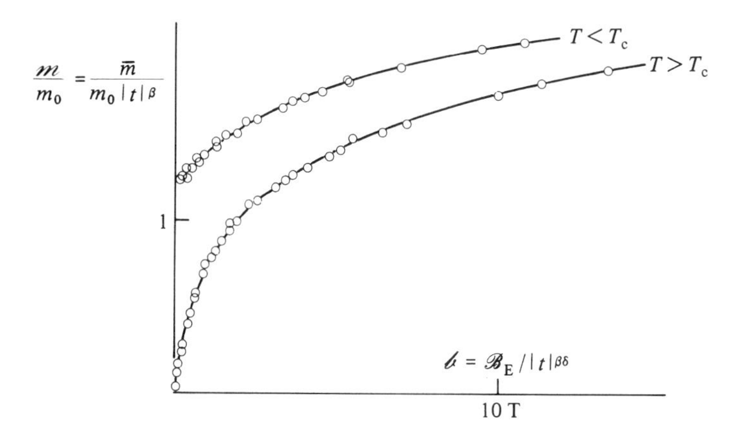

that, in the course of fluctuations, a given site is known to be spin up. Then, although the *mean* internal field is zero, the *actual* internal field acting on the immediately neighbouring sites is quite large. Now just above T_c the magnetisability is extremely large, so the local internal field due to the site of interest is almost certain to line up the neighbouring sites. But then, by the same argument, the neighbouring sites are almost certain to line up the *second*-nearest sites, and so on. The probability of being spin up will only die away slowly with distance – we say that the system has a large *coherence length*, ξ. Thus just above T_c, although the mean magnetisation is zero, the *local* magnetisation is not. There will be regions of size ξ having substantial magnetisation in a definite direction, although the boundaries and direction of magnetisation of these regions will change in the course of thermal fluctuations (Fig. 18.13). This explains the large heat capacity which persists above T_c, even though the *mean* magnetisation is zero. A similar phenomenon occurs at the liquid–vapour critical point. The system contains relatively large regions of liquid-like and vapour-like density, which fluctuate continuously. (These regions scatter light, and the system has a milky appearance, known as *critical opalescence.*) In the same way, the ferromagnetic just *below* T_c has relatively large fluctuating up spin and down spin regions, as it does above T_c.

Local order near T_c and the coherence length

Fig. 18.13. Illustration of the concept of decimation. The spins are all identical, and interact with their near neighbours. The aim of decimation is to retain only the spins shown as bold arrows in the formal expression for F, the other spins behaving as a medium which modifies the effective interactions between the bold spins. At a critical point, repeated decimations leave the expression for F unchanged in form. (The orientations shown are typical of situations near a critical point: notice the large degree of short-range order.)

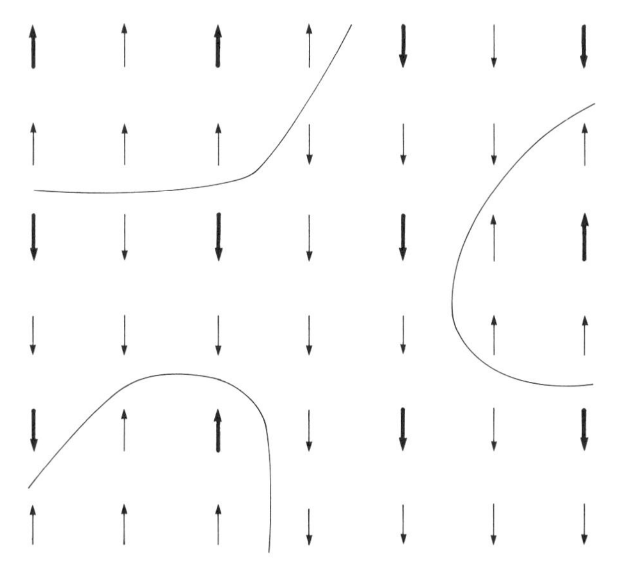

The only difference is that there are now slightly more of one type than the other.

The key to understanding the critical behaviour is evidently to study models in which the interactions are of short range, and to take proper account of the effects of fluctuations. A good deal of progress has been made in this direction. One point to notice is the significance of the number of dimensions d in which the system is extended. In a one-dimensional system such as a line of spins it is known that there is *never* true long-range order, and hence no phase transition. The reason is that a change from up spin to down spin at some point on the line has a finite probability, so that on a long line there will be many such reversal points, and the system must always contain both spin up and down spin regions. In one dimension neither up spin nor down spin will predominate, because there is no way in which a dominant region will cause smaller regions to shrink. In two or more dimensions, by contrast, the dominant region tends to *surround* the minority region, and since the interface has a positive energy, the minority regions will tend to shrink, and we have feedback leading to a phase instability. In fact, it has been shown that in a hypothetical system in four or more dimensions the pressure from the dominant phase would be so strong that even near T_c the fluctuations would be negligible and the mean field theory would become valid. Two- and three-dimensional systems lie between these two limits, but they have proved extremely difficult to analyse exactly. A landmark was Onsager's exact solution of the two-dimensional Ising model published in 1940. The Ising model is an idealised ferromagnet with up and down spin states and a nearest-neighbour interaction of the form $-J\sigma_i\sigma_j$, where σ_i is a spin index equal to ± 1 which indicates the spin state of the ith site. The critical exponents for Onsager's solution are shown in Table 18.2. They are quite different from those for the mean field theories, showing that in two dimensions the effects of fluctuations are very strong. Little progress has been made on exact solutions for realistic models in three dimensions.

Importance of the number of dimensions

Onsager's solution of the 2-d Ising model

More interesting than exact solutions of particular models is the substantial progress made recently in understanding the general physical reasons for the appearance of scaling laws near T_c, and universal sets of critical constants. This work is based on a simple and penetrating idea. At the critical point the coherence length becomes *infinite*. Thus there is nothing (apart from the crystalline lattice and the size of the sample) to characterise the scale of the fluctuations. It was suggested by Kadanoff that at the critical point (provided we choose a scale large compared with the lattice spacing and small compared with the sample size) *the appearance of the fluctuations would not be altered if we changed our length scale*, apart from suitable scalings of our units of $\bar{m}$, time, etc. Such an assumption provided good prospects of explaining the scaling law for the free energy.

Kadanoff's criterion

A number of ingenious techniques have been developed by Kadanoff, Wilson and others which allow us to identify the critical point by using this criterion, to verify the scaling law, and to determine the critical exponents, *without* obtaining an exact solution or describing in detail the complex structure of the fluctuations near the critical point. One of these techniques is *decimation*: we shall explore it in outline only. We start with a set of spins on a lattice (Fig. 18.13). The Hamiltonian takes some suitable form, such as

$$\mathscr{H}_0 = -J_0 \sum \sigma_i \sigma_j - \sum_i \sigma_i \mathscr{B}_0, \tag{18.47}$$

where the first sum is taken over all nearest-neighbour pairs, and the second over all spins. The first term represents the nearest-neighbour coupling and the second the effect of the external field, $\mathscr{B}_0$. The corresponding free energy may be expressed by the relation

$$e^{-F/kT} = \sum_{C_0} e^{-\mathscr{H}_0/kT} \times e^{-F_{R0}/kT}, \tag{18.48}$$

where F_{R0} is the non-singular part of the free energy which is of non-magnetic origin. The sum in (18.48) is taken over all possible spin configurations C_0. We now attempt, in a formal sense, to *double the lattice size* by rewriting (18.48) as a sum over the configurations C_1 of only those spins which are shown in bold in the diagram. The idea is that the remaining *residual* spins act as a medium which transmits an *effective* interaction between the bold spins, and under the action of an external field applies an *effective* external field to the bold spins. This formal transition of viewpoint is made quite simply as follows. We separate the sum $\sum_{C_0}$ over all configurations into a double sum $\sum_{C_1,r}$ over the configurations C_1 of the bold spins, and the configurations r of the residual spins:

Outline of the technique of decimation

$$e^{-F/kT} = \sum_{C_1} \left[\sum_r e^{-\mathscr{H}_0/kT} \right] \times e^{-F_{R0}/kT}. \tag{18.49}$$

Once the sum over the residual configurations has been performed the term in square brackets is a function of the configuration of bold spins only, and it turns out that it may be written as $\exp(-\mathscr{H}_1/kT - R_0/kT)$, where $\mathscr{H}_1$ is an *effective* Hamiltonian acting between the bold spins, and R_0 is a constant. Thus

$$e^{-F/kT} = \sum_{C_1} e^{-\mathscr{H}_1/kT} \times e^{-F_{R1}/kT}, \tag{18.50}$$

where $F_{R1} = F_{R0} + R_0$ is a new *effective* regular part of the free energy, which now includes some terms of magnetic origin.

The new effective Hamiltonian $\mathscr{H}_1$, which is a function of the bold spin co-ordinates only, in general has a more complicated form than (18.47): it includes effective interactions between bold spins other than nearest neighbours. In a serious decimation calculation these more distant interactions are important, and have to be taken into account. Here, where our aim is only to describe the essence of the method, I shall ignore them, and assume for simplicity that the new Hamiltonian $\mathscr{H}_1$ has the same *form* as (18.47), the difference being that the new Hamiltonian refers to the bold spins only, and has a new *effective coupling constant* J_1 and a new *effective applied field* $\mathscr{B}_1$. Using reduced variables $K=J/kT$ and $b=\mathscr{B}/kT$, we therefore have a new reduced coupling constant K_1 and a new reduced field b_1 which are definite functions of the old ones,

$$K_1=f(K_0), \quad b_1=g(b_0). \tag{18.51}$$

(In practice, b_1 depends on K_0 also, but, for simplicity again, I shall ignore this.) Once we have established the functions f and g, we may clearly repeat the process, obtaining a sequence of effective reduced coupling constants $K_0, K_1, K_2, \ldots$, effective reduced fields $b_0, b_1, b_2, \ldots$, and effective regular parts of the free energy $F_{R0}, F_{R1}, F_{R2}, \ldots$, as we repeatedly double the scale on which we view the spin lattice. It is this process of doubling the scale and finding what happens to K, b and F_R in the process which is known as *decimation*. The operation of finding the new values of K, b and F_R from the old is known in the literature as *renormalisation*, and the set of all such renormalisation operations is known as the *renormalisation group*.

We may now identify the critical point very simply. In general we find that for a given coupling constant J_0 there are special values of temperature T and applied field $\mathscr{B}_0$ for which the sequences for K_n and b_n tend to *definite limits* K^* and b^*, where

Identification of the critical point

$$K^*=f(K^*), \quad b^*=g(b^*). \tag{18.52}$$

When this happens then, clearly, after a few decimations, if we change the length scale by a further decimation we do not change the effective Hamiltonian, and the fluctuations therefore appear unchanged at our new scale. But that is precisely Kadanoff's criterion for the critical point. In practice such limiting behaviour occurs at $\mathscr{B}_0=0$, and, for a given value of J_0, at some definite temperature, which we may now identify as the critical temperature: we have here a systematic method for finding T_c. Such calculations are successful when based on realistic models, and predict values of T_c a little below the mean field prediction, in agreement with measurement.

To predict the critical exponents we observe that when we are close to the critical point for which $K=K^*$, $b=b^*=0$, we have by differentiating (18.51)

$$\delta K_n \simeq f'\delta K_{n-1}, \quad b_n \simeq g' b_{n-1}, \tag{18.53}$$

where $\delta K_n = K^* - K_n$, so that after n decimations we have

Determination of the critical exponents

$$K_n = K^* - \delta K_n \simeq (J_0/kT_c)[1-(f')^n t],$$

and

$$b_n \simeq (g')^n b_0 \simeq (g')^n \mathscr{B}_0/kT_c, \tag{18.54}$$

where we have written K^* as J_0/kT_c. Now the singular part of the free energy after n decimations F_{Sn} has a value per spin which is a function of K_n and b_n. In other words, since J_0 and T_c are constants, we may regard the value per spin as a function of $(f')^n t$ and $(g')^n \mathscr{B}_0$. After n decimations the number of spins is $N/(2^d)^n$, where d is the number of dimensions. Thus

$$F_{S_n} 2^{dn} = F[(f')^n t, (g')^n \mathscr{B}_0]. \tag{18.55}$$

Writing $2^{dn} = \lambda$, and expressing f' and g' as $(2^d)^p$ and $(2^d)^q$, respectively, we have

$$\lambda F = F(\lambda^p t, \lambda^q \mathscr{B}_0). \tag{18.56}$$

This has the form required by the scaling hypothesis, and the critical exponents may be written down in terms of f' and g', using (18.43).

We have thus seen, in outline, how the renormalisation technique provides a means of calculating T_c, a justification of the scaling hypothesis and a method for finding the critical exponents. It has also proved successful in explaining universality: it turns out that the details of the original Hamiltonian cease to matter after the first few decimations, and that the values of the critical exponents depend only on the symmetry and dimensionality of the system.

At present much theoretical and experimental investigation of the details of critical point phenomena continues. Nevertheless we seem to be able to say with some confidence that the scaling hypothesis and the theoretical explanations of it give an essentially correct picture of the behaviour very close to the critical point. Though our theories of phase transitions are necessarily as incomplete in detail as those of the equation of state it does seem that mean field theory modified near the critical point by a suitable scaling theory to take account of critical fluctuations gives a reasonably good picture of the essential physics in most cases.

Problems

18.1 *The triple point of water* — Sketch the phase diagram of water near 0 °C, using the following information. At 1 atmosphere ice melts at 0 °C. At this temperature its latent heat of melting is 3.4×10^5 J kg^{-1} and its density is 0.92×10^3 kg m^{-3}, while the latent heat of evaporation of water is 2.5×10^6 J kg^{-1} and its vapour pressure is 6.03×10^2 N m^{-2}. Estimate the triple point temperature and pressure. [Use the Clausius–Clapeyron equation (8.9).] *§8.3.1*

18.2 *The phase diagram of ^{4}He* — Sketch the rather unusual phase diagram of ^{4}He, given that $T_c = 5.2$ K, $P_c = 2.3$ atm, that the solid is denser than the liquid, that the solidification pressure is 24 atm at 1 K, and that above 1 K the solid has less entropy than the liquid while below 1 K the heat capacities of both solid and liquid (due mainly to phonons) are extremely small.

18.3 *The van der Waals critical point* — Expand van der Waals' expression (18.10) for P to third order in powers of $\delta V = V - V_c$ on the critical isotherm. By considering the values of $(\partial P/\partial V)_T$ and $(\partial^2 P/\partial V^2)_T$ at the critical point obtain expressions (18.11) for V_c, T_c and P_c, and show that the critical isotherm has the form $\delta P/P_c = -\frac{3}{2}(\delta V/V_c)^3$. Show also that just below T_c the isotherms take the form $\delta P/P_c = -\frac{3}{2}(\delta V/V_c)^3 + 2(\delta T/T_c)(2 - 3\delta V/V_c)$. With the help of the equal area rule deduce that on the coexistence curve $\delta P/P_c = 4\delta T/T_c = -(\delta V/V_c)^2$ (see Fig. 18.11).

18.4 Show that, according to van der Waals' equation (18.10), a liquid may exist in a metastable state of *negative pressure* when $T < 27T_c/32$. Estimate the lowest such pressure which can exist in water at 20 °C, assuming that bubble nucleation is completely inhibited. [For water, $T_c = 647$ K, $P_c = 218$ atm. Such negative pressures sometimes occur near ships' propellors. If nucleation does occur the water boils explosively, and the resulting *cavitation* produces shock waves which may damage the propellor.]

Negative pressures in liquids

18.5 At very low temperatures ^{4}He will dissolve 6% of ^{3}He, but is almost insoluble in ^{3}He. Above 0.87 K the two liquids are completely miscible. Sketch the binary phase diagram for the liquid and vapour phases (i) at a pressure of 30 torr, where the boiling points are 2.08 K for ^{4}He and 1.32 K for ^{3}He, and (ii) at a pressure of 0.01 torr, where they are 0.79 K and 0.36 K respectively. [In the second case there is a triple point just above 0.36 K at which the two liquids are in equilibrium with almost pure ^{3}He vapour.]

Binary phase diagram for ^{3}He/^{4}He mixtures

In a *dilution refrigerator*, almost pure liquid ^{3}He enters a *mixing chamber* containing liquid ^{4}He at the operating temperature of 10 mK, and cools it by dilution (why?). The diluted ^{3}He diffuses up a heat exchanger tube through stationary liquid ^{4}He to a *still*, where there is a free liquid surface at about 0.5 K. At this temperature the vapour pressure of the ^{4}He component remains extremely small, and almost pure ^{3}He vapour is pumped off the surface at 0.01 torr, helped by a small heater. This vapour is compressed to 30 torr by a diffusion pump, and delivered to a *condenser*, cooled to 1.32 K in an ordinary pumped liquid ^{4}He cryostat. The liquid ^{3}He then passes in turn through a coil in the still (where it cools to 0.5 K), a constriction where the pressure drops to the region of 0.01 torr and the heat exchanger, before finally reentering the mixing chamber. Identify the stages of this process on the two phase diagrams.

The dilution refrigerator

18.6 In § 18.8 we discussed the phase separation which occurs in alloys when E_2 is negative. When E_2 is positive *mixing* is favoured. In this situation we may observe an *order–disorder transformation*, in which, on cooling, the system changes from a random solid solution to a *superlattice structure*, detectable by X-ray diffraction, in which the A atoms tend to lie on the sites of the a-sublattice, and the B atoms on the sites of the b-sublattice, each a-site being surrounded by [illegible] b-sites and *vice-versa*.

The order–disorder transformation

Consider the case in which there are equal numbers of A and B atoms and $v_A = v_B$, using as order parameter $\sigma = 1 - 2f^b_A$, where f^b_A is the proportion of A atoms on b-sites. Show that, if the atoms on each sublattice are arranged at random,

The Bragg–Williams theory or order–disorder

$$G = -NE_2\sigma^2 + \tfrac{1}{2}NkT[(1+\sigma)\ln(1+\sigma) + (1-\sigma)\ln(1-\sigma)] + \text{const.}$$

By minimising G with respect to σ, deduce that $\sigma = \tanh(2E_2\sigma/kT)$, and find the critical temperature. Show that the forms predicted for the order parameter, energy and heat capacity as functions of temperature are the same as those for the Weiss theory shown in Fig. 18.2.

18.7 The wave drag D on a ship of given hull form and length l travelling at velocity v has the form $D = \rho g l^3 W(v^2/gl)$, where ρ is the density of water, g is the acceleration due to gravity and W is the drag function. Show that the relation between D, l and v obeys a scaling law of the form $\lambda D = f(\lambda^p l, \lambda^q v)$, where $p = \frac{1}{3}$ and $q = \frac{1}{6}$.

The idea of a scaling law, and the mean field critical exponents

Using (18.39) find the mean field forms near T_c for $\bar{m}$ and $(\partial\bar{m}/\mathscr{B}_E)_T$ as functions of t in zero field and the form of $\bar{m}$ as a function of $\mathscr{B}_E$ on the critical isotherm, and so obtain the mean field critical exponents β, γ, γ' and δ, shown in Table 18.2. Show that the scaling results quoted after (18.39) are consistent with a scaling $F'_S = \mu^4 F_S$ for the singular part $F - F_c$ of the free energy. On this assumption find the corresponding scaling of $S_S = S - S_c$, and obtain the mean field values of the critical exponents α, α' in Table 18.2 and n, m in (18.40). [It was shown by Landau that the same exponents will be found for *any* theory which assumes that F is a regular, expandable function of $\bar{m}$, $\mathscr{B}_E$ and T near the critical point. It follows that the appearance of other critical exponents is an indication that the critical instability has given F an essential singularity at the critical point.]

Relations between the critical exponents

18.8 Use the scaling hypothesis (18.40) to find a relation similar to (18.41) which shows how the singular parts of $\bar{m}$ and $(\partial\bar{m}/\partial\mathscr{B}_E)_T$ depend on t and $\mathscr{B}_E$ near the critical point. Deduce the forms of the singular parts of $\bar{m}$ and $(\partial\bar{m}/\partial\mathscr{B}_E)_T$ as functions of t in zero field and the form of $\bar{m}$ as a function of $\mathscr{B}_E$ on the critical isotherm. Hence confirm the relations (18.43) and (18.44).

19

The fundamental assumptions reviewed

The theory of heat has two parts: the identification of states of thermal equilibrium, and the determination of how a given system *approaches* thermal equilibrium. In this chapter we review the statistical assumptions which we have used in tackling these two problems. We shall find that our methods of identifying the standard equilibrium state are soundly based, but that it is much harder to determine rigorously how rapidly systems approach equilibrium, or even to demonstrate that they necessarily converge onto the standard equilibrium state at all. These difficulties concerning the approach to equilibrium involve questions of principle which have fascinated some of the best minds of physics. I must emphasise that the issues concerned here are primarily questions of philosophy and rigour. Very few physicists doubt that the results of statistical thermodynamics are correct. Indeed, Einstein thought it the most securely founded branch of physics, because it is in essence no more than an elaboration of probabilities, and does not depend on a particular physical model in the way that other branches of the subject do. The difficulties of principle are interesting, however. They are particularly concerned with the question of what we mean when we say that a thermodynamic system has been *randomised*. Because this question is most easily discussed for classical systems, we deal with them first, in the first three sections. We shall then turn to quantum systems. In § 19.4 we shall introduce the *density matrix*, an improved representation of the statistical state, which we then use to discuss the approach to equilibrium from the quantum point of view, and the master equation in particular. We shall end the book with a discussion of the behaviour of entropy in fluctuations.

19.1 Accessibility and the ergodic assumption

In classical physics we view the approach to equilibrium as a process of randomisation which occurs because representative points originally in a given neighbourhood of phase space are gradually dispersed as they follow their individual trajectories: *questions about randomisation are, in classical physics, questions about the nature of trajectories in phase space.* We begin our

§ 13.2

discussion of this randomising process by returning to the *accessibility assumption*: we have generally assumed that for given constraints an isolated system is able to pass from the neighbourhood of any one state to the neighbourhood of any other having the same energy. This accessibility assumption seems very natural at first sight. It is, however, easy to find special cases where it fails, so it certainly cannot be proved rigorously in general. In this section we shall explore some of these cases.

§ 13.4

First, situations do arise in which there are obvious abnormal limitations on the parts of phase space accessible to the system. For instance, in the thermodynamics of an isolated galaxy it is obvious that the six components of linear and angular momentum are, like the energy, *constants of the motion*, and this will certainly limit the degree of randomisation of the motion which can occur. In a sense, such limitations are a form of external constraint. Momentum conservation, like energy conservation, does not depend on the detailed internal properties of the system but simply on how we permit it to be influenced by the outside world. It is probably best to regard such cases (which do not arise very often) as part of the general problem, always present in thermodynamics, of deciding how to characterise the constraints. Situations also arise, as we have seen, in which the phase space has two or more parts which are disconnected, or almost disconnected. We have noted, for instance, that a thermodynamic phase in metastable equilibrium may be regarded as a system stuck in a relatively small part of its phase space, and it may take the system a very long time to find its way out and reach true thermal equilibrium. We have also observed, in connection with the third law, that some systems such as glasses at low temperatures are able to explore only the very restricted part of phase space which corresponds to thermal vibration, atomic rearrangement being either impossible or extremely unlikely. Such failures of the accessibility assumption are *internal* properties of the system concerned, and are therefore harder to anticipate than the first type. They may have important practical consequences. All we can do about them is to be alert to such possibilities, and treat them individually as they occur. They raise no special difficulties of principle.

Externally imposed constants of the motion

Disconnected parts of phase space

§ 18.6

§ 4.5

Apart from abnormal restrictions of the sort just discussed, the accessibility assumption does not cause any *practical* difficulties, but it remains very difficult to justify. The nature of the difficulty becomes apparent if we consider a few counter-examples (Fig. 19.1). In a perfectly reflecting, perfectly rectangular box, the trajectory of a single particle is never randomised, for $|p_x|$, $|p_y|$ and $|p_z|$ are conserved as well as the kinetic energy. A planet subject to a perfect inverse square law gravitational field moves in a closed elliptical orbit which never explores the whole of the space available to it for a given energy and angular momentum. In an ideal harmonic solid, the phonon modes are independent and the energies stored in them never come into thermal equilibrium. Clearly, in each of these cases there is some ideal and in practice unobtainable property which is preventing the system from exploring

The problem of internal constants of the motion

the whole of its phase space. We need only roughen slightly the walls of the box, or make the Sun slightly non-spherical, or introduce a small anharmonic force into the lattice, and the motions will change slowly so that they *do* explore the whole of the phase space. It is therefore tempting to dismiss such special counter-examples as abnormal cases which would never arise in practice. Further consideration suggests, however, that this might be wrong. For a system of N particles Hamilton's equations consist of $6N$ coupled first-order differential equations. Such a set of equations may be converted into $6N-1$ first-order equations in the canonical variables, and one equation connecting these variables with time. It follows that in general there will be as many as $6N-1$ constants of the motion (corresponding to the $6N-1$ constants of integration), and one function of the canonical variables which increases steadily with time, corresponding to motion along the trajectory in phase space. Thus for *any* closed system, the trajectory is subject to $6N-2$

§ 13.3

Fig. 19.1. Two counter-examples to the accessibility assumption. In a perfectly rectangular, perfectly elastic box the direction of motion of a single molecule is never randomised. In an exact inverse square law the orbit of a planet is closed, and it does not explore the shaded area to which it should have access if we take only energy and angular momentum conservation into account. However, roughening the walls of the box, or flattening the sun so as to distort the inverse square law field will perturb these trajectories so that they do explore the whole of the available phase space.

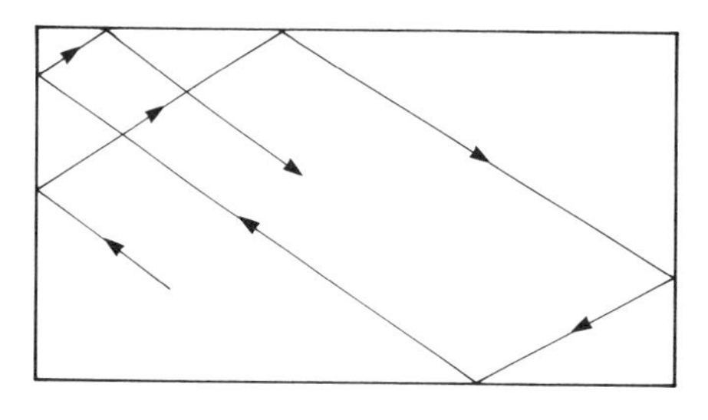

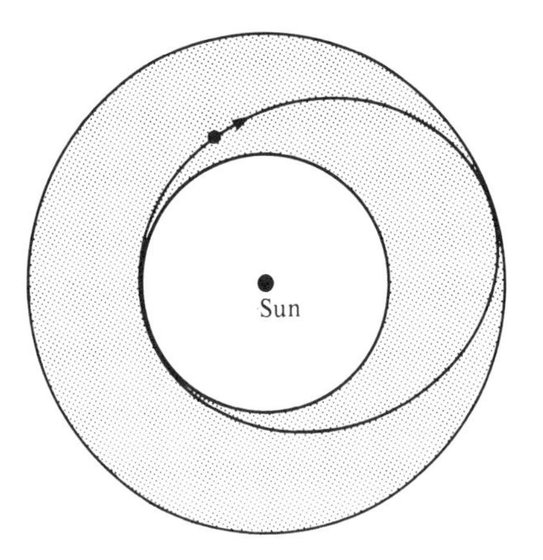

restrictions in addition to the energy restriction. This makes it seem exceedingly surprising that the accessibility assumption should work at all for any system.

Quasi-ergodic trajectories

The way round this difficulty becomes apparent when we consider that the trajectories might be *quasi-ergodic*. An *ergodic* trajectory would be one which actually passes through all points in the relevant part of phase space. The existence of the $6N-2$ restrictions shows that the trajectories cannot be ergodic. They may, however, be quasi-ergodic, in the sense that they pass *infinitely close* to all points in the relevant part. Consider, for instance, the planetary orbit shown in Fig. 19.1. For an exact inverse square law the orbit is an ellipse given by

$$1/r=(1/r_0)[1-e\cos(\theta-\theta_0)]. \qquad (19.1)$$

In this orbit the constants r_0 and e are fixed by the energy and angular momentum, and the constant θ_0 is a constant of integration. The orbit is closed, and the accessibility assumption fails. If, however, we add a very small inverse cube term to the inverse square force, the orbit becomes

$$1/r=(1/r_0)[1-e\cos\{\alpha(\theta-\theta_0)\}], \qquad (19.2)$$

where α is close to but not equal to unity. The orbit now slowly precesses, and if α is an irrational number it eventually covers the plane for $1-e<r_0/r<1+e$ infinitely densely, in spite of the fact that θ_0 is a constant of the motion. It does not actually pass through all points (for $\theta_0=0$, for instance, it does not pass through the points $r=r_0$, $\theta=2\pi n$) but it passes infinitely close to all points.

Classical statistical thermodynamics is in fact based on the idea that, apart from a very few special cases of the sorts discussed above, the trajectories are always quasi-ergodic, and therefore always *appear*, on our scale of interest, to become randomised in spite of the fact that they are subject to $6N-2$ *hidden constants of the motion*. It turns out that this is actually a very reasonable assumption. The question of whether the trajectories should be expected to be quasi-ergodic has been investigated using *measure theory* by Birkhoff and others. Their results show in effect that for systems containing many interacting particles the chance that any appreciable fraction of the trajectories will not be quasi-ergodic in any given representation is extremely small. For such systems we may therefore adopt the accessibility assumption with great confidence, but we cannot prove that it is valid. (The measure theory does not apply to quantities like energy and momentum which may be held fixed from outside the system.)

The importance of practical canonical variables

There is a feature of this argument which we must emphasise now and return to later. A given system may be described by many alternative sets of canonical variables. We normally employ some set of *practical variables*, in terms of which the observables of interest are simple and relatively slowly varying functions of position in phase space. When this is so we do not need to know the *exact* positions of the representative points of the ensemble: it is enough to know their mean density in a given neighbourhood, and significant changes in the system correspond to macroscopic changes in this density. Birkhoff's argument assumes that we are using some such *given* set of practical variables, and he concludes that the trajectories are then exceedingly likely to be quasi-ergodic. But this does not mean that we cannot *choose* a perverse and abnormal dynamical representation in which the trajectories are not quasi-ergodic. We may, for instance, choose to use as our dynamical variables the $6N-1$ constants of the motion and the remaining function which increases steadily with time. This would have the effect of straightening out all the trajectories so that the representative points run parallel and at the same speed. In this representation no mixing of representative points occurs at all. Does this invalidate the results of statistical thermodynamics? The answer is no: in choosing an abnormal representation which straightens out the trajectories we simultaneously make the *observables* into exceedingly involuted functions whose values fluctuate exceedingly rapidly with position in the representation space. In such a perverse representation the fact that no macroscopic mixing occurs is of no significance, for the important information about the state of the system is now contained in the microstructure of the ensemble: to find the mean values of the observables we should need to know *exactly* where the representative points were, and no longer just their coarse-grained probability distribution. Such an abnormal representation would be quite unusable in practice. In what follows we shall take it for granted that we are using practical canonical variables, for which the observables are simple and relatively slowly varying functions, that the important information about the state of the system lies in the coarse-grained distribution of representative points, and that the trajectories are extremely complex and quasi-ergodic.

Abnormal representations

To summarise, it is clear that, if we use practical canonical variables, the only restrictions on accessibility likely to be important in practice are restrictions imposed from outside the system, and occasionally, the splitting of phase space into isolated or almost isolated pieces. Apart from such limitations, measure theory shows that for large systems it is extremely unlikely that the accessibility assumption will fail.

19.2 The problem of time reversal symmetry

We already know how to identify the equilibrium states of an isolated system in classical statistics. In classical theory, using practical variables, we take the system to be in equilibrium when the local ensemble density is independent of time. We have already seen that this occurs if, and only if, the density $\tilde{\rho}$ is the same for all mutually accessible parts of phase space – and when the accessibility assumption holds, this means that $\tilde{\rho}$ must be the same for all states of the same energy.

§ 13.4

In trying to see how the ensemble *approaches* such an equilibrium state, however, we encounter a special difficulty. Imagine that we could take a ciné film of the behaviour of the Universe which shows the details of the motions of all its constituent particles, and we then project the film *backwards* so that the direction of time is reversed. This has the effect of reversing all the velocities and momenta, but the accelerations $\mathbf{a}$ and rates of change of momenta are unchanged (because $\mathbf{a} = \mathrm{d}\mathbf{u}/\mathrm{d}t$, and the signs of $\mathrm{d}\mathbf{u}$ and $\mathrm{d}t$ are both reversed). Now it turns out that in the reversed motion the forces predicted by the laws of physics at each point are also unchanged. (This is obvious for a simple potential force. For a velocity-dependent force such as the force $e\mathbf{v} \wedge \mathscr{B}$ acting on a moving charged particle we observe that, since all the electric currents in the Universe are reversed, the velocity $\mathbf{v}$ and the magnetic field $\mathscr{B}$ are *both* reversed also, so this force, too, is unaltered.) Since both the accelerations and the forces are unchanged when we run the film backwards the particles still appear to obey the laws of physics. We may say that *the laws of physics are time-reversal invariant*: they do not distinguish between the positive and negative directions of the time axis. We shall see later that the same is true of Schrödinger's equation in quantum theory.

The meaning of time-reversal symmetry

This is a startling result for thermodynamics. In both the quantum and the classical treatments we have assumed that as time proceeds the probability of finding the system in a given state becomes steadily more evenly spread over the mutually accessible states, which corresponds to an increase of entropy. According to thermodynamics, then, if we run time backwards the entropy should *decrease*. But if the system still obeys the same laws when time runs backwards, why should there be any difference in the behaviour of the entropy? It seems that in deducing the law of increase of entropy from the laws of mechanics we must either have made some error in logic or introduced some hidden further assumption. A little thought shows that we have, in fact, made a significant further assumption. To see what it is, consider again Fig. 13.2, in which we showed how an ensemble of representative points originally close together in phase space becomes spread out over the constant energy surface as time proceeds. Observe carefully a special feature of this argument. *We assumed that within the original cell the probability distribution was uniform*, corresponding to a random distribution of representative points. It is easy to see that if, instead of distributing the points at random in the original cell we had arranged them in certain narrow strips (Fig. 19.2) then, although they would have spread out at first, eventually they would all have met again in the original cell, so that $\tilde{\rho}$ would have first become more even, but subsequently less even, and the entropy would have first increased but subsequently decreased. It is clearly always possible to do this: we may select from the original ensemble just those representative points which at some later time happen to arrive in some particular area of phase space. For such a specially chosen ensemble the entropy will at some later stage *decrease* with time.

p. 141

Entropy decreasing ensembles do exist

This makes it clear that the law of increase of entropy in an isolated system is not a consequence of the laws of mechanics alone (which, as their time-reversal symmetry makes obvious, are *neutral* so far as entropy increase or decrease are concerned) but involves also our assumption that we may take the starting probability to be *uniform* within a sufficiently small cell in phase space. Experience shows that this assumption works well, and we may, indeed, tentatively suggest it as a new empirical law of physics, the *principle of uniform local probability*:

The principle of uniform local probability

> When using practical canonical variables we may always treat the distribution of representative points within any sufficiently small cell of phase space as uniform for the purpose of calculating the probabilities of future events.

(The reason for the limitation to practical variables is obvious: in an abnormal representation in which the values of the observables varied rapidly with position inside the cell, we could not use a smoothed distribution to calculate their averages, either in the present or in the future.)

§ 19.1

It is, in fact, this empirical principle which distinguishes the future from the past, and such an important feature of physics deserves careful examination. We must remember that statistical mechanics claims to predict not what will happen in any given experiment, but what will happen *on average* if an experiment is repeated many times. Thus our starting ensemble should represent a very large number of repetitions of the same experiment, set up as far as possible in the same way each time. There are three conceivable ways in which significant non-uniformities on a fine scale, or *microstructure*, might

Fig. 19.2. An entropy-decreasing ensemble. In Fig. 13.2 we showed how representative points originally in a box at A become spread over the constant energy surface. Here we show that, if we choose points within A which lie close to four special lines, then, although in the first 15 cycles the points appear to spread out, after 57 cycles they all meet again in box A, and the entropy appears to decrease towards the end of this process.

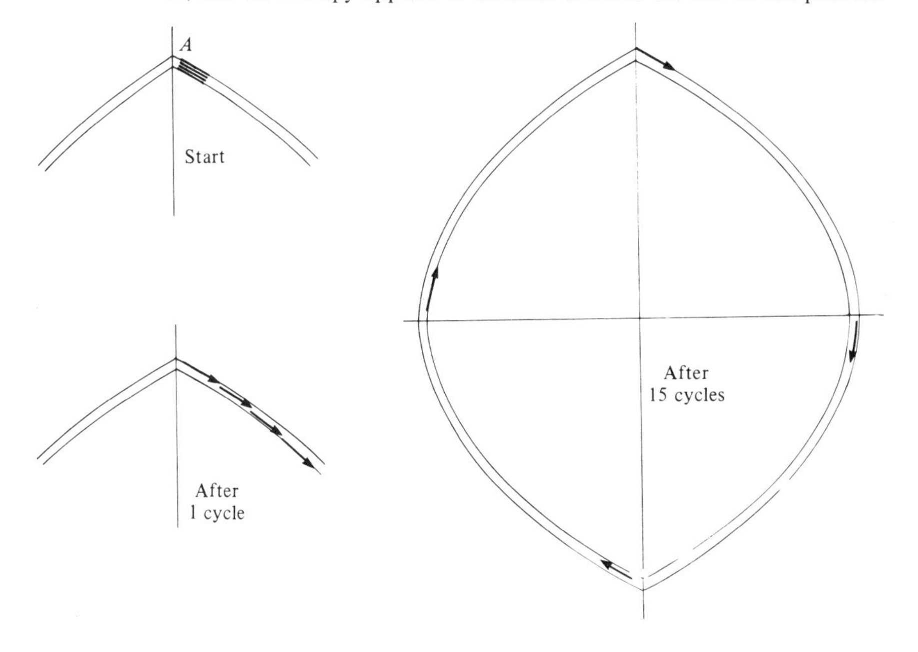

appear in such an ensemble, and thus vitiate the principle of uniform local probability and possibly lead to a decrease of entropy. First, significant microstructure might arise by *chance*. However, it is easy to show that this has a negligible probability. If we examine all the possible ensembles which are consistent with our limited knowledge of the starting state, the proportion of such ensembles which have the right microstructure to make the entropy decrease at later times is *exceedingly* small. Thus we shall in practice only see the entropy decrease at later times if there is some *very* powerful influence which tends to select these special ensembles at the outset. Secondly, we must remember that, once the experiment has started there *will* be microstructure in the ensemble, associated with the past behaviour of the system. If we return to Fig. 13.2 yet again, we see that although, some time after the start, the representative points will have been spread over the whole energy shell in phase space, they will not cover it *uniformly*. The representative points originally in region A will have been drawn out into a very long and narrow filament which is wound many times around the energy shell. Within any given cell of the shell, if we look on a fine enough scale, we shall see that the filament itself contains points *at the same density as the original region* (for its area has not changed), but that the turns of the filament are separated by much wider empty regions. Thus each cell will contain an exceedingly complicated microstructure which represents, in a sense, the *past* history of the experiment. In the particular example shown in the diagram it is obvious that as time goes on the microstructure will become finer and finer, so that we may safely ignore it in predicting *future* events. However, it is not entirely obvious that, in some more complex system, the process could not reverse itself, so that the microstructure associated with the past history was reconverted into a *macroscopically* non-uniform distribution at some time in the future. Some authors have attempted to prove that such a system is exceedingly unlikely to arise. Alternatively, we may rely on a different argument, emphasised by Landau, who notes that no experiment can be perfectly isolated from its surroundings. Thus in each member of the ensemble there will always be different small, random perturbations due to the surroundings which cause the phase space trajectory to wander slightly from its ideal path. At a sufficiently fine level these random perturbations will always rapidly blur over any microstructure which the ensemble may have, thus justifying the principle of uniform local probability. We therefore have good reason to believe that any microstructure due to the *past* history of the experiment may be ignored for the purposes of predicting its *future* behaviour. There remains a third bizarre possibility that the *whole Universe* (the experiments and their surroundings) has been elaborately preprogrammed in some special starting state so that when we repeat a set of experiments the starting states of the experiments frequently combine with the apparently random small effects of the surroundings to generate just that very small fraction of possible outcomes which correspond to a decrease of entropy. Of course, we have no means of proving

p. 141

Microstructure as a record of the past

Landau's blurring argument

by deduction that this is not the case. It is simply a matter of observation that it is not, and in this sense we must regard the principle of uniform local probability as a purely empirical law.

It is worth noting that this last aspect of the principle is closely related to our notions of *cause* and *effect*, and of *memory*. Since the laws of physics are time-reversal invariant, they do not themselves distinguish between past and future, and on the microscopic scale we may as well say that future events 'cause' present events as the other way round. On the macroscopic level, however, matters look very different. Bodies affect one another by force fields which are transmitted through space at a finite velocity. We observe, for instance, that a macroscopic accelerating charge acts as a source of electromagnetic waves which spread out from it to influence neighbouring macroscopic charges. As the wave spreads out it gradually becomes weaker and less recognisable until we lose it in the microscopic fluctuations of the distant electromagnetic field. Under these circumstances it is natural to think of the accelerating charge as the *cause* and the wave as the *effect*, because we can recognise the cause as a single, obvious, macroscopic, visible event. To argue the other way around we should have to regard the future 'cause' of the past 'effect' as part of the fluctuating electromagnetic field at infinity, which we cannot see or identify in any simple way. It is characteristic of our way of thinking that we look for a *small* number of readily identifiable causes which we then link with a *larger* number of effects which become less obvious and significant but more numerous as we distance ourselves from the cause. Since, as we have seen, macrostructure is always being converted into microstructure as time proceeds *forwards*, we naturally regard the future as caused by the past, and not the other way round. In the same way, since 'memories' are essentially relatively microscopic and widespread records of single macroscopic events, we have memories of the past, but no memories of the future. It could, in principle, be otherwise. The equations of electromagnetism are time-reversible, and there is a second possible solution of the equations for the electromagnetic field, the *advanced potential* solution, in which an electromagnetic wave gathers itself together from the distant fluctuations in the field and *converges* onto the charge, being absorbed at the moment when the charge accelerates. This is not contrary to the laws of electromagnetism; but in practice such waves do not appear, except at the exceedingly low level expected for random fluctuations of the distant field. This is in accordance with the principle of uniform local probability, which tells us that there is no microstructure in the distant field which is significant for *future* macroscopic events. If, contrary to ordinary experience, we frequently observed converging waves correlated with the accelerations of nearby particles, we should presumably have to revise our notions of cause and effect, and of direction in time. We should presumably begin to speak of converging waves as 'caused' by the accelerating charge in the future. This would not be so strange as it sounds, for we might in such a universe also discover patterns in our heads which were correlated with

Cause and effect, and memory

macroscopic events in the *future*: we might, in fact, have 'memory' of the future, and be well aware that the charge would be accelerating at a later time. Such a universe would not be at variance with the laws of classical physics. The principle of uniform local probability, however, implies that the *real* Universe does not contain the very special microstructure needed to generate such events.

19.3 Coarse graining and the classical H-theorem

We are now ready to discuss the law of increase of entropy in classical statistics, and at once meet a difficulty. We saw in the last section that if we consider the ensemble for a strictly isolated system there is a sense in which the representative points never become completely uniformly spread over the 'accessible' phase space, because the ensemble retains a microstructure corresponding to its past history. Indeed, since any group of moving representative points continues to occupy the same dimensionless hyper-volume $d\Omega$ at the same density $\tilde{\rho}_f$ as it moves, it is clear that the integral $-k\int \tilde{\rho}_f \ln \tilde{\rho}_f \, d\Omega$ for the *exact* entropy $\tilde{S}_f$ *never changes*, if by $\tilde{\rho}_f$ we mean the exact, fine-grained density which takes account of the microstructure. In practice, however, we are not interested in $\tilde{\rho}_f$. When we are using practical canonical variables the values of the observables depend only on the *coarse-grained density* $\tilde{\rho}$, which is the value of $\tilde{\rho}_f$ averaged over a suitable small cell in phase space. Moreover, we never have any information which would allow us to calculate $\tilde{\rho}_f$, because we know that in an ensemble of real systems Landau blurring will all the time be carrying $\tilde{\rho}_f$ in the direction of $\tilde{\rho}$; and the principle of uniform local probability allows us to replace $\tilde{\rho}_f$ by $\tilde{\rho}$ for the purpose of calculating future probabilities. It was therefore suggested by the Ehrenfests that we should always work with $\tilde{\rho}$ and with the corresponding *coarse-grained entropy* $\tilde{S} = -k\int \tilde{\rho} \ln \tilde{\rho} \, d\Omega$ as variables. It is this coarse-grained entropy which obeys the H-theorem, as we now show.

The fine-grained entropy is constant in a strictly isolated system

Coarse-graining

We consider the ensemble at time 0, and again at a considerably later time t. The principle of uniform local probability tells us that we may discuss the distribution at time t by treating the distribution at time 0 as though it had no microstructure, so that $\tilde{\rho}_f = \tilde{\rho}$ at $t = 0$. We may then write the change in coarse-grained entropy as

The classical H-theorem

$$\begin{aligned}\Delta\tilde{S} &= \tilde{S}(t) - \tilde{S}(0)\\ &= \tilde{S}(t) - \tilde{S}_f(t)\\ &= -k\int [\tilde{\rho} \ln \tilde{\rho} - \tilde{\rho}_f \ln \tilde{\rho}_f]_t \, d\Omega. \end{aligned} \tag{19.3}$$

where we have used the facts that $\tilde{S}(0) = \tilde{S}_f(0)$, and that $\tilde{S}_f$ is independent of time. To see that $\Delta\tilde{S}$ is positive imagine that, within a given cell at time t, we start with distribution $\tilde{\rho}_f$, and then change $\tilde{\rho}_f$ into $\tilde{\rho}$ by transferring probability from subcells where $\tilde{\rho}_f$ is greater than $\tilde{\rho}$ to subcells where it is smaller, until the whole cell is at the same probability $\tilde{\rho}$. In such a small transfer from subcell 1 to

Compare §4.2

subcell 2 the change of entropy is equal to

$$-k[-\mathrm{d}\tilde{\rho}_{f1} \ln \tilde{\rho}_{f1} + \mathrm{d}\tilde{\rho}_{f2} \ln \tilde{\rho}_{f2}] \, \mathrm{d}\Omega_0 = k \ln (\tilde{\rho}_{f1}/\tilde{\rho}_{f2}) \, \mathrm{d}\tilde{\rho}_{f2} \, \mathrm{d}\Omega_0, \qquad (19.4)$$

where $\mathrm{d}\Omega_0$ is the common volume of the two subcells. Since we chose $\tilde{\rho}_{f1} > \tilde{\rho}_{f2}$ and $\mathrm{d}\tilde{\rho}_{f2} > 0$, this term is positive. By summing such contributions we see that the whole process of changing from $\tilde{\rho}_f$ to $\tilde{\rho}$ within each cell can only increase the entropy, and $\Delta\tilde{S}$ must be positive. Thus the coarse-grained entropy $\tilde{S}$ can only increase, so long as we may always safely treat the initial distribution as though it were uniform on a sufficiently fine scale, in accordance with the principle of uniform local probability.

Clearly, it is the coarse-grained rather than the fine-grained entropy which corresponds to the non-equilibrium entropy discussed in Chapter 4. The question then arises of whether its value depends in any significant way on the size of cell over which the coarse graining is performed, or on our choice of practical canonical variables. In answer, we note first that, so long as the cell used is small compared with the range in phase space over which $\tilde{\rho}$ varies appreciably, $\tilde{\rho}$ and hence $\tilde{S}$ are both independent of cell size. Moreover, both $\tilde{\rho}$ and $\mathrm{d}\Omega$ are invariant under a canonical transformation from one set of practical variables to another. Thus $\tilde{S} = -k \int \tilde{\rho} \ln \tilde{\rho} \, \mathrm{d}\Omega$ is independent of both cell size and representation, for practical representations. Evidently the coarse-grained entropy is well defined.

Coarse-grained entropy is independent of cell size and representation

19.4 Matrix mechanics and the density matrix

We now turn again to quantum systems. Up to this point we have described the statistical state of a quantum system by quoting the ensemble probabilities $\tilde{p}_i$ of its approximate energy states, ψ_i. This description has two apparent defects. It is *representation dependent*: it is related to our particular choice of approximate energy states. It also provides no means of describing *linearly combined states* of the form $\sum c_i\psi_i$, which appear frequently in quantum theory. It will turn out that these deficiencies are more apparent than real: it is only rarely that the description in terms of the $\tilde{p}_i$ values proves to be inadequate. Nevertheless it will be helpful to our discussion to introduce a more complete statistical description, in terms of a quantity known as the *density matrix*. The density matrix may be expressed in various quantum representations, but the most natural for our present purposes is *matrix mechanics*, which I shall assume to be well known, and describe only briefly.

In matrix mechanics we represent any state function of interest ψ as an expansion $\sum a_\alpha \phi_\alpha$ in terms of some complete, orthonormal set of *basis functions* ϕ_α, which may be chosen in any convenient way. We then think of ψ as represented by the set of coefficients a_α, known as the *state vector*. When an operator $\hat{B}$ acts on state ψ it converts it to some new state ψ'. It is easy to show that the corresponding new state vector a' may be obtained from the old state vector a by an operation of *matrix multiplication*

Matrix mechanics

$$a'_\beta = \sum_\alpha B_{\beta\alpha} a_\alpha, \qquad (19.5)$$

where $B_{\beta\alpha}$ is the *matrix representation* of the operator $\hat{B}$ and is given by

$$B_{\beta\alpha}=\int \phi_\beta^* \hat{B} \phi_\alpha \, d\tau. \tag{19.6}$$

If we measure some observable B while the system is in state ψ then, unless the state function ψ happens to be an eigenstate of the operator $\hat{B}$, quantum theory does not predict with certainty the value which we shall obtain: we shall obtain one of the eigenvalues of the operator $\hat{B}$, but we cannot say which. However, the theory does predict what value we shall obtain on average. This value is known as the *expectation value* $\langle B \rangle$, and is equal to $\int \psi^* \hat{B} \psi \, d\tau$. In terms of the state vector this relation may be expressed as

Expectation values

$$\langle B \rangle = \sum_{\beta,\alpha} B_{\beta\alpha} a_\alpha a_\beta^*. \tag{19.7}$$

In statistical thermodynamics we do not know the state function with certainty. We have to represent our limited knowledge by imagining an ensemble of systems, each in some definite quantum state, the weights of the various states in the ensemble representing our assessment of the probability of finding the system in the state concerned. In this situation the value of B which we expect to find if we measure the system involves a *double* average. For each state we must average over the *quantum uncertainty* in B: this is the averaging which is represented in the expectation value (19.7). But we must also average over the systems in the ensemble, to allow for the *statistical uncertainty* represented in the ensemble. The result is the *statistical expectation value*,

Statistical expectation values and the density matrix

$$\langle \bar{B} \rangle = \sum_{\beta,\alpha} B_{\beta\alpha} \overline{a_\alpha a_\beta^*}$$

$$= \sum_{\beta,\alpha} B_{\beta\alpha} \rho_{\alpha\beta}. \tag{19.8}$$

It is the quantity $\rho_{\alpha\beta}$, the ensemble average of $a_\alpha a_\beta^*$, which is known as the *density matrix*.

To illustrate the working of the density matrix, consider a system which consists of a single spin of magnitude $\frac{1}{2}$. We use as basis functions the spin up state $\chi_\uparrow$ ($\sigma_z = \frac{1}{2}$) and the spin down state $\chi_\downarrow$ ($\sigma_z = -\frac{1}{2}$), which in this case are also good approximate energy states. Consider the following five possible density matrices:

Significance of the diagonal and off-diagonal terms

$$\begin{pmatrix} 1 & 0 \\ 0 & 0 \end{pmatrix}, \begin{pmatrix} 0 & 0 \\ 0 & 1 \end{pmatrix}, \frac{1}{2}\begin{pmatrix} 1 & 1 \\ 1 & 1 \end{pmatrix}, \frac{1}{2}\begin{pmatrix} 1 & -1 \\ -1 & 1 \end{pmatrix}, \frac{1}{2}\begin{pmatrix} 1 & 0 \\ 0 & 1 \end{pmatrix}. \tag{19.9}$$

For the spin up state the state vector is (1, 0), and it is easy to see that the first matrix represents an ensemble in which every system is spin up. In the same way the second matrix represents an ensemble in which every system is spin down. If we recall that the linearly combined state $\psi_+ = (1/\sqrt{2})(\chi_\uparrow + \chi_\downarrow)$ with state vector $(1/\sqrt{2})(1, 1)$ represents a spin lying in the positive y direction

($\sigma_y = \frac{1}{2}$), we see that the third matrix corresponds to an ensemble in which every system is in ψ_+. The fourth matrix represents in the same way an ensemble in which every system is in state $\psi_- = (1/\sqrt{2})(\chi_\uparrow - \chi_\downarrow)$, for which $\sigma_y = -\frac{1}{2}$. Note that in any density matrix a diagonal term such as $\rho_{\alpha\alpha}$ is the ensemble average of $a_\alpha a_\alpha^*$, which is the ensemble probability of finding the system in basis state α. If, as here, the basis states are the approximate energy states, *the diagonal terms of the density matrix are identical with the probabilities* $\tilde{p}_i$. In the third and fourth matrices, for instance, we note that $\tilde{p}_\uparrow = \tilde{p}_\downarrow = \frac{1}{2}$, as one would expect for linearly combined states like ψ_+ and ψ_-. The *off-diagonal* terms measure *phase coherence*, a concept which has no classical analogue. For instance, in ψ_+ the spin up and spin down terms appear with the same phase, and in the third matrix the corresponding off-diagonal terms are positive. In ψ_-, on the other hand, the spin up and spin down terms are in anti-phase, and the corresponding off-diagonal terms of the fourth matrix are negative. Recalling that the y spin operator is the Pauli matrix

$$\hat{\sigma}_y = \frac{1}{2}\begin{pmatrix} 0 & 1 \\ 1 & 0 \end{pmatrix}, \tag{19.10}$$

and using (19.8) we confirm that the statistical expectation value for σ_y is zero for the first and second matrices, but $\pm\frac{1}{2}$ for the third and fourth matrices, respectively, as expected. The fifth matrix represents an ensemble without phase coherence. It might, for instance, contain equal numbers of spin up and spin down systems, or alternatively equal numbers of systems in states ψ_+ and ψ_-. The statistical expectation value of σ_y is zero for this matrix, as, again, one would expect.

Off-diagonal operators

This illustration shows that the density matrix is in some ways superior to our description in terms of $\tilde{p}_i$ values. We notice how the density matrix can handle without difficulty the last three types of ensemble, which have different degrees of phase coherence, and different values of the observable σ_y. The $\tilde{p}_i$ description does not distinguish between these three ensembles, and therefore cannot handle observables such as σ_y for which the approximate energy states are not eigenstates.

Unitary transformations and the invariant trace

The density matrix is also superior in the sense that it may be used with *any* convenient set of basis functions. In matrix mechanics we may change from one such set to another by applying a *unitary transformation* to the state vectors and to the operator matrices. The density matrix transforms in the same way, and is thus not tied to any particular basis set. Note in particular that, if we write the matrix product $\sum_\alpha B_{\beta\alpha}\rho_{\alpha\gamma}$ as $(\hat{B}\hat{\rho})_{\beta\gamma}$, we see using (19.8) that the statistical expectation value $\langle \bar{B} \rangle$ is just the sum of the diagonal terms of this matrix. Now the sum of the diagonal terms of a matrix, known as its *trace*,

$$\mathrm{Tr}\,(\hat{A}) = \sum_\alpha A_{\alpha\alpha}, \tag{19.11}$$

is *invariant* under a unitary transformation. Thus, as we would anticipate, the value of $\langle \bar{B} \rangle$ is the same whatever basis we use.

19.5 The equilibrium density matrix

Our next step is to find the form of the density matrix in thermal equilibrium. To do so we must first obtain its equation of motion in an isolated system. We may easily write down $(\mathrm{d}/\mathrm{d}t)(a_\alpha a_\beta^*)$ for such a system with the help of the time-dependent Schrödinger equation, which in matrix mechanics takes the form

$$\dot{a}_\alpha = (-\mathrm{i}/\hbar)\sum_\gamma H_{\alpha\gamma} a_\gamma, \tag{19.12}$$

Equation of motion of the density matrix

where $H_{\alpha\gamma}$ is the matrix for the Hamiltonian operator. On taking the ensemble average of the result we find that

▶ $$\dot{\rho}_{\alpha\beta} = (-\mathrm{i}/\hbar)\sum_\gamma [H_{\alpha\gamma}\rho_{\gamma\beta} - \rho_{\alpha\gamma}H_{\gamma\beta}], \tag{19.13}$$

which is the required equation of motion.

The equilibrium density matrix is one in which $\dot{\rho}_{\alpha\beta} = 0$ for all α, β. It is most easily identified when we use as basis functions the approximate energy states ψ_i. In this basis H_{ij} is almost, but not exactly, diagonal. We may therefore take the off-diagonal terms of H_{ij} to be small quantities. On this understanding we may identify the equilibrium state as being, to a good approximation, one in which $\dot{\rho}_{ij}$ is zero to the first order of small quantities. This state has a simple form. Its *diagonal* terms ρ_{ii}, which are in fact the $\tilde{p}_i$ values, are a smooth function of the energy E_i, and the *off-diagonal* terms take the form

Form of the equilibrium density matrix

$$\rho_{ij} = \frac{\tilde{p}_j - \tilde{p}_i}{E_j - E_i} H_{ij}, \tag{19.14}$$

which, since $\tilde{p}$ is a smooth function of E, is always first-order small, even if E_j approaches E_i. To check that this is a valid solution, we write $\dot{\rho}_{ij}$ retaining first-order terms only as

$$\dot{\rho}_{ij} \simeq (-\mathrm{i}/\hbar)[(E_i - E_j)\rho_{ij} + (\tilde{p}_j - \tilde{p}_i)H_{ij}], \tag{19.15}$$

using (19.13). For the state just described we easily confirm that $\dot{\rho}_{ij} = 0$ to first order for both diagonal and off-diagonal terms.

On examining the nature of the equilibrium density matrix which we have discovered, we find that we must take account of a new principle. The fact that the diagonal terms $\tilde{p}_i$ are a function of energy only was to be expected: this is the *principle of equal equilibrium probability* as it applies to the density matrix. But we have *also* discovered that the off-diagonal terms of the density matrix are very small in thermal equilibrium. In other words, in thermal equilibrium there is negligible phase coherence between the approximate energy states: this is the *principle of random equilibrium phases*. This new

The principle of random equilibrium phases

principle has a number of important consequences. One of them is that for any observable B we have, in thermal equilibrium,

$$\langle \bar{B} \rangle = \sum_{j,i} B_{ji}\rho_{ij} = \sum_i B_i \tilde{p}_i, \tag{19.16}$$

where B_i stands for B_{ii}, which is the expectation value of B in state i. But this is exactly how we would have expected to write the ensemble average of B in the $\tilde{p}_i$ description. Thus the principle of random equilibrium phases justifies our use of the $\tilde{p}_i$ description for systems in equilibrium, even for non-diagonal operators: evidently the $\tilde{p}_i$ description is not so bad, after all. In particular for a system in the canonical state at temperature T we have

Justification of the usual expressions for off-diagonal observables in equilibrium

$$\langle \bar{B} \rangle = \sum_i B_i p_i = \sum_i B_i \, e^{-E_i/kT} / \sum_i e^{-E_i/kT}, \tag{19.17}$$

as usual.

§9.1

This result is written in the approximate energy representation. It is sometimes useful to rewrite it in a form which is independent of representation. To do so, we use the idea of an *operator exponent*, by introducing the operator $\hat{T} = \exp(-\hat{H}/kT)$ where $\hat{H}$ is the Hamiltonian operator, on the understanding that it is to be interpreted as the series $1 - \hat{H}/kT + (\hat{H}/kT)^2/2! \ldots$. In matrix mechanics a term such as $\hat{H}^3$ takes the form of a matrix product $\sum_{kl} E_{ik}E_{kl}E_{lj}$, etc., and $\hat{T}$ must be interpreted as a matrix T_{ij} which may be written out as a series of matrix products of this type. Now, in the approximate energy representation used above, E_{ij} is effectively diagonal. A little investigation shows that the partition function $\sum \exp(-E_i/kT)$ may therefore be written as the sum of diagonal terms $\sum T_{ii} = \mathrm{Tr}\,(\hat{T})$. In the same way $\sum B_i \exp(-E_i/kT)$ may be written as the sum of diagonal terms of the product matrix $\sum_i (\sum_j B_{ij}T_{ji}) = \sum_i (\hat{B}\hat{T})_{ii} = \mathrm{Tr}\,(\hat{B}\hat{T})$. We may therefore rewrite $\langle \bar{B} \rangle$ in the form

The invariant expression for $\langle \bar{B} \rangle$ at temperature T

$$\langle \bar{B} \rangle = \frac{\mathrm{Tr}\,(\hat{B}\hat{T})}{\mathrm{Tr}\,(\hat{T})} = \frac{\mathrm{Tr}\,(\hat{B}\, e^{-\hat{H}/kT})}{\mathrm{Tr}\,(e^{-\hat{H}/kT})}. \tag{19.18}$$

Because the traces are invariant this formula for $\langle \bar{B} \rangle$ holds in any representation and may be used even when we have no idea what the energy states look like. It provides the starting point for much modern formal statistical mechanics, such as the Nambu formalism.

19.6 The approach to equilibrium in quantum theory

All the difficulties concerning the approach to equilibrium which we met in classical statistics have their counterparts in quantum theory. For instance, to show that the equilibrium density matrix discussed in the previous section is the *only* possible type of equilibrium matrix we need to prove an *accessibility assumption*, that all approximate energy states of the same energy are linked, directly or indirectly, by off-diagonal terms of the Hamiltonian. The problems of trying to demonstrate this for any practical set of states that we are likely to use are similar to the problems of trying to prove the quasi-ergodic

The accessibility assumption

§ 19.1

assumption in the classical case, and I shall not elaborate them. We simply note that if we use practical approximate energy states, then, as in the classical case, only quite abnormal external restrictions or the occasional splitting of the accessible states into separate restricted groups are likely to cause difficulties. Apart from such special cases the accessibility assumption is very unlikely to fail in large systems.

As in classical theory, however, one can *make* the accessibility assumption fail by choosing a sufficiently perverse abnormal set of basis states. Consider, for instance, the *exact energy states*, which certainly exist, although we have no hope of finding them for a large system. In more practical representations, as we shall see, the important information is carried by the diagonal terms $\tilde{p}_i$ of the density matrix, and we may usually ignore the off-diagonal terms, just as we ignore the microstructure in the classical case. But in the exact energy representation there are no off-diagonal terms in the Hamiltonian, no quantum jumps, and the $\tilde{p}_i$ values never change. This does not invalidate the results of quantum statistics, however. In this very abnormal representation the *observables* are far from diagonal, and the measurable changes which occur in the system are described not by the $\tilde{p}_i$ values, which are constant, but by the off-diagonal terms of the density matrix ρ_{ij}, which oscillate at the frequencies $(E_i - E_j)/\hbar$, as may be seen from (19.13). Like the abnormal representation discussed in § 19.1, to which it is in fact closely related, the exact energy representation would be quite unusable in practice, for, even if we could find the exact energy states, it would be much too difficult to obtain expressions for the observable behaviour of the system in terms of the oscillations of the individual off-diagonal terms of the density matrix.

The exact energy representation

The problem of time-reversal symmetry is as evident in quantum theory as it was in classical theory. To reverse the velocities of all the particles we must replace each wave function ψ by its complex conjugate ψ^*. It is convenient to replace the basis functions ϕ_α by their conjugates ϕ_α^* at the same time, so that the state vectors a_α are also replaced by their complex conjugates, and the density matrix $\rho_{\alpha\beta}$ likewise. We then find (remembering that $H_{\alpha\beta}^* = H_{\beta\alpha}$) that $\rho_{\alpha\beta}^*$ obeys the usual equation of motion (19.13) if t is replaced by $-t$. Thus the motion of the density matrix has time-reversal symmetry.

Time-reversal symmetry for the density matrix

In classical theory we found that the fine-grained entropy did not in fact change with time, and a similar difficulty arises in quantum theory. The natural way to define the fine-grained entropy when using the density matrix is to find the representation in which the density matrix is diagonal, and to define $\tilde{S}_\mathrm{f}$ as $-k \sum \tilde{p}_\mathrm{f} \ln \tilde{p}_\mathrm{f}$ in this representation. Treating the matrix $\rho_{\alpha\beta}$ as an operator $\hat{\rho}_\mathrm{f}$ we may then identify the fine-grained entropy as

The fine-grained entropy is independent of time

$$\tilde{S}_\mathrm{f} = -k \operatorname{Tr} (\hat{\rho}_\mathrm{f} \ln \hat{\rho}_\mathrm{f}), \tag{19.19}$$

since $(\ln \hat{\rho}_\mathrm{f})_{ii} = \ln (\hat{\rho}_\mathrm{f})_{ij}$ when $\hat{\rho}_\mathrm{f}$ is diagonal. Now the operation of converting $\hat{\rho}_\mathrm{f}(0)$ at $t=0$ into $\hat{\rho}_\mathrm{f}(t)$ at a later time is in fact a unitary transformation. Since the trace is invariant under a unitary transformation it is evident that the fine-grained entropy $\tilde{S}_f$ is independent of time, as in classical theory.

Evidently, as in classical theory, we shall only be able to show that $\tilde{S}$ increases with time if we employ coarse-graining, which in quantum theory means that we must replace ρ_{ij} by its double average over a suitably sized group of similar states i and a second group of similar states j. We shall find that when we do this in a suitable practical representation, the smoothed density matrix is already almost diagonal, so that we may define $\tilde{S}$ as $-k\sum \tilde{p}_i \ln \tilde{p}_i$, as usual. There are three requirements which such a practical representation must meet, in quantum theory. First, as we have already noted, we must use approximate energy basis states which are mutually accessible, being coupled by weak off-diagonal terms of the Hamiltonian. Secondly, we must limit ourselves to representations in which all the observables of interest are diagonal, or nearly so. The reason is that, if non-diagonal observables are important, the *initial* density matrix may have appreciable off-diagonal terms. (For our spin-$\frac{1}{2}$ particle, for instance, if σ_y is initially positive the off-diagonal parts of its density matrix will be large and positive also.) In such situations the master equation would not be valid, and, moreover, the statistical state of the system could not be described by the $\tilde{p}_i$ values alone, as we have seen. At first sight this seems a severe restriction, but it turns out that for *large* systems this is not so. von Neumann has shown how, in principle, for *any* large system we may always find a representation in which, to an excellent approximation, all the quantities which may in practice be observed, the *macroscopic observables*, are diagonal. It is this von Neumann representation which corresponds most closely to the classical description. For instance, in handling the statistics of nuclear magnetic resonance, in which we have a strong magnetic field in the z direction, and the nuclear spins have a transverse component which precesses around the z axis, we normally treat the magnetisation of the whole sample as a classical object obeying a classical equation of motion. In quantum terms this means that we are using basis states in which the z component of magnetic moment has been made slightly uncertain, but the x and y components have become almost certain. The energy is now slightly uncertain also, but on the scale of the complete system all of these uncertainties are negligible. In a practical representation of this particular type we can have no information on the initial values of the off-diagonal terms of the density matrix. We must therefore assume that their phases are incoherent, and on coarse-graining it follows that we must take the initial values of the off-diagonal part of the density matrix to be zero. The *third* requirement of the practical representation will appear later.

The coarse-grained density matrix

Practical representations

§ 19.4

The method of macroscopic observables

The initial state of the density matrix

We now take the important step of obtaining the master equation for a practical representation which meets these requirements. Because the off-diagonal part of H_{ij} is small, the off-diagonal parts of ρ_{ij} never become large, and we may treat them as solutions of the first-order Equation (19.15). Moreover, since the jump rates are small we know that the diagonal terms $\rho_{ii}=\tilde{p}_i$ are slowly varying. Thus the driving term $(\tilde{p}_j-\tilde{p}_i)H_{ij}$ in (19.15) is

Derivation of the master equation

approximately independent of time. Its exact time dependence is not important, but to treat it as having been constant since $t=-\infty$ raises difficulties when $E_i=E_j$. We therefore write it as $(\tilde{p}_j-\tilde{p}_i)_0 H_{ij}\,\mathrm{e}^{st}$, where s is a small positive quantity, as an indication that it has been acting for a long but finite time (a device commonly used in quantum theory). The solution to (19.15) then takes the form

$$\rho_{ij}=\frac{\tilde{p}_j-\tilde{p}_i}{E_j-E_i+\mathrm{i}\hbar s}H_{ij}. \tag{19.20}$$

This result is similar to (19.14). Here, however, $\tilde{p}_j-\tilde{p}_i$ will not necessarily tend to zero when E_j-E_i tends to zero, because the system is not in equilibrium: notice how the introduction of s prevents ρ_{ij} from diverging at this point. Using this value of ρ_{ij} we may write down $\mathrm{d}\tilde{p}_i/\mathrm{d}T=\dot{\rho}_{ii}$ with the help of (19.13). We find that

$$\mathrm{d}\tilde{p}_i/\mathrm{d}t=(\mathrm{i}/\hbar)\sum_j |H_{ij}|^2\frac{\tilde{p}_j-\tilde{p}_i}{E_j-E_i+\mathrm{i}\hbar s}+\text{c.c.} \tag{19.21}$$

We next make the important assumption, which we shall discuss shortly, that it is possible to find groups of similar final states j within which $|H_{ij}|^2$ and $\tilde{p}_j$ are slowly varying functions of E_j. We may then replace the sum over j by an integral over the density of states within each group, and find that

$$\mathrm{d}\tilde{p}_i/\mathrm{d}t=\sum_{\text{groups}}(\mathrm{i}/\hbar)|H_{ij}|^2(\tilde{p}_j-\tilde{p}_i)g_j\int_{-\infty}^{\infty}\left[\frac{\mathrm{d}E_j}{E_j-E_i+\mathrm{i}\hbar s}-\text{c.c.}\right], \tag{19.22}$$

where $\tilde{p}_j$ and g_j are evaluated for a given group at energy E_i, since the integral is dominated by contributions at that energy. Contour integration shows that the value of the integral is $-2\pi\mathrm{i}$, independent of the exact value of s. Thus

$$\mathrm{d}\tilde{p}_i/\mathrm{d}t=\sum_{\text{groups}}(2\pi|H_{ij}|^2 g_j/\hbar)(\tilde{p}_j-\tilde{p}_i)$$

▶ $$=\sum_j(2\pi|H_{ij}|^2/\hbar\,\delta E)(\tilde{p}_j-\tilde{p}_i)$$

$$=\sum_j \nu_{ij}(\tilde{p}_j-\tilde{p}_i), \tag{19.23}$$

§2.4

where we have used the accessibility convention in the second line. This is the master equation.

The origins of irreversible behaviour

In obtaining this result we have made the transition from a time-reversible equation of motion for ρ_{ij} to an irreversible equation of motion for $\tilde{p}_i$. It is important to see how this has come about. It does so for two reasons. First, as in the classical case we have *coarse-grained*: we have ignored any initial microstructure in the off-diagonal parts of the density matrix, and have treated $\tilde{p}_j$ as a smooth function of E_j within each group of final states. As in classical theory it is conceivable but unlikely that this procedure is invalid, that the real ensembles encountered in experiments contain microstructure which could reverse the tendency of the $\tilde{p}_i$ values to become equalised. We can only assert, in accordance with the principle of uniform local probability, that such significant microstructure is not apparent in practice. Secondly, we have assumed that it is possible to find substantial groups of similar final states j within each of which $|H_{ij}|^2$ is a slowly varying function of energy. Now, we noted earlier that the master equation fails in the presence of certain *non-randomising perturbations*, such as a uniform electric field. Another such example is a set of identical scattering centres arranged on an extended regular crystal lattice. The master equation fails in these cases because the perturbations concerned can scatter a particle of given momentum into only a single final state, or into a small number of very different final states. In such cases there is no substantial group of final states over which the integral (19.22) may be performed, and our derivation fails. For a set of scattering centres arranged at random, however, on writing H_{ij} as a sum $\sum H_{ij}^{(n)}$ over perturbations associated with the individual centres n, we expect to be able to write $|H_{ij}|^2$ as $\sum_n |H_{ij}^{(n)}|^2$, the cross-terms cancelling out on average because they have random phases. In this situation the overall scattering rate will be the sum of the scattering rates for the individual centres, and, since each centre can scatter a particle of given momentum into a large number of similar final states of slightly different momentum, our derivation of the master equation is valid. This, then, is our third requirement for the practical representation: we must choose a representation in which all non-randomising perturbations are included in the approximate Hamiltonian, so that the off-diagonal parts are associated only with randomising perturbations, such as scattering centres in random positions.

§ 19.2

§ 2.4

The requirement of randomising perturbations

To summarise our discussion of quantum statistics, we have seen that the density matrix approach has justified our early conclusions about how to calculate equilibrium thermal averages as expressed in (9.1), and provided us with a formula for them (19.18) which is independent of representation. We have also been able to justify the master equation approach, provided we use a practical approximate energy representation in which the macroscopic observables of interest are diagonal, or nearly so, and the residual scattering corresponds to a randomising perturbation only. We may then give a unique definition of the non-equilibrium entropy $\tilde{S}$ as $-k \sum \tilde{p}_i \ln \tilde{p}_i$, where $\tilde{p}_i$ is the coarse-grained probability in the practical representation, and this coarse-grained entropy obeys the H-theorem. In other representations $\tilde{S}$ may be written, if required, as $-k \operatorname{Tr}(\hat{\rho} \ln \hat{\rho})$, where $\hat{\rho}$ is the coarse-grained density matrix.

§ 9.1

19.7 The behaviour of entropy during fluctuations

You may well have noticed an apparent discrepancy in our treatment of entropy earlier in the book. Once a system has reached equilibrium it fluctuates randomly around its equilibrium state, moving sometimes towards equilibrium and sometimes away from it. It seems that in this process the entropy must rise and fall. How is this to be reconciled with the H-theorem,

§§4.2, 19.3

which, in both its quantum and classical forms, appears to demonstrate that the entropy *never* falls? Moreover, if the entropy *does* fall in a fluctuation, might it not be possible to take advantage in some way of favourable fluctuations and so defeat the second law? This idea was taken up early in the development of statistical thermodynamics by Maxwell, who liked to imagine a small *demon*, an intelligent creature small enough to observe the individual atoms. He supposed that the demon stationed himself by a small trap-door in the wall separating two containers of gas at the same temperature and pressure. If the demon opens the door whenever he sees a fast molecule approaching from the right or a slow molecule approaching from the left it seems that with patience he can establish a temperature difference, making the left-hand gas hotter than the right-hand gas, without expending any work. Thus the demon appears to provide a mechanism for making heat flow from a colder to a hotter body while leaving the rest of the world unchanged, contrary to the Clausius statement of the second law.

Maxwell's demon

Let us begin our discussion of this issue by commenting on the absolute nature of the H-theorem. We noted when we first discussed the concept of entropy that there are several possible attitudes to what it is. In this book, however, I have firmly taken the view that the entropy is a function of the state probabilities, $\tilde{S} = -k \sum \tilde{p}_i \ln \tilde{p}_i$. This means that the entropy, so defined, is not really a function of what the system is *actually* doing, but a function of *how much we know* of what the system is doing. From this point of view the H-theorem just reflects the fact that our knowledge of what the system is doing is continually becoming less precise, and this is indeed true *so long as we make no new measurements on the system.* Using this point of view, we have to accept that it *would* be possible, by taking a very large number of measurements, to decrease the entropy of a system. For instance, if we have 10^{23} spin-$\frac{1}{2}$ particles in zero magnetic field we have $2^{10^{23}}$ possible states, all with the same energy. If we know nothing of which way up the spins are we say that we have a configurational entropy equal to $10^{23}k \ln 2$, a substantial entropy. If, by some means, we could measure the state of every spin, we could then say that the system was definitely in a single quantum state, and we should conclude that its entropy was reduced to *zero* at the moment of measurement. (This would, however, take a great deal of measurement. For comparison, observe that the disc store of a very large computer can store about 10^{11} bits of information. The corresponding amount of measurement could be used to reduce the entropy of some object by $10^{11}k \ln 2$, which is 10^{-12} J K^{-1} – hardly a very striking entropy reduction.) From this point of view, if a fluctuation occurs but we do not measure it, our idea of the probabilities and hence our value for the entropy is unchanged. But if, like the demon, we *measure* the microscopic details of what the system is doing, we *can* take advantage of the corresponding entropy reduction. Thus it seems that Maxwell was right, and that the demon can defeat the second law, in principle.

The effect of measurement on entropy

This attitude was accepted until 1951, when Brillouin pointed out that the argument assumed, wrongly, that the demon could make the necessary measurements without generating any entropy in the process. In fact the molecules will be bathed in equilibrium temperature radiation and therefore invisible to the demon unless he uses a flashlight. On calculating the minimum entropy generated by the flashlight, Brillouin discovered that it was *greater* than the entropy reduction engineered by opening and shutting the trap door, so the second law was saved after all. One can make the same point in a somewhat different way by observing that a measurement corresponding to one bit of information may be used to reduce the entropy of a system at best by $k \ln 2$. But to make use of the information we must first record it, and it is easy to see that the recording process will necessarily generate an entropy greater than $k \ln 2$. For instance, a computer memory for one bit will consist of some bistable element, having two positions of minimum energy separated by an energy barrier ϕ which, if the memory is to be reliable, must be much greater than kT, where T is the temperature of the element. To switch the element from one state to the opposite state generates a heat of about ϕ corresponding to an entropy increase of ϕ/T, which is much greater than k. Since on average we shall have to switch the element on half the occasions when we store information in it, it is clear that the process of storing the information alone wastes much more entropy than the $k \ln 2$ which it saves.

The measuring process wastes more entropy than it saves

One can see in yet another way that we cannot beat the second law by taking measurements, *if we remember to include the entropy of the observer himself.* There is nothing special in this context about a human observer. Any scheme which a human could operate to beat the second law could clearly be automated. But having automated it we may treat the automation system itself as just another thermodynamic system. We ourselves, standing outside it, are taking no measurements, so we may validly apply the H-theorem to the joint assembly of thermodynamic system plus automated observer. Since the H-theorem applies to the joint assembly it is clear that we do not expect the joint assembly, treated as a whole, to be able to beat the second law.

To summarise, then, it is true in principle that we *can* reduce the entropy of an isolated system, as defined in this book, by taking measurements of the system (though the number of measurements needed to produce even a tiny reduction in entropy is formidable). But we do not reduce the entropy of the whole world in this process, for there is a compensating increase of entropy associated with the process of observation. There is therefore, alas, no prospect of providing for the energy needs of mankind by suitably marshalling a small horde of Maxwell demons.

Answers to problems

Chapter 3

3.1 (ii) $Q_A = 10^{23}$; $Q_B = 2\times 10^{23}$; $T = 1045$ K.

3.2 2.86×10^{-2} J K^{-1}; $(\overline{\Delta E^2}/E^2)^{\frac{1}{2}} = 4.4\times 10^{-11}$.

3.3 $p_2 = 0.2, 0.4$; $p_1 = 0.8, 0.6$; $T = 5.23$ K, 17.9 K.

3.4 $T = \pm\infty$, -17.9 K, -0. This system could only be in equilibrium with a reservoir at these temperatures if the reservoir itself had an upper bound on its energies.

Chapter 4

4.1 $S = 0.50k$; $0.67k$.

4.2 $S = 2.86\times 10^{-2}$ J K^{-1}.

4.3 5.27 J K^{-1}; 5.74 J K^{-1}.

4.4 1.3×10^{30}; $\Delta S_A = -0.56\times 10^{-21}$ J K^{-1}; $\Delta S_B = 1.52\times 10^{-21}$ J K^{-1}.

4.5 1.90×10^{-25} J; 2.23×10^{-25} J.

Chapter 5

5.2 Flux of atoms $= \frac{1}{4}An\bar{c}$; flux of energy $= (\frac{1}{4}An\bar{c})\,2kT$. The rate at which a molecule of speed c hits the wall is proportional to c, so the faster molecules have a greater chance of escaping.

5.4 One would not expect it to be possible to transmit a signal through the gas faster than the molecules are moving. 320 m s^{-1}; 1010 m s^{-1}. Because the *temperature* falls with height.

5.5 227 K; 440 K.

Chapter 6

6.1 From 6.31×10^{-3} m^3 at 1 atm to 2.52×10^{-3} m^3 at 2.5 atm. 925 J; 583 J; 37%.

6.2 -1 °C; 31 °C.

6.3 64%. For little fuel, $r_e \to r_c$ and $\eta \to (T_2 - T_1)/T_2$. Larger compression ratios increase T_2, and hence η.

6.4 5.0×10^{10} J; 2.6×10^{10} J.

6.5 (a) 6.3 J K^{-1}; (b) 1.39 J K^{-1}; (c), (d) 1.83×10^{-5} J K^{-1}: when charging, the cell does work QV, but the energy stored is only $\frac{1}{2}QV$, so $\frac{1}{2}QV$ must be dissipated in the leads; (e) nil: reversible process; (f) 5.76 J K^{-1}: see derivation of (5.14).

6.6 49 000 J mol^{-1}; more if irreversible; see problem 7.4.

Chapter 7

7.1 $F = -NkT \ln [\alpha V T^{\frac{3}{2}} e^{-\frac{3}{2}}]$; $P = -(\partial F/\partial V)_T = RT/V$
$S = Nk \ln [\alpha V (2E/3Nk)^{\frac{3}{2}}]$; $P/T = (\partial S/\partial V)_E = R/V$
$E = \frac{3}{2}Nk[e^{S/Nk}/\alpha V]^{\frac{2}{3}}$; $P = -(\partial E/\partial V)_S = 2E/3V$.

7.2 To minimise F, the surface area must be minimised at a given volume.

7.3 1.57 $m^3 s^{-1}$.

7.4 $\Delta G = -2.38 \times 10^5$ J mol^{-1}; $\phi_{min} = 1.24$ V.

7.5 $\Delta A = RT_R[\ln (P_R T_1^{\frac{5}{2}}/P_1 T_R^{\frac{5}{2}}) + \frac{5}{2} - (T_1/T_R)(\frac{3}{2} + P_R/P_1)]$; 3010 J.

Chapter 8

8.4 $(\partial T/\partial V)_S = (-T/C_V)(\partial P/\partial T)_V$.

8.5 $2.5k$, if we treat the argon as a cubic lattice for simplicity.

8.6 $(\partial E/\partial \mathscr{A})_T = \Gamma - T d\Gamma/dT$; 0.121 J m^{-2}; because heat enters: work per unit area $= \Gamma =$ surface *free* energy density.

8.7 About 0.09 K if the skater is equivalent to a uniform pressure of 10^6 N m^{-2}.

Chapter 9

9.2 3.9×10^{-24} J, using the *high* temperature limit of the expression for $\bar{\varepsilon}$.

9.3 1.03×10^{-21} J; 8.61×10^{-20} J; about 15 K and 600 K, but see also problem 12.9.

9.4 $F_r = -kT \ln (2IkT/\hbar^2)$; $S_r = k \ln (2\,eIkT/\hbar^2)$, and, as expected in the classical limit, $C_r = k$ and $\bar{\varepsilon}_r = kT$. The interlevel spacing must be small compared with kT for states having appreciable occupation.

9.6 $f = (NkT/2L_{max}) \ln [(1+x)/(1-x)]$; $+1.2$ K.

Chapter 10

10.1 3000 K; 0.5 m; 1.4×10^{-5} m; about 2 W.

10.2 Raised by about 3×10^{-6} K; never had any significant effect.

10.3 6 weeks (useful for space probes).

10.4 About half.

10.5 -3.8×10^{-5} N m^{-1}, an appreciable contribution to the total Γ. This type of calculation will fail at higher T (i) because Γ will fall with T, and (ii) because the low temperature limit ceases to be valid.

Chapter 11

11.2 μ has an extra term mgh.

11.3 P stays constant, $\Phi = \int dw = -PV$, $G = \int \mu\, dN = \mu N$, $\Phi_G = 0$.

Chapter 12

12.1 $\mu=\varepsilon$; $C=\frac{1}{2}k(\varepsilon/kT)^2 \operatorname{sech}^2(\varepsilon/2kT)$.

12.2 $Z=\frac{1}{2}\sum_{n=0}^{\infty}\sum_{m=0}^{\infty}\exp[-(n+m)h\nu/kT]+\frac{1}{2}\sum_{n=0}^{\infty}\exp(-2nh\nu/kT)$. At high T the chance of both being in the same state is negligible, and the number of states available is essentially half the number for two separate oscillators: so, apart from a constant, S is the same as for two oscillators. At low temperatures there is only a single important excited state at energy $h\nu$, as for one oscillator.

12.4 3.3×10^{-19} J, 2.8×10^{42} J^{-1} mol^{-1}; 3.3×10^{42} J^{-1} mol^{-1}, 91 K. Potassium is not an ideal free electron metal.

12.5 $\lambda_T=3\times10^{-11}$ m for N_2; 8×10^{-10} m for He; $\mu=-2.7\times10^{-21}$ J – just below the conduction band, $T_c=3.0$ K.

12.6 $\mu=-kT\ln[(mkT/2\pi\hbar^2)A/N_A]-\phi$. Valid when spacing on plane $\gg\lambda_T$.

12.7 13.4 J; heat into surroundings; isothermal expansion of O_2 to correct partial pressure, then mix with air reversibly through a membrane.

12.8 Fractions ionised are 3.7×10^{-5} and 0.12.

12.9 (i) p-H_2, 4.1×10^{-4} R, (ii) 3:1 mixture, 1.0×10^{-4} R, (iii) p-H_2 changing to o-H_2, 1.6 R. Note that K must be even for p-H_2 and odd for o-H_2. In each case only the two lowest states are significant; allow for both degeneracies.

Chapter 13

13.1 Boundaries remain parallel to x axis, height and width remain constant. Filament wrapped 541 times around the space between trajectories.

13.2 47% N_2, 5% O_2, 48% H_2. (Not a reliable calculation: the stratosphere is not very close to thermal equilibrium.)

13.3 Saturates in same way, but low field susceptibility is three times smaller.

13.4 (i) k, $3k/2$: CO_2 has 2 rotations and 4 vibrations, H_2O has 3 rotations and 3 vibrations; (ii) 0.52°; (iii) 20 μV.

13.5 The classical description treats the particles as distinguishable: if indistinguishable, divide γ by $N!$, since in the classical limit the particles are all in different states.

Chapter 14

14.1 Each atom has 12 neighbours at distance r, 6 at $\sqrt{2}r$, 24 at $\sqrt{3}r$, 12 at $\sqrt{4}r$, etc.; $\alpha=1+(0.5)/2^3+2/3^3+1/4^3+2/5^3\ldots\simeq1.17$; $r_s/r^*=0.97$. The latent heat should be the binding energy of the lattice at rest $=1.35\times10^{-20}$ J atom^{-1} less the zero point energy of vibration $=\frac{9}{8}N\hbar\omega_D=0.14\times10^{-20}$ J atom^{-1}. (Kinetic energy in vapour and PV terms in the enthalpy are negligible at low T.)

14.4 $T_B=25$ K, $T_i=49$ K, $\Delta T=-0.13$ K.

14.5 11 300 torr = 14.8 atm.

14.6 Extrapolation suggests $T_c=3.8$ K. (Observed $T_c=3.34$ K.)

Chapter 15

15.1 Rotation of a molecule involves moving the nuclei, and molecular rotations are in the classical limit at room temperature (§9.1.2).

15.2 $+2.1\times10^{-9}$ K.

15.5 Assuming $kT_{max} \approx$ level splitting gives $\mathscr{B}_{res} \approx 6 \times 10^{-2}$ T. Cooling below 0.04 K takes us to a region of low heat capacity where the salt cannot effectively cool other objects. Field needed is about 1.8*T*.

15.7 Fraction $= 3.0 \times 10^{-10}$. Potential energy change for electron moving from A into B is $[(3+1.2)-0.2-(4+1.0)]$ eV $= -1.0$ eV. Kinetic energy change for electrons at Fermi level $= +1.0$ eV. No heat enters or leaves. Since μ is the same in both metals, the flow is reversible.

Chapter 16

16.7 $T_{eff} = (T_1 R_1 + T_2 R_2)/(R_1 + R_2)$.

16.8 Spectral density is $4kTR/(1+R^2C^2\omega^2)$. We find $\frac{1}{2}C\overline{V^2} = \frac{1}{2}kT$, in accordance with the equipartition theorem.

16.10 5.4×10^{-20} J.

Chapter 17

17.3 7.7×10^{-12} s; 1.2×10^{-5} m; about 10^{10} Hz. But in fact Ohm's law would break down for a different reason at only 10^5 Hz, the frequency at which the rf skin depth is equal to l. At higher frequencies we can no longer treat the local current density as a simple function of the local electric field: the *anomalous skin effect*.

17.4 When $\omega \sim 1/\tau$, the molecules rotate through large angles against the viscous drag of the medium in each cycle. The relaxation time could be estimated by treating the rest of the liquid as a viscous medium. A discussion in terms of molecular collisions would be more accurate, but much more difficult.

17.6 Prediction implies that $\alpha/\beta = 273$ K. This fails at room temperature because the scattering is then dominated by temperature varying processes. $\alpha = 40.9$ μV K^{-1}; $\beta = -0.078$ μV K^{-2}; 524 °C; 1.23×10^{-2} J C^{-1}; 2.9×10^{-5} J C^{-1} K^{-1}.

Chapter 18

18.1 $T_{trip} = 0.007$ °C; $P_{trip} = 6.03 \times 10^2$ N m^{-2}.

18.2 There is no triple point. The solid–liquid line is level below 1 K, and rises at higher temperatures. Below 24 atm ^{4}He is *liquid* at $T=0$ (because the zero-point energy of motion of the light ^{4}He atom is lower in the liquid than in the solid).

18.4 At the limit of local stability $P = -1040$ atm.

18.5 The ^{3}He behaves like a Fermi gas within the ^{4}He. Its entropy per particle is proportional to $n^{-\frac{2}{3}}$, so there is a substantial increase in entropy and hence latent heat on dilution.

18.6 $T_c = 2E_2/k$.

18.7 $n = \frac{1}{2}$, $m = \frac{3}{4}$.

Suggestions for further reading

The references range from elementary to graduate level, and are given under each heading in order of increasing difficulty.

General texts on statistical thermodynamics

Kittel & Kroemer (1980), Reif (1965), Rushbrooke (1949), Wannier (1966), Hill (1960), Reichl (1980), Landau & Lifshitz (1980).

Low temperature physics and techniques

Rosenberg (1963), White (1968).

Chapter 2

Probability: Meyer (1970), Weatherburn (1952).
Statistical foundations: Reif (1965). See also references for Chapter 19.

Chapters 3 and 4

Temperature and entropy: for a similar approach see Kittel & Kroemer (1980), Reif (1965). For a different but widely used approach, Gibbs' method of maximum likelihood in the canonical ensemble, see Schrödinger (1946), Rushbrooke (1949), Hill (1960).
The third law: Pippard (1964), Rushbrooke (1949), Wilks (1961).
Techniques for reaching negative temperatures: Andrew (1955).

Chapter 5

Elementary kinetic theory of ideal gases: Flowers & Mendoza (1970), Collie (1982), Jeans (1959). See also references for Chapter 17.

Chapters 6, 7 and 8

Partial differentiation: Riley (1974).
Classical thermodynamics: Adkins (1983), Zemansky & Dittman (1981), Pippard (1964), Guggenheim (1949), Callen (1960).
Engineering thermodynamics: Van Wylen & Sonntag (1978), Roger & Mayhew (1980).
Chemical thermodynamics: McClelland (1973).

Chapter 10

Blackbody radiation: Reif (1965).
Phonons: Ashcroft & Mermin (1976), Ziman (1964), Born & Huang (1954).

Chapter 12

Electron gas in metals: Kittel (1971), Ashcroft & Mermin (1976). See aiso references for Chapter 17.
Bose condensation: London (1954).
Chemical equilibrium constants: Rushbrooke (1949), Hill (1960).

Chapter 13

Classical statistical mechanics: Rushbrooke (1949), Tolman (1938).

Chapter 14

Equation of state for gases and liquids: Flowers & Mendoza (1970), Frenkel (1946), Croxton (1974), Hill (1960), Mayer & Mayer (1940), Reichl (1980).

Chapter 15

Electromagnetic field energies: Bleaney & Bleaney (1976).
Thermal behaviour of spin systems: Kittel & Kroemer (1980), Garrett (1954), White (1970).
Superconductivity: Reichl (1980), Ashcroft & Mermin (1976), Tinkham (1975).

Chapter 16

Fluctuations: Reif (1965), Landau & Lifshitz (1980).
Noise and stochastic processes: Reif (1965), Robinson (1974), MacDonald (1962), Papoulis (1965).
Rates of chemical reaction: Hill (1960).
Quantum theory of the dependence of scattering rate on occupation: Dirac (1958).

Chapter 17

Advanced kinetic theory: Kennard (1938), Chapman & Cowling (1970), Lifshitz & Pitaevskii (1981).
Relaxation time and the Boltzmann transport equation: Reif (1965), Ashcroft & Mermin (1976), Reichl (1980).
Transport properties of metals: Ashcroft & Mermin (1976), Ziman (1964), Pippard (1965).
Non-equilibrium thermodynamics: Reif (1965), Becker (1967), de Groot & Mazur (1962), Prigogine (1961).

Chapter 18

Phase transitions and critical point behaviour: Kittel & Kroemer (1980), Reichl (1980), Stanley (1971), Pfeuty & Toulouse (1977), Patashinskii & Pokrovskii (1979).
Ferromagnetism: Ashcroft & Mermin (1976), White (1970).
Liquid–vapour transition: see references for Chapter 14.
Phase nucleation: Frenkel (1946), Becker (1967).

Chapter 19

Fundamentals of classical and quantum statistics: Reichl (1980), ter Haar (1954), Tolman (1938), Jancel (1969), Farquhar (1964).
Field-theory techniques in statistical physics (Nambu formalism)*:* Kadanoff & Baym (1962), Lifshitz & Pitaevskii (1980), Abrikosov, Gorkov & Dzyaloshinskii (1963).
Entropy in fluctuations: Brillouin (1951).

References

Abrikosov, A. A., Gorkov, L. P. & Dzyaloshinskii, I. E. (1963). *Methods of Quantum Field Theory in Statistical Physics.* Engelwood Cliffs, New Jersey: Prentice Hall.

Adkins, C. J. (1983). *Equilibrium Thermodynamics*, 3rd edn, Cambridge: University Press.

Alder, B. J. & Wainwright, T. E. (1959). 'Studies in molecular dynamics'. *Journal of Chemical Physics*, **31**, 459–66.

Andrew, E. R. (1955). *Nuclear Magnetic Resonance.* Cambridge: University Press.

Ashcroft, N. W. & Mermin, N. D. (1976). *Solid State Physics.* New York: Holt, Rinehart & Winston.

Becker, R. (1967). *Theory of Heat.* Berlin: Springer Verlag.

Berman, R. & MacDonald, D. K. C. (1952). 'Thermal and electrical conductivities of copper at low temperatures'. *Proceedings of the Royal Society*, **A211**, 122–8.

Bleaney, B. I. & Bleaney, B. (1976). *Electricity and Magnetism*, 3rd edn, Oxford: University Press.

Born, M. & Huang, K. (1954). *Dynamical Theory of Crystal Lattices.* Oxford: Clarendon Press.

Brillouin, L. (1951). 'Maxwell's demon cannot operate'. *Journal of Applied Physics*, **22**, 334–7.

Broyles, A. A., Chung, S. U. & Sahlin, H. L. (1962). 'Comparisons of radial distribution functions'. *Journal of Chemical Physics*, **37**, 2462–9.

Callen, H. B. (1960). *Thermodynamics.* New York: Wiley.

Chapman, S. & Cowling, T. G. (1970). *Mathematical Theory of Non-Uniform Gases.* Cambridge: University Press.

Collie, C. H. (1982). *Kinetic Theory and Entropy.* London: Longman.

Croxton, C. A. (1974). *Liquid State Physics.* Cambridge: University Press.

de Groot, S. R. & Mazur, P. (1962). *Non-equilibrium Thermodynamics.* Amsterdam: North Holland.

Dirac, P. A. M. (1958). *Principles of Quantum Mechanics*, 4th edn, Oxford: Clarendon Press.

Ehrenfest, P. & Ehrenfest, T. (1959). *Conceptual Foundations of the Statistical Approach in Mechanics.* Ithaca, New York: Cornell University Press.

Farquhar, I. E. (1964). *Ergodic Theory in Statistical Mechanics.* London: Interscience.

Flowers, B. H. & Mendoza, E. (1970). *Properties of Matter.* London: Wiley.

Frenkel, J. (1946). *Kinetic Theory of Liquids.* Oxford: University Press.

Garrett, C. G. B. (1954). *Magnetic Cooling.* Cambridge, Massachusetts: Harvard University Press.

Guggenheim, E. A. (1945). 'The principle of corresponding states'. *Journal of Chemical Physics*, **13**, 253–61.

Guggenheim, E. A. (1949). *Thermodynamics.* Amsterdam: North Holland.

Henry, W. E. (1952). 'Spin paramagnetism of Cr^{3+}, Fe^{3+} and Gd^{3+} at liquid helium temperatures'. *Physical Review*, **88**, 559–62.

Hill, T. L. (1960). *Statistical Thermodynamics.* Reading, Massachusetts: Addison-Wesley.

Hirschfelder, J. O., Curtis, C. F. & Bird, R. B. (1967). *Molecular Theory of Gases and Liquids.* New York: Wiley.
Ho, J. T. & Litster, J. D. (1969). 'Magnetic equation of state of $CrBr_3$ near the critical point'. *Physical Review Letters*, **22**, 603–6.
Jancel, R. (1969). *Foundations of Classical and Quantum Statistical Mechanics.* London: Pergamon.
Jeans, J. (1959). *Kinetic Theory of Gases.* Cambridge: University Press.
Kadanoff, L. P. & Baym, G. (1962). *Quantum Statistical Mechanics.* New York: Benjamin.
Kannuluik, W. G. & Carman, E. H. (1952). 'Thermal conductivity of rare gases'. *Proceedings of the Physical Society*, **65B**, 701–9.
Kennard, E. H. (1938). *Kinetic Theory of Gases.* New York: McGraw Hill.
Kittel, C. & Kroemer, H. (1980). *Thermal Physics*, 2nd edn, San Francisco: Freeman.
Kittel, C. (1971). *Solid State Physics*, 4th edn, New York: Wiley.
Kreuzer, H. J. (1981). *Non-equilibrium Thermodynamics and its Statistical Foundations.* Oxford: Clarendon.
Landau, L. D. & Lifshitz, E. M. (1980). *Statistical Physics*, 3rd edn, part I revised and enlarged by E. M. Lifshitz and L. D. Pitaevskii. Oxford: Pergamon.
Lifshitz, E. M. & Pitaevskii, L. D. (1980). *Statistical Physics*, part 2. Oxford: Pergamon.
Lifshitz, E. M. & Pitaevskii, L. D. (1981). *Physical Kinetics.* Oxford: Pergamon.
London, F. (1954). *Superfluids*, vol. 2. New York: Wiley.
Ma, S.-K. (1976). *Modern Theory of Critical Phenomena.* Reading, Massachusetts: Benjamin.
MacDonald, D. K. C. (1962). *Noise and Fluctuations.* New York: Wiley.
McClelland, B. J. (1973). *Statistical Thermodynamics.* London: Chapman & Hall.
Mayer, J. E. & Mayer, M. G. (1940). *Statistical Mechanics.* New York: Wiley.
Meyer, P. L. (1970). *Probability and Statistical Applications*, 2nd edn, Reading, Massachusetts: Addison-Wesley.
Michels, A., Botzen, A., Friedman, A. S. & Sengers, J. V. (1956). 'The thermal conductivity of argon'. *Physica*, **22**, 121–8.
Papoulis, A. (1965). *Probability, Random Variables and Stochastic Processes.* New York: McGraw Hill.
Patashinskii, A. Z. & Pokrovskii, V. L. (1979). *Fluctuation Theory of Phase Transitions.* Oxford: Pergamon.
Pfeuty, P. & Toulouse, G. (1977). *The Renormalisation Group and Critical Phenomena.* Chichester: Wiley.
Pippard, A. B. (1964). *Classical Thermodynamics.* Cambridge: University Press.
Pippard, A. B. (1965). *Dynamics of Conduction Electrons.* London: Blackie.
Prigogine, I. (1961). *Thermodynamics of Irreversible Processes.* New York, London: Interscience.
Reichl, L. E. (1980). *Modern Course in Statistical Physics.* Austin: University of Texas Press.
Reif, F. (1965). *Statistical and Thermal Physics.* New York: McGraw Hill.
Ridley, B. A., Schulz, W. R. & Le Roy, D. J. (1966). 'Kinetics of the reaction $D+H_2 = HD+H$'. *Journal of Chemical Physics*, **44**, 3344–7.
Riley, K. F. (1974). *Mathematical Methods for the Physical Sciences.* Cambridge: University Press.
Robinson, F. N. H. (1974). *Noise and Fluctuations in Electronic Devices and Circuits.* Oxford: Clarendon Press.
Roger, G. F. C. & Mayhew, Y. R. (1980). *Engineering Thermodynamics and Heat Transfer*, 3rd edn, London: Longman.
Rosenberg, H. M. (1963). *Low Temperature Solid State Physics.* Oxford: Clarendon Press.
Rushbrooke, G. S. (1949). *Statistical Mechanics.* Oxford: Clarendon Press.
Schrödinger, E. (1946). *Statistical Thermodynamics.* Cambridge: University Press.
Stanley, H. E. (1971). *Phase Transitions and Critical Phenomena.* Oxford: Clarendon Press.
ter Haar, D. (1954). *Statistical Mechanics.* New York: Rinehart.
Tinkham, M. (1975). *Superconductivity.* New York: McGraw Hill.
Tolman, R. C. (1938). *Principles of Statistical Mechanics.* Oxford: University Press.
Van Wylen, G. & Sonntag, R. E. (1978). *Classical Thermodynamics*, 2nd edn, New York: Wiley.

Waelbroeck, F. G. & Zuckerbrodt, P. (1958). 'Thermal conductivities of gases at low pressures'. *Journal of Chemical Physics*, **28**, 523–4.
Walker, C. B. (1956). 'X-ray study of lattice vibrations in aluminium'. *Physical Review*, **103**, 547–57.
Wallace, D. J. & Zia, R. K. P. (1978). 'The renormalisation group approach to scaling in physics'. *Reports on Progress in Physics*, **41**, 1–85.
Wannier, G. H. (1966). *Statistical Physics*. New York: Wiley.
Weatherburn, C. E. (1952). *Mathematical Statistics*. Cambridge: University Press.
Weiss, P. & Forrer, R. (1926). 'Aimantation et Phénomène Magnétocalorique du Nickel'. *Annales de Physique*, **5**, 153–213.
Weyl, H. (1911). 'Das asymptotische Verteilungsgesetz der Eigenwerte linearer partieller Differentialgleichungen'. *Mathematische Annalen*, **71**, 441–79.
White, G. K. (1968). *Experimental Techniques in Low Temperature Physics*, 2nd edn, Oxford: Clarendon Press.
White, R. M. (1970). *Quantum Theory of Magnetism*. New York: McGraw Hill.
Wilks, J. (1961). *The Third Law of Thermodynamics*. Oxford: University Press.
Wood, W. W. (1968). 'Monte Carlo studies of simple liquid models'. In *Physics of Simple Liquids*, eds. H. N. V. Temperley, G. S. Rowlinson & G. S. Rushbrooke, pp.115–230. Amsterdam: North Holland.
Yourgrau, W., van der Merwe, A. & Raw, G. (1982). *Irreversible and Statistical Thermophysics*. New York: Dover.
Zemansky, M. W. & Dittman, R. H. (1981). *Heat and Thermodynamics*, 6th edn, New York: McGraw Hill.
Ziman, J. M. (1964). *Theory of Solids*. Cambridge: University Press.

Index

including tables of notation and short courses

Greek letters

Table of constants

Boltzmann's constant	k	1.38066×10^{-23} J K^{-1}
Triple point of water		273.16 K
Origin of Celsius temperature scale	T_0	273.15 K
Velocity of light	c	2.997925×10^{8} m s^{-1}
Planck's constant	h	6.62618×10^{-34} J s
	$\hbar = h/2\pi$	1.05459×10^{-34} J s
Proton charge	e	1.60219×10^{-19} C
Proton mass	m_p	1.67265×10^{-27} kg
Electron mass	m_e	9.10953×10^{-31} kg
Permittivity of free space	ε_0	8.8542×10^{-12} F m^{-1}
	$4\pi\varepsilon_0$	1.11265×10^{-10} F m^{-1}
Permeability of free space	μ_0	1.2566×10^{-6} H m^{-1}
		$= 4\pi \times 10^{-7}$ H m^{-1}
		$= 1/\varepsilon_0 c^2$
Bohr radius	$4\pi\varepsilon_0\hbar^2/m_e e^2$	5.29177×10^{-11} m
Bohr energy	$\frac{1}{8}m_e(e^2/\varepsilon_0 h)^2$	2.17991×10^{-18} J
		$= 13.6058$ eV
Bohr magneton	$\mu_B = e\hbar/2m_e$	9.27408×10^{-24} J T^{-1}
Stefan–Boltzmann constant	$\sigma = \pi^2 k^4/60\hbar^3 c^2$	5.6703×10^{-8} W m^{-2} K^{-4}
Avogadro's number	N_A	6.02205×10^{23} mol^{-1}
Molar gas constant	$R = N_A k$	8.31441 J K^{-1} mol^{-1}
Standard atmosphere	P_0	1.01325×10^{5} N m^{-2}
Molar volume at *STP*	$V_0 = RT_0/P_0$	22.41383×10^{-3} m^3 mol^{-1}

Rough values:

kT at 300 K	4.1×10^{-21} J $\simeq \frac{1}{40}$ eV
Acceleration of gravity, g	9.8 m s^{-1}